DATE DUE

THE

EVOLVING

LANDSCAPE

The independence
of the geographic
discipline will be
sustained not by its
isolated exclusiveness
but by how well it
performs its task
of explaining the
landscape.

HOMER

ASCHMANN

THE EVOLVING LANDSCAPE

Homer Aschmann's Geography

Edited by Martin J. Pasqualetti

With a Foreword by John Brinckerhoff Jackson

THE JOHNS HOPKINS UNIVERSITY PRESS

Baltimore and London

TO ERIKA BRITT

Published in cooperation with the Center for American Places,
Harrisonburg, Virginia

The Johns Hopkins University Press
2715 North Charles Street
Baltimore, Maryland 21218-4319
The Johns Hopkins Press Ltd., London

Library of Congress Cataloging-in-Publication Data
will be found at the end of this book.
A catalog record for this book is available
from the British Library.

ISBN 0-8018-5310-9

Frontispiece: Photograph of Homer Aschmann
by Brigham Arnold, 18 March 1949, South of El Mármol
in Baja California. Reproduced by permission.

All photographs by Homer Aschmann, except as noted.

Contents

Foreword

John Brinckerhoff Jackson

HOMER ASCHMANN and I belonged to the generation that came to maturity in the years following World War II. That was the time when cars were again available and inexpensive and when travel cross-country became popular. I was older than Aschmann, and though I was always an admirer of his writings and proud to have published two early contributions of his in *Landscape,* it was not until a few years ago that we met and became friends. I had only recently given up publishing a small magazine devoted to what I inaccurately called "human geography" and was living in retirement in Santa Fe. On the other hand, Aschmann was already widely recognized as a cultural geographer of distinction with a sizable following; he had, in fact, already produced most of the papers included in this volume. But we at once took a liking to each other. Aschmann was an engaging conversationalist and a world traveler, and we shared an absorbing interest in the various ways humans had come to terms with their environment, specifically how we created landscapes as places where we could live and work and celebrate together. Both of us, that is to say, deeply respected the impulse to change and improve the immediate world as our home—a favorable reaction by no means popular with environmentalists. In any case, it was that affection we felt for the labors of humankind that determined our two very different careers.

Except for one very pleasant course with Derwent Whittlesey at Harvard, I never had training in geography or geographical theory. The years I spent wandering, usually by motorcycle, through the landscapes and cities of this country and western Europe, were without plan and were entirely devoted to the pursuit of excitement and pleasure. I took an immense number of photographs, not with the hope of producing works of art, but simply to record places and buildings I thought were typically vernacular. The unusual, I am ashamed to say, had no appeal; the geometrical patterns of open fields in southern Germany, the crossroads' country store in Arkansas, Main Street, whether in Spain or Nebraska or Alberta—those were what I wanted to remember, and when, on my visits to campuses in this country, I showed those pictures and explained them, I was surprised to have them well received. There were in fact many young Americans who were glad to hear about everyday landscapes and everyday life, for it was fashionable in certain circles at the time to describe America as a society in total decay. I seemed to offer evidence to the contrary.

While I was busily vindicating the American way of life, Aschmann, already an established scholar and cultural geographer, was traveling systematically and exhaustively throughout Mesoamerica, searching for the origins of certain landscape features and the manner in which primitive societies had often adapted to forbidding environments with ingenuity and skill. What was the reason for a certain pattern of field and settlements in the mountainous regions of Mexico? How to account for the changes in the vegetation cover when a region was settled by a new population? The answers to these geographical riddles were not easy to find, but Aschmann was enough at home in geology and ethnobotany and anthropology to decipher them, and incidentally, to contribute to the prehistory of many landscapes. He succeeded precisely because he studied not merely the terrain but also the culture of its inhabitants; it was the human aspect of the problem that engaged his attention. Few things exasperated him more than the cult of the Sierra Club and other conservation groups and their assumption that the only area worth exploring was the untouched wilderness. Granted that wilderness offered rest and solitude, why could those blessings not be found in the open countryside?

It was this deliberately commonsensical reaction on Aschmann's part that attracted me, and one of his criticisms of the wilderness cult was published in *Landscape*. It is true that at first the Sierra Club seemed in harmony with my liking for the vernacular rural landscape and its traditional way of life. But it was not long before the well-intentioned suburban nature-lovers lost control of the movement and were succeeded by tenured eco-fanatics determined to frighten the public by publicizing the approaching cosmic doom: the destruction of the ozone layer, the pollution of Antarctica, the disappearance of the Amazon rain forests and of countless invaluable species. The solution was, of course, the elimination of technology, the city, civilization, and ultimately humans themselves, and it was suggested by the radical environmentalist leaders that we would be wise to worship Gaia, the reputed earth goddess. What was most shameless in this crusade was its perverse dislike of humanity, and one result was that many persons, including myself, came to loathe the radical environmentalists. In fact, they have so willfully distorted the sensible aims of the true environmental movement that the movement now has more enemies than friends.

As a native of Southern California, Aschmann was deeply offended by the chaotic urbanization that overtook his homeland. We could certainly have used his geographical expertise and human judgment in explaining urban sprawl, but his silence on the subject may well have reflected his belief that the eco-fanatics would eventually be silenced by the technocrats, and would (to use an elegant 19th-century phrase) collapse into innocuous desuetude; this belief turned out to be an academic folly.

Let us hope that happens. Nevertheless, I wish Aschmann had had more to say about cities and highways and population growth. I wish he could have recognized—without necessarily endorsing—the religious impulse in back of our relationship with the natural environment, and most of all I wish he were present in our computerized world to remind us that our most innocent joys, as well as our deepest emotions, derive from our contact with the earth, for that is what Aschmann knew; and if we read this book carefully, that is what we learn.

Preface and Acknowledgments

As you will note if you read even a small portion of this book, some people consider Homer Aschmann to be one of the shining lights of the last half of 20th-century American geography. Many others have never heard of him. It is with a particular eye to the latter group that this book is addressed, but Aschmann's aficionados should find a special enjoyment delving again into the work of this particularly vibrant and unique personality.

Aschmann's relatively low profile was the result of several factors. He worked in a small department on a fairly small university campus, and his publications often appeared in appropriate if obscure places (in over fifty different journals, in several languages, and in at least seven countries). Furthermore, he was not a self-promoter, preferring to let his work stand on its own. Given these circumstances, I felt that without this book it was likely that the contributions of an extraordinary observer and commentator would remain scattered and hard to find, possibly lost, despite all the merits with which he endowed them.

Inevitably, Aschmann's strong personal traits and views influenced what he put into print and how it was received. For this reason, even though the principal emphasis here is on how landscapes evolve, I felt it appropriate to capture a flavor of the Aschmann idiosyncrasies that helped produce the folklore existing within certain geography circles.

In preparing this book, I applied a light editorial touch to Aschmann's papers out of respect for his meticulous attention to what ended up in print. Stylistic changes were minimal—capitalization of terms such as place-names were made consistent within individual articles, but a consistent style of spelling was not imposed for the entire book. Spanish accent marks were added, however, wherever missing in the original manuscript (with the exception of the Vizcaíno translation) in the spirit of Aschmann's concern for accurate rendering of foreign languages. Substantive deletions are always indicated with ellipses, occasional inserted comments with brackets. I retained as many of his maps as possible, and I also left the units of measurement as they were in the original versions.

The only unavoidable change to content has been with the photographs. It has been impossible to locate the originals, except in the case of "The Evolution of a Wild Landscape and Its Persistence in Southern California." I borrowed substitutes from Aschmann's extensive and well-indexed slide collection. In a few places, I have supplemented these with my own.

All the papers included here, with the exception of the one on the Canary Islands Lanzarote and Fuerteventura, have been published elsewhere. Looking over the list of Aschmann's publications will make it obvious to readers that I did not include everything Aschmann published. It will also be apparent that his interests were actually broader than is here demonstrated. I did not, for example, include his papers on remote sensing, soils, archaeology, aviation, climatology, geographical education, Australia, or Madeira. His curiosity and his ability to bring it to bear on every question he had about the land deserve a strong salute.

I organized the papers around several landscape themes of interest to Aschmann, although the borders between them are not rigid. Several of the papers, for example, could have been placed in more than one section.

I have several people to thank. First, I wish to express my special gratitude to John Brinckerhoff Jackson for providing the foreword. The *New York Times* once described Jackson as "the premier recorder of the human environment," and Aschmann held him in high regard.

I also express a special thank-you to Elliot McIntire, Professor of Geography, California State University, Northridge, for assembling the publications list, for gathering anecdotes about Aschmann, and for helping in many other ways. Thanks are also due to Barbara Trapido-Lurie for her help in redoing and refining certain maps. Others who have contributed anecdotes, suggestions, photographs, and support include Brigham Arnold, Jack Bale, Karl Butzer, Richard A. Crooker, William M. Denevan, James R. Drake, Gary Dunbar, Herb Eder, Robert E. Ford, Richard Francaviglia, John Fraser Hart, Edwin H. Hammond, Carl Hansen, Leslie Hewes, Virginia (Coleman) Janzig, Carl L. Johannessen, Robert Kelly, Melvin G. Marcus, Marvin W. Mikesell, John A. Milbauer, Thomas Pagenhart, Tom Patterson, Leland R. Pederson, Robert M. Petersen, Edward Price, John Q. Ressler, Jonathan Sauer, Steven G. Spear, James Sullivan, Yi-Fu Tuan, H. Jesse Walker, Bret Wallach, Robert C. West, Wilbur Zelinsky.

Warm appreciation also goes to George F. Thompson, President of the Center for American Places, for supporting the publication of this book and for making many helpful suggestions throughout. And lastly, I wish to express thanks to Louise Aschmann, who Homer once told me was smarter than he. That, he firmly believed, was quite a compliment.

An Iconoclast on the Loose

Martin J. Pasqualetti

WHEN COLUMBUS stumbled upon the vast Western Hemisphere at the close of the 15th century, the dispersed and lightly equipped inhabitants who lived there had only minor dominion over the land. Their impact on landscapes, especially small when compared with the scale of changes in large parts of Europe, was, however, soon to soar. Those who followed Columbus brought with them such a mighty combination of animals, technology, and avarice that the power to alter landscapes increased drastically and would eventually have monumental consequences. The legacy of this period in history, now all about us, fueled the craft and career of Homer Aschmann, a gifted geographer who combined insatiable curiosity, broad knowledge, rare insight, and genuine sensitivity with an irascibility, eccentricity, determination, and toughness appropriate to the task of exploring and explaining the physical and cultural mechanics of the evolving landscape.

By chance and choice Aschmann lived most of his life within the Mediterranean climate along western North America. He used the lens of history and archaeology to study the first contact of the Europeans with the gentle coastal people who occupied this agreeable environment. These people led lives unstressed by weather or shortages of sustenance, blessed not only with adequate supplies of mountain water but also with food easily obtained by hunting and collecting on the rich land or by fishing in the bountiful sea. In few other places were these early Europeans to encounter such a welcoming, comfortable, unthreatening, familiar, and salubrious place as that which we now call Southern California. Aschmann began early to study these and other landscapes, examining how they had evolved from the form found by the earliest Europeans to what he encountered during his lifetime.

Soon after the arrival of the Spanish, the native peoples were subjugated to foreign ways and decimated by exotic diseases. Long-held cultural traditions were quickly swept aside and largely forgotten. In the Southern California area, a Spanish flavor began to replace the influence of the altered native cultures. Population increased slowly at first, but by the 20th century the power of exponential growth was in full display, and Southern California bulged with fresh arrivals. Members of each new group, unconnected to the past and bent on change, contributed to the process of reshaping and redistributing the land to suit their needs and ideals.

As a geographer, Aschmann was a professional observer of landscapes, and for 40 years he had a perfect observation post on the eastern edge of this mild region, at the Riverside campus of the University of California. Even there, 50 miles from the center of activity around Los Angeles, the impact of technology was obvious and relentless. He saw the orange groves and the calmly paced life of his early career submerged by the changes rippling out from the Los Angeles epicenter. He looked on as Southern California confronted the economic rapids of agriculture, oil, aerospace, and service, noting that each activity was accompanied by residential surges and signature landscapes.

Aschmann was amazed as, year after year, migration into the area continued unchecked. Undaunted by the growing angst within the populations who lived there, migrants kept arriving, drawn by the residual reputation of a lifestyle of purported ease and opportunity but ignorant of the local landscape history or what they might find once they arrived. With increasing frequency they found a place whose natural assets had been buried, burned, paved, camouflaged, crowded, neutralized, or removed. It was no longer home to a few thousand Indians but to 12 million mechanized souls who chased the good life through a layer of air pollution so dense it obscured the picturesque landscapes that had helped foster the Southern California cliché of panoramic vistas, wide beaches, palm trees, blue skies, and snow-capped mountains.

The most startling characteristic of the changes in the Southern California landscape was the speed with which they occurred. That this feature took Aschmann aback is reflected in his review of a book about the more stable landscapes of Maine: "The tiny amount of change in the [Maine] countryside over a 60-year lifetime comes as a shock to one who has seen Southern California modified in a far shorter time."[1]

In addition to studying the intense and quick changes of land and life that engulfed him—and perhaps in response to them—Aschmann spent much of his time in areas primitive and unspoiled by comparison. He focused early on the depopulated Central Desert of Baja California, the Guajira Peninsula of Colombia, aboriginal peoples of the American Southwest, the dry Atacama and Norte Chico of Chile, the Canary Islands, and the northwestern reaches of Australia. But his attention always returned to Southern California, where the backdoor transformations of the throngs who lived there served as counterpoint to the sparsely inhabited locations he visited elsewhere.

A powerless and reluctant witness to the effects of the forces of change, Aschmann (in 1981) described their likely consequences with death-knell clarity: "They are frightening. I suspect we have twenty, twenty-five years in which time to come into some sort of ecological balance with the environment. If not, the possibilities are not attractive."[2] Contributing to his reputation as a sometimes outrageous commentator, he went on to suggest one possible solution: "I think that a good nuclear war that knocked out ninety percent of the species might lengthen life by several tens of thousands of years."[3]

Not a man to extrapolate a hopeful future from present trends, Aschmann nevertheless dedicated his working life to the goal of bringing the elements of landscape evolution into focus by studying how culture and nature are woven into what we now see on the earth's surface. He considered this a noble, even expected, aspiration: "An awareness of the intricate relationships between man and his environment is a major realm for scholarly investigation and informed concern on the part of all men who profess to be educated."[4]

There is probably no name more apt for a geographer than *Homer* and perhaps the subconscious reason he preferred Homer over his given first name of Harold was that it reflected his unquenchable thirst for excursion and exploration. He must have considered it a functional coincidence to have been born into the age of mobility, a windfall of good timing that facilitated his travels and increased his opportunities to visit varied landscapes, even if—he would surely acknowledge—such freedoms also caused some of the environmental problems he lamented.

Aschmann projected vividly in his writings the understanding he had accumulated in these places. He believed all geographers should possess, as he did, comfortable command of myriad facts about many places from the work of others, but he preferred to form his own opinions, if possible through extended and repeated personal visits. He thought of landscapes as palimpsests whose underlying nature could remain hidden to those in a hurry. He would, nevertheless, opt for a brief time at a field site rather than no time at all. And he loved to wander. So when people asked him what profession he would have pursued had he not become a geographer, he would reply without hesitation and quite seriously, "A truck driver."[5] Indeed, there was a believable ring to it: being paid to travel.

As a university professor, Aschmann considered himself a member of the intellectual elite ("What good are we if we are not?" he remarked), but his demeanor was anything but pretentious. He dressed casually, leaving his tie on the doorknob between classes, always the same tie. When in the field and not sleeping on a cot, he would share rooms, often with his students, in simple hotels and join them at nearby gathering spots to soak up local culture.

Aschmann "had an empathy for Indians,"[6] said a colleague, and this affinity benefited his studies in the rural areas he liked to frequent. He regarded them as people of the earth and thus naturally good geographers with more to teach us about the land than we could learn anywhere else. He would seek out these people for casual conversation, gently probing them for their knowledge and wisdom about the environment in which they lived. He held equal esteem for other nonprofessional observers: "I am impressed, that it is the amateur who can spend decades visiting and revisiting . . . a region . . . who gets the valuable insights. The professional . . . may be forced to rush to print too early."[7]

One of Aschmann's goals, especially later in life, was to retard geography's slide toward limited, impersonal, sedentary research. He mourned the increasing reluctance among young geographers to leave the comforts of their offices, and insisted that those who worked with him get outside to look at the land and sniff the air. His message was not just for advanced students, but for beginning undergraduates as well: after all, "these are real, not imaginary landscapes; some exposure to field observation belongs in the first course."[8]

Again and again Aschmann demonstrated a willingness and uncommon ability to follow the scent of explanation wherever it led, with the aim of a more complete understanding of the land. For this reason he regularly consulted and wrote for a wide variety of journals, including those in the fields of linguistics, ethnography, archaeology, botany, soil science, history, geomorphology, aerial photography, anthropology, and demography. Always on the lookout for hints and suggestions that others less perceptive would not find, he noted that, during his lifetime, geographers who frequented the hallways of other disciplines had become a rarity.

Recognizing that subject specialization was common in many disciplines, Aschmann resisted the trend in his own work, believing it a particularly inappropriate approach for those studying the complexities of landscape origin and function. He was especially adamant in dismissing the increasing incentives to find *practical* application in what he did: "Our rich society can afford a modest number of scholars impelled merely by curiosity."[9]

Geography's range and freedom were the elements most attractive to Aschmann; he saw them as its strength, its value, and its future: "I expect and hope that many of us would answer the question of why we entered geography first with the statement that it was interesting and congenial, and then with the follow-up that the discipline was non-restrictive. It permitted, even encouraged, me to follow my interests wherever they led. It is absolutely vital that the subject be kept that way."[10] He believed that the increasing compartmentalization of the discipline would diminish its appeal to the very type of broadly talented people who contributed most strongly to building its reputation.

Aschmann's belief in the necessity of unbridled intellectual freedom reflected the close disciplinary links he came to appreciate most formally in graduate school, particularly those between geography and social sciences such as anthropology. He felt that melding anthropology with geography was especially valuable when studying aboriginal peoples in their natural settings: "Geographers along with anthropologists have consistently made contributions in the form of community and local regional studies among the sedentary Indians descended from the high civilizations of Western Latin America."[11]

It was his emphasis on and understanding of landscapes that explains why Aschmann drew so readily from several disciplines. He realized that the history of every landscape is written in many languages and that translating just one of them can only provide a partial understanding of how the landscape has evolved.

Aschmann felt that thorough familiarity with the cultural details of local inhabitants was essential to understanding how landscapes develop. This was particularly true in areas once inhabited by aboriginal cultures: "Dietary preferences, agricultural practices, land holding and inheritance patterns, and familial and village level social organization are the sorts of culture elements that may survive from the Indian past."[12] He argued that all this information provided the building blocks important to understanding human use of the earth, and he was steadfast in believing that such an approach had but one natural home: "Only geography endeavors to maintain this perspective."[13]

Aschmann could process multiple signals from the landscape and develop a quick and often remarkably perceptive interpretation. His natural preference, however, was for unhurried study, and in this he reflected the advice of Carl Sauer, his mentor at Berkeley, who encouraged those interested in landscapes to take the time to "sit on vantage points and stop at question marks."[14] Such reflection, Aschmann maintained, helped avoid the type of baseless conjecture that was potentially damaging to geography.

Although Aschmann had an unshakable faith in the value of the geographical perspective, only rarely did he write pedagogical papers, believing doing so took time from important long-term projects. When he did yield to the temptation, it was with contrition: "too much of our limited research energy is invested in minor descriptive studies or Denkschrifts and apologias of at best only transient interest, of which this paper is an example. The academic pressure to publish regularly is admitted, and in my case responded to, but palms should go to him who spends a decade on the substantial monograph."[15]

Aschmann was unusually attentive to students, perhaps a somewhat surprising trait for such a widely respected and productive scholar. He did not hide in his office behind closed doors as academics are increasingly pressured to do, but made himself available without restriction to those with serious questions or points to discuss, whether late at night, on weekends, during holidays, or early in the morning. That he always seemed to be in his office was an example of his unusual work ethic and an admission of the realities of his profession: "Academics have about as much time to publish as letter carriers"; nights and weekends were work time. He thought that being a good geographer took more self-discipline, both in the classroom and in the library, than most people could muster.

When people entered his office, they were likely to find him reading yet another book he had acquired in exchange for a written review. He always felt this arrangement advantaged him because he wanted to read these books anyway, and this way he got them free. In addition, reviews provided a convenient forum for personal views he felt would be inappropriate to express within more formal papers. But he felt that reading books, whether free or not, could never replace being in the field.

If books and fieldwork were two wells he used to quench his thirst for knowledge, a watertight memory was a third; it was the talent many remembered about him most vividly. He would, with little prompting, recite at length from a Gilbert and Sullivan operetta, quote obscure Spanish or German poetry, or tell earthy limericks late into the night at a bar or at his home, where he would invite friends and students to eat and talk and share a bottle.

Believing that unhurried contemplation is the basis of the most valuable landscape insights, Aschmann did not rush students through their graduate work, even when others might have thought some prodding was in order. "It is not unreasonable to hope," he noted, "that continuing observation and study will be cumulative, making possible the recognition of patterns and trends that one less experienced could not see."[16] Recalling the amount of time Aschmann himself had spent in graduate school, one of his former classmates said: "Aschmann used to say that the 14 years he spent gaining the Ph.D. degree at Berkeley was about the right amount of time for such an undertaking, which, he asserted, should be leisurely."[17]

This preference for unhurried study also helps explain why Aschmann, like Sauer, decided to become a geographer rather than a mathematician or physicist: "In the latter fields at the frontier of knowledge, each problem is a unit. A brilliant solution allows you to tackle another problem from essentially the same base. Typically the greatest achievements are made by persons in their youth. In geography, one faces a world so varied that a lifetime is too short to even see it, let alone learn much about how each place got to be the way it is."[18]

Aschmann's character embodied the concept of *juxtaposition*. He had the enviable facility to be engrossed in a 16th-century Spanish Jesuit diary one minute and the next minute be reciting Chaucer's "The Miller's Tale." He could judge an undergraduate research paper with the latitude appropriate to its neophyte author, but become furious when a graduate student missed a key fact through sloth. Somewhat socially shy, he would unhesitatingly introduce himself to strangers at professional meetings and in the field. Although some considered him the epitome of a curmudgeon, he also demonstrated a sensitivity, even a nostalgia, when he expressed—as he did in the mid-1960s—his worry about the degree to which we changed the shape and use of the land: "Until we know better what we want, I can only regard it as fortunate that the moon rather than the places closer to home is receiving our attention."[19]

Aschmann was particularly attracted to empty or nearly empty quarters and was especially pleased when such places turned up unexpectedly. He often demonstrated a flair for approaching these sites from their least unexplored frontier, and delighted in flipping conventional wisdom on its head with an individualist's twist in perspective. Perhaps the best example of this is the paper he wrote pointing out that wild landscapes persisted even in heavily urbanized Southern California.[20]

Despite Aschmann's strong attraction to wildlands, he had no tolerance for environmental artifice, once calling the authors of a Sierra Club book on wilderness a "bunch of fanatics."[21] Neither an activist nor a promoter of activism, he nonetheless admired people who stood firm and supported a solid argument with facts and experience. He even admitted, in a particularly gracious moment: "I am glad that fanatics devote themselves to protecting a few areas from spoilation for commercial gain."[22] But he saw this as an enormous task: "Making, by trial and error or by ingenious insight, our humanized world the closest approximation to paradise would seem a task capable of employing human creativity into the indefinite future."[23] He could not resist pointing out, however, that such devotion alone will not satisfy "even the present population's need for a chance to experience non-artificial landscapes."[24]

During his four-decade tenure at UC Riverside, Aschmann managed to experience—often under less than comfortable circumstances—a greater assortment of landscapes than most of us will ever encounter, all the time collecting and cataloging what he saw and learned. Judging from the history of his field experience, he seemed to take perverse pleasure in hardships, danger, and tests of physical endurance. In Baja, for example, he overcame great thirst, hunger, and a ruptured appendix; in the Guajira he was several times held at gunpoint; in Upper Volta and elsewhere he overcame the discomforts of scorching heat. He survived these and many other trials without complaint or reluctance, compensated by the hard-won insights others more pampered would tend to miss.

His appreciation of the element of danger bolstered a larger-than-life reputation, as did his attraction to arm wrestling, drinking in dangerous bars, or—as if to demonstrate his intent to always see the world from a different perspective than anyone else—performing flawless handstands on steep pinnacles at Bryce Canyon or at geography meetings. He had a fondness for smoke and drink (habits that added to his legend), admired courage and physical prowess, and remained silent in the face of discomfort. The description "a man's man" seemed, for him, a good fit. Although his gruff surface was part of what made him intriguing, everyone remembers that beneath this dressing hid a soft heart, unbending intellectual honesty, and tireless energy to pursue facts.

Like many geographers, Aschmann enjoyed the aerial perspective. One of his earliest opportunities for observations of this type was at the controls of "heavys," B-24s, during World War II. This was obviously not the safest way to see new places. Once, with his aircraft irreparably damaged, he maintained control until he believed he was over friendly territory. After he reached the ground he was captured nevertheless because "they had moved the border."

Never one to waste an opportunity, Aschmann used the year of captivity to "polish up" his German and to research and write a paper ("Kriegie Talk") on the jargon that developed among American airmen in German prison camps. He never lost his interest in the aerial view of the Earth, later publishing papers on commercial aviation and high-altitude remote sensing,[25] and he also continued his fascination for the origins of names, especially place-names, as several papers in this book demonstrate.

Over a beer or beside a campfire, Aschmann would talk about these and other wartime experiences to emphasize that his definition of a good geographer included being alert for opportunities to increase understanding of one's surroundings. To his chagrin, and perhaps understandably, the message his students tended to hear with their cynical ears was that his POW experience helped explain both his proficiency in German and the high standards he enforced for their own language study.

Such standards were in store for his students throughout their graduate educa-

7

tion. Realizing what was ahead, some of them decided they did not want to run the Aschmann gauntlet, and left; others managed to make their way through his two lan-guage exams, five written exams, and extensive oral exams, all administered with a rigor appropriate to his view of geography.[26]

He knew he was being stringent: "The standards of preparation and literacy I am suggesting for doctoral candidates in geography are admittedly revolutionary, even brutal."[27] But there was in his mind no acceptable alternative; he saw his job as one of attracting the best students and getting the most out of them. He also mourned the continued relaxation of standards in geography, in particular because he knew that landscapes, though often appearing simple, were like intricate tapestries. He knew that unraveling them and understanding their complex arrangement was often a gradual process. This approach helps explain why, although receiving his B.A. at 19 and his M.A. at 22, Aschmann did not finish his Ph.D. until he was 34. Only part of that long period was attributable to the interruptions of war and the need to sup-port his growing family. He felt compelled to follow every lead, walk every trail, talk with every person, and read every book before he would feel comfortable assembling the facts, thinking through what they meant, and writing down his observations. The step that usually took the most time—he might tell you if you were lucky—was thinking and reflecting, mentally sifting and sorting. He called it "staring at the wall."

As a university professor, Aschmann assumed the special duty of identifying those who would make good geographers:

> Students are admitted to most courses with diverse backgrounds, and the instructor must re-present elementary material if he is not to lose a large fraction of the class. For the abler major or graduate student, completing a long series of requirements will be dull and discouraging, and the consci-entious but uninspired student will be attracted and comforted. Such stu-dents may make fine citizens, but if they come to constitute the bulk of geographers, and there are too many symptoms that this is occurring, ge-ography has no future in the better liberal arts colleges.[28]

Aschmann also believed graduate education in geography shortchanged students and the discipline when it did not require proficiency in related fields of study:

> The other threat [to the future of geography] is the denial of the opportu-nity for a graduate student to do serious work in an ancillary discipline. This is not a matter of taking a random general course, but a program of study that ultimately will permit him to take a seminar in the other field, to ex-perience directly, and in competition with that field's graduate students, how research is prosecuted on another frontier of knowledge. An occasional fine graduate student may thus be lost to another discipline, but its breadth of interest gives geography a singular advantage in competing for minds of wide-ranging curiosity.[29]

8

As examples of this point, he offered: "The net reproductive rate in demography, the theory of rent and interest in economics, the concept of culture in anthropology, and functional versus evolutionary models for societies are the sorts of ideas referred to, and the geographer must be prepared to give a clear and accurate exposition of those he feels are relevant."[30] Within the same context, he presaged a particularly visible societal emphasis that was to find broad sponsorship three decades later: "Some reactionaries like me even feel noble in exposing students to the notion that there is aesthetic merit in a culturally diverse and less than perfectly efficient world."[31]

Aschmann's devotion to repeated and painstaking observation began early in his career. He chose the Central Desert of Baja California as the site of his early field-work after first visiting there with Carl Sauer and two fellow students, Thomas Pagenhart and Brigham Arnold, in the late 1940s. Baja California seemed to suit Aschmann's fondness for little-known places. There, he honed his talents for field observation and interview, widened even further his competency in ancillary disciplines, and developed the ability to read Spanish archival materials. Together, these various skills helped prepare him to tackle landscape complexities elsewhere as well.

It is a predisposition of the natural geographer to enjoy reading well-written and perceptive landscape description. Aschmann relished such activity, and though he spent most of his time reading scholarly publications, he always sought out the work of experienced amateur observers. Illustrating the reason for this pursuit, he sometimes recalled Sauer's reference to the book *Kosmos*, Alexander von Humboldt's attempt to comprehend the physical universe. Though this tome was considered the marvel of its age, Sauer had pointed out that it is now largely unread: "On the other hand, any serious student of Mexico today or in the future must read Humboldt's less pretentious essay on the Kingdom of New Spain. Put another way, the greatest discoveries of science are destined to be superseded. Informal and insightful observations at a given place in time find a permanent place in human knowledge."[32] Moreover, locating good sources of these observations requires devoted detective work, a requirement Aschmann enforced upon himself and those he tutored.

Following his own recommendations about the benefits of cumulative knowledge, Aschmann returned to Baja many times, tapping local wisdom, uncovering just one more fact, producing over the years a stack of articles and books on any Baja theme he found pertinent in explaining landscape changes, including ecology, demography, archaeology, ethnography, cave paintings, and the construction of the Baja highway. His most comprehensive and enduring work, *The Central Desert of Baja California*, set a standard for scholarship and dogged fieldwork that is by now spawning a third generation of research in several disciplines.

Whether examining Baja California's depopulated Central Desert, the subsistence economy of the Guajira Peninsula of Colombia, the impact of aboriginal lifestyles on the southwestern United States, the use of offshore islands for settlement, vegetative change in arid northern Chile, or the pockets of habitation in Australia's Northern Territory, he seemed most at home where few lived.

He often discussed teaching cultural geography, which he claimed said, "is terribly hard to teach in a coherent, interesting, and intellectually challenging manner."[33] His approach was "to convince the student that the data of cultural geography are derived from field observation and are only as valid as those observations."[34] Landscapes, he explained, always had more to tell those who were perceptive and patient enough to hear: "One could continue to become a wiser geographer."[35]

When geographers began adopting statistical techniques and modern technological aids, Aschmann argued that it was a Faustian bargain for cultural geographers to favor the use of numerical data and inanimate equipment over hard-won insight and inspiration. Such skill might help a student land a job, but it did not bode well for the discipline:

> The rise in employment opportunities for geographers in the non-academic world, as in various planning agencies, is generally considered to be a favorable development, but it . . . presents a dangerous temptation to the major graduate departments. Such employers always prefer to hire people with semi-professional skills; in drafting, making detailed surveys, or doing statistical tabulations. They are immediately useful and later they can be introduced into more responsible positions where their real geographic knowledge may be applicable. The department that emphasizes turning graduate students into technicians is competing with undergraduate engineering schools on their home grounds, and will probably be less successful. The short term placement opportunities may make administrators happy but are likely to prove a mess of pottage for which breadth of vision, our academic birthright, is squandered. Any good graduate department must at least keep open a path of study that permits the bright student to avoid the detour into acquiring high competence as a low-level technician.[36]

He never felt fully comfortable when office-bound or when staring at papers full of numbers; rather, he preferred to draw his understanding of landscapes on-site, usually beginning "by worrying about subsistence problems."[37] Being on the land allowed him to integrate information and nuance in ways he felt only properly trained geographers were disposed and qualified to do. As a cultural geographer "concerned with the diversity of landscapes over the earth . . . not accounted for by physical differentiation,"[38] he saw no other way to proceed than to exercise direct and personal involvement in the process. Those who followed his advice invariably began to understand "the limitations of modern statistical data."[39]

One of the inherent limitations Aschmann viewed in such data was their inability to measure or integrate landscape complexity; he thought a concentration on raw numbers missed the essence of the many interrelated themes that only the human brain is designed to blend and interpret. This explains why he promoted acquiring a wide and varied range of knowledge of a place before suggesting explanations of its origins, its developmental history, and its present use. He knew that developing the necessary skills was a tall order and not for everyone: "the cultural geographer is . . . likely to need a very considerable store of . . . information, and to be self-selected as one who doesn't mind acquiring it."[40] One of the reasons that so few are cut out for Homer Aschmann's brand of geography is because it requires an uncommonly large investment of time to develop the necessary tools, and this is especially daunting for those reluctant to leave the conveniences of the office. It also runs counter to present university tendencies to reward productivity over reflection.

Academic geography in the United States has suffered many blows in recent years, and several notable departments (such as at the University of Chicago) have been disbanded. Understandably, this trend worried Aschmann, and he believed that part of the problem stemmed from geographers not paying adequate attention to communicating outside the discipline:

> Geographers have not managed to impress the academic and intellectual world with the relevance of their investigations to the welfare of society; nor that their approach is fundamental, based on massive data, and capable of contributing to general understanding. Had we done so . . . colleges would find us indispensable; tomorrow's prospective leaders would be widely exposed to an appreciation of man and his environment they will need, for aesthetic if not for actual survival reasons, as they remold the world; from the able . . . students, not burdened with immediate economic demands, we might recruit scholars especially capable of advancing our intellectual frontiers. In the long run we must.[41]

In the final analysis, Aschmann felt his professional mission was to understand all that the land could reveal to him. With an ability on par with that of Aldo Leopold and J. B. Jackson, this master observer regularly broke into clearings of perception about the land few could find for themselves. All the while he prodded his fellow geographers to remember that "Geography's potential contributions to the welfare of the society . . . are major. Their realization is our challenge."[42] In his 72 years, Homer Aschmann worked tirelessly to meet this challenge and, as one admirer phrased it, "to place the treasures he discovered on earth onto the printed page."[43] This act of generosity started and ended, as does this book, in California.

SOUTHERN

CALIFORNIA

Introduction

Daniel D. Arreola

Homer Aschmann was a Southern Californian. With the exception of a few years away at Berkeley and in the military, he lived there his whole life, an unusually long period for observation. And he watched it change. In the early years of Aschmann's life, Southern California was a landscape of barley fields, orange groves, and scattered settlements bordered by beaches and snow-capped mountains. Over the years he witnessed first its gradual and then its accelerated transformation to a sprawling urban landscape of 12 million people, characterized by unprecedented cultural diversity.

Aschmann understood the temptation and drawbacks of superficial landscape observation. He knew that one of the most important prerequisites to explaining landscapes is time enough to observe slowly, carefully, and repeatedly over a long frame of reference. Directing a sharp eye toward both the general and the peculiar, he saw beneath the surface image that the popular media often project to the world. He knew, as his fellow geographer David Sopher expressed it, that "the evidence of regional distinction is there for the watcher to note and to put into proper perspective."[1] Thus, he observed Southern California, this "Island on the Land," as both nurturing nest and medium of professional curiosity.[2]

The landscapes of Southern California accommodated but also sustained him because his observations were never locationally or visually indifferent. His long attachment and association with this land gave him the luxury to observe, reflect, and comment critically upon small parts of the mosaic that evolved during his lifetime into the Southern California landscape we now see.

Aschmann's varied interests usually found an outlet through his regional focus on Southern California landscapes, whether about place-names, climate, rock art, wildlands, vegetative change, history, or ethnography. This is why his writings on the region are found not only in this section but in several other places in this book as well.

Here, we are attracted to some of these titles that explore Southern California: "Proprietary Rights to Fruit on Trees Growing on Residential Property," "Purpose in the Southern California Landscape," and "A Late Recounting of the Vizcaíno Expedition and Plans for the Settlement of California." In these and many other articles and even in book reviews, Aschmann always had something interesting to say. These pieces further reveal the now uncommon plurality of academic interests that marked his writings.

Aschmann seemed equally at home discoursing about any number of things cultural or environmental, historical or prehistorical. He had a special affinity for odd, intriguing, and culturally tantalizing topics. For example, in the "Proprietary Rights" paper he examines one of the commonest landscapes of the Southern California residential scene: ornamental fruit trees on urban and suburban lots.

Recounting his own boyhood experiences of stripping fruit from neighborhood trees in East Los Angeles during the 1930s, Aschmann advances an unusual "fugitive-child culture" hypothesis, which states that the type of fruit tree planted can influence the degree to which fruit will be "harvested" by wandering youth. Loquat trees, for example, which yield a small but juicy fruit, were widely planted in the Southland. They, like apricots and plums, were generally considered public property even when found on private land because the fruit was not usually harvested for commercial sale in the region; invariably, they were certain to be picked. Avocado trees, on the other hand, were also extremely common in parts of Southern California, yet they were usually considered private property because their fruit had commercial value; they therefore enjoyed better security from eager hands. Besides, as Aschmann recalls, avocados "were not suitable for eating on the run" and "taking fruit was for sport not nutrition." Aschmann reports that these picking activities were clearly illegal but at the same time given full social acceptance in the neighborhood. His positing of the moral question "whether formal laws and their enforcement constitute an aid or a hindrance to the maintenance of satisfactory relations within social groups" demonstrates his concern with linking human activity and the landscape to larger social issues.

When this paper was reprinted in an anthology about the proprietary dimensions of forestry in 1988, the editors of that volume noted that Aschmann's account "establishes that this [tree raiding] is not an anarchic activity but rather is a highly socialized activity conforming to rules of tree tenure found across a number of societies."[3] Aschmann's curiosity about ordinary domestic landscapes preceded by several decades the work of others who did not begin exploring this theme until the 1970s and 1980s, when the study of landscape elements such as lawns and yard space came into vogue.

Aschmann's curiosity followed a completely different tack when he looked into the origin of the expression Miracle Mile, a section of Wilshire Boulevard in midtown Los Angeles. In little more than one page (see the "Linguistics" section, later in this volume), he outlines the initial oral use of the term, its first appearance in commercial advertising when the boulevard extended west of downtown Los Angeles amidst open fields of barley and beans. Aschmann notes the irony of such a flamboyant name for a street developed in the middle of the Great Depression; he might have added the minor paradox of such an opulent and exclusive shopping strip named after an avowed socialist, Cincinnati-born H. Gaylord Wilshire (1861–1927). Miracle Mile was a pioneer development for Los Angeles because it was one of the first commercial strips designed especially to appeal to the automobile.

Of all his papers, comments, and reviews—including a half dozen on the peculiarities of the Mediterranean climate and vegetation peculiarities—Aschmann reflects most insightfully on his homeland in "Purpose in the Southern California Landscape." In a remarkably restrained argument, he makes known through poignant observations his feelings about a land and people he had witnessed over several decades. Almost from the first, he confronts the oft-repeated view of the Southland as secondary to the San Francisco Bay area as the perceived cultural hearth of the Golden State. "As long as I can remember, the boorish provincialism of Los Angeles has been contrasted deprecatingly with the urbane sophistication of San Francisco, but it was to Los Angeles that people chose to come."

And so they came, and continue to come, and some would say, as Aschmann did, to a fault. But Aschmann correctly noted that it was not just a benign climate and spectacular scenery that drew folk to the Southland landscapes they would eventually re-form. It was the *culture* of the region "that served as a powerful attraction to further immigration." According to Aschmann, three "desires" shaped Southern California's cultural landscape: a demand for privacy, particularly the desire for an individual house set on a separate lot and the disinclination to travel by any means other than the automobile; a freedom from social constraint in behavior, facilitated by a pluralist migration that meant "there never was a dominant cultural group or social class"; and direct access to the environment, whether beach, mountain, or desert, via an extensive road network keyed to the automobile. In the end, Aschmann's prescription for the Southland chimes his prescient disappointment: "The distinctive Southern California landscape and living pattern will fade, and the landscape will come to be like that of other American urban areas."

Some three centuries before Aschmann penned his views on the future of Southern California, a Carmelite brother, writing in Puebla de Los Angeles of New Spain, recounted the aspirations of the Vizcaíno expedition of 1602–3. As Aschmann's very translation of this colonial document makes plain, Fr. Antonio de la Ascención saw great promise in the coastal lands of California and may have been one of the state's first boosters when he advised the royal court of the great opportunity to add this territory to the "Kingdom." Had Spain acted with immediate dispatch in this matter, instead of sleeping for 150 years before effectively settling California, certainly the Southland and, perhaps, much of the rest of the Golden State, might have evolved to become an independent Latin American state.[4] Given the rapid demographic change that has transpired in Southern California, where Hispanic Americans now constitute almost one-third of the region's population, the Southland may yet, in fact, evolve to become a place quite different than even Homer Aschmann might have imagined.

Purpose in the Southern California Landscape

The Southern California style of living is too attractive. . . . Individual decisions about the kind of environment one creates to live in, when multiplied by a sufficient number, become disastrously damaging.

It is hard to view the humanized world of the Southern California coastal plain as a whole as anyone's concept of heaven.
—"People, Recreation, Wild Lands, and Wilderness"

Most Americans have heard a good deal about Southern California and a considerable fraction of them have at least visited the region. Even the bulk of the resident population as well as those who have toured the area extensively, however, know it only in the last two decades. The relatively small group of people who have observed Southern California more or less continuously since before World War II can scarcely evade recognition that its landscapes have undergone profound modification. The reader is asked to accept on faith without photographic documentation the assertion that the look of the place has been changed phenomenally in less than a generation.

The observer of this region finds that it has features quite distinct from those of other metropolitan centers within the United States, and even more variant from those in the rest of the world. Some are distinctions in kind and arise largely from the peculiar physical environment. The special features of the cultural landscape generally originate from intense development in particular directions. Trends began here and have been carried farther than elsewhere: thus there are more cars and freeways and more examples of isolated single-family, one-story, ranch-style residences.

Immigration to Southern California

As a place in which to live as opposed to a picturesque spot to visit or tour, Southern California has long had enough appeal to draw a virtual flood of immigrants. As long as I can remember, the boorish provincialism of Los Angeles has been contrasted deprecatingly with the urbane sophistication of San Francisco, but it was to Los Angeles that people chose to come. In the process, however, perhaps inevitably, some of the appeal has been lost. Even as Los Angeles develops art and music centers surpassed in the United States only by those of New York, the statement that "it is a nice place to visit but I wouldn't want to live there" can no longer be laughed off as the hopeless jealousy of someone who invested in property in the Bay Area.

The distinctive climate of Southern California, mild but far from uniform, is well known. The varied and essentially interesting, if occasionally dangerously unstable, topography is similarly notable and real. It is pertinent to note, however, that for more than two and a quarter centuries after the area was first visited and publicly reported it did not attract even one permanent European resident. When Europeans settlement finally occurred in 1769, it was for missionary and strategic or geopolitical purposes. Until 1870 the total population of Southern California had not grown at all; the fading Indians were no more than replaced by immigrants. Then, in less than a century the population has increased, largely by immigration, roughly two hundredfold. The point of this historical aside is that the natural landscape, as it had been modified by long-term Indian occupance, was of no particular appeal to immigrants of European origin. Only after the modern cultural landscape had been implanted to some degree and the Southern California way of life had begun to evolve did its appeal to the immigrant express itself with such overwhelming and self-invigorating force.

The more peculiarly Southern Californian the region became, the more people wanted to come and be a part of it. There was always a disgruntled minority who grieved at the costs of development and complained of the loss of open space and the increased crowding of certain recreational amenities such as beaches. Only within the past decade or so has it become apparent to many that this paradise on earth has deteriorated. Both the local residents and the inhabitants of other parts of the United States and the world may do well to contemplate what caused Southern California's present predicament. The residents must also look for a solution which will preserve and restore as much as possible the advantages which originally brought them here.

In his perceptive essay on the future of Southern California Edward Price noted that, more than in almost any other major population concentration, people came to the region to consume the environment rather than to produce from it.[1] An unusually large fraction brought with them moderate or considerable resources obtained elsewhere. These were invested locally, and there were no complaints if they yielded great returns. But in many instances obtaining the good life was a major consideration. It is only in this frame of reference that the enormous sums invested in developing and maintaining orange groves between 1880 and 1910 become understandable. If he did not go broke, the orange grower came close to achieving his objective of living a rural life in a handsome setting and in an area so densely settled that he could enjoy most of the urban amenities. Further, he created a cultural landscape, one now tragically reduced to tatters, that served as a powerful attraction to further immigration. The retired Iowa farmer who came to Long Beach between 1910 and 1930 had fewer resources, but he endeavored to make his residence and neighborhood resemble a small Iowa town while he enjoyed the more salubrious climate and sought casual employment if he needed it.

People did work, and with growing population and markets the service, construction, and other industries have developed apace. It was only with World War II and the rise of the air-frame industry, however, that the producer or job seeker became a dominant element among the immigrants. For a bit more than two decades people have come to Southern California because it was a high-wage, labor-scarce locale. Projected in expanded form into the future, this latter type of immigration promises little but trouble.

Modification of the Landscape

The past 90 years of phenomenal growth occurred in a landscape very little modified by previous works of man. The immigrants brought enough exotic wealth to give them great potency in developing the landscape. At the same time, technological capacity, just beginning to burgeon as America industrialized by exploiting inanimate energy, has expanded at an accelerating rate. It was to Southern California that man first brought water by aqueduct for more than 300 miles over mountain and valley. Except for the cost involved in reworking already developed property, our capacity to rework the physical landscape is essentially unlimited. Deserts bloom far from any natural water source. Harbors and marinas have been cut into straight coasts. Hills are planed into terraces suitable for mass-designed, though opulent, houses. In the words of the hymnal, "the mountains shall be made low" if any land developer sees an opportunity for profit.

Goals in Developing Southern California's Cultural Landscape

This is a landscape of desire. J. B. Jackson expressed the idea that cultural landscapes all fall into this category.[2] They are the product of the resident society's effort to construct heaven. Their variant forms derive from the need to gain a living from the given environment, the differing technological capabilities of the various societies, and those societies' diverse ideas of what heaven should be like. In Southern California the first two limitations have had little and progressively less potency. The cultural landscape is as it is because men have deliberately chosen to make it that way. Should one find it less than his own ideal, and many do, there are several probable lines of explanation. One's ideal may be sharply different from that of those most responsible for shaping the cultural landscape. Rigid governmental controls and legally established private property rights may have interfered to such a degree that neither agency was able to approach its goal. A set of individual decisions on how to make a private heaven, which worked so brilliantly toward a public heaven when affluent people chose to develop orange groves, are producing disastrous results now that population densities have increased by an order of magnitude. The last sort of explanation seems to be the most important, and its understanding requires an effort to recognize the modalities of the private heavens sought so enthusiastically by those who shaped the cultural landscape of Southern California.

Privacy, freedom from social constraint in behavior, and direct access to the environment at home, at the beaches, and in the mountains are three major goals sought by the Southern Californian. It seems quite feasible to identify many of the most characteristic features of the Southern California cultural landscape as outgrowths of an intense striving for these goals, modified of course by the prior cultural experiences of the populace.

*   *   *

1. The demand for privacy is expressed in the individual house set on a separate lot and the almost complete unwillingness of a Southern Californian to travel except in his own automobile. The demand for privacy in housing might have been satisfied with less cost in space by the Mediterranean enclosed patio or atrium. The prevailing pattern was certainly influenced by the fact that so many of the immigrants came from the small-town Midwest with its separate houses and yards. As opposed to the heartland, however, the demand for privacy requires a high, vision-blocking fence around each backyard. Once this mode for private housing is accepted, extremely low urban population density is inevitable, and the low density induced further landscape molding features. The collapse of the already established and capitalized public transportation system, which in the 1930s proceeded so far that it could never be stemmed, is attributed by most students of this grim history, whether businessmen or social scientists, to decreasing demand. Blithely ignoring the expense, each Southern Californian who possibly could, acquired a car and drove it to work. The feedback effects of low-density settlement that meant infrequent public service and also space for a garage for each household are not to be ignored, but the private automobiles' takeover needed a psychological impetus. The Southern Californian was willing to pay dearly for his right to seal himself off in his car from unsought social contact as he made his daily journey to and from his place of employment.

2. The flood of immigrants after 1870 had diverse sources. In the larger, growing communities, of which Los Angeles is the extreme example, there never was a dominant cultural group or social class. Retired Iowa farmers might predominate in Long Beach, retired navy officers in Coronado, and recently Jews in Beverly Hills, but no group was especially strong in Southern California as a whole. No proper, traditional way of life ever established itself as a model for the whole community. Cultural pluralism, of course, is well recognized as characteristic of most American metropolitan centers, but most also have their basic model or image. Southern California has none. Whether it be religious expression, house types, dress, or mode of entertaining, Southern Californians prize only their right to do as they please. Their free and easy manners are less abrasive to neighbors when practiced in low-density urban settlements and fenced backyards. The especially intense search for privacy can be justified in part by social sensitivity. Perhaps it should be reiterated that the goals in living identified as Southern Californian seem much less distinctive now than they

did 10 years ago. Other parts of the country, even of the world, have made them their own. But it is in Southern California that the full impact of these goals on a nearly pristine landscape has had its maximum effect.

3. Having come to consume the environment, that is, the climate and scenery, the modal Southern Californian makes a vigorous effort to do so. Around his separate house he does have a good deal of outdoor living, cooking, and entertaining in a fenced and private backyard, abetted by a climate which gives him a long season in which such activities are comfortable. Here there is further encouragement for low-density residential development. On weekends he has long sought to sample in person the remarkably diverse environments reasonably proximate to him. The beach, the mountains, and the high and low deserts are regularly visited by swarms of families, each in its own automobile. To extend the range of these weekend excursions and to make rugged places accessible, an extraordinary road network was demanded and obtained. Excellent highways lace the San Gabriel Mountains on what are perhaps the steepest and most unstable slopes in the world. The cultural focus on the automobile as a recreational instrument and the road to make it effective has been emulated elsewhere in this country. Southern California gave leadership and intensity to the development of the focus.

*   *   *

It is reasonable and appropriate to add that these personal goals in creating and enjoying the cultural landscape are essentially humane, even noble in themselves. They enrich the human spirit and are to be valid for the whole populace, not restricted to an aristocracy of any sort. It is the expectable but unplanned concomitants of such goals that have created a difficult, even dangerous, situation.

Results of Too Much Success

The most grievous difficulty is the one least subject to amelioration. The Southern California style of living is too attractive. It could be exported, but it is enjoyed most fully in its hearthland or core area. The continuing flood of immigration is now placing unbearable pressure on the finite environment. The beaches and accessible mountain resorts are suffering more intensely each year, and, though the radius of the weekend excursion has risen to 300 miles, its object is already likely to be found too crowded for full enjoyment. Areas where resort cabins can be placed show higher densities than the cities themselves.

The catalog of ills arising from the demand for low-density residential settlement needs only to be listed: destruction of agricultural land and the replacement of orange groves by endless and uninspiring suburbs. As the suburbs spread out the distance to work grew greater; the necessity for automobiles increased and public transportation essentially disappeared. Commuting 50 or more miles in Southern California is not confined to wealthy exurbanites; anyone may do it. The cost in time, automobile maintenance, and freeway construction is so enormous that residents like to avoid thinking about it. The by-product of traffic jams is not alleviated by new freeways; they extend the system and increase its total burden. Total air pollution rises, and the appealing climate of the coastal basins, with their summer inversion over marine air, has become a trap and a plague.

Failure to Develop Urban Amenities

Conversely, the truly urban amenities specific to older high density cities have been neglected. Diversity of option within short distance is lacking. There is no real theater district; Southern California has many good restaurants, but they are so spread out that a separate expedition must be planned to each; the opportunities to procure specialized and distinctive goods and services at even the largest of the shopping centers are comparable to those available in an unsophisticated town of 50,000 people, and at the same time central Los Angeles is steadily losing its commercial richness. Each center offers an almost identical selection, and it is faddish and limited. Petula Clark is contradicted; everything is not waiting for you in downtown L.A.

Problems have certainly been aggravated by unimaginative and repetitive residential tract development as well as taxation policies which encourage subdivision of agricultural land. Once the ideal of the individual house on a fenced lot was identified, developers provided the home-seeking immigrant with nothing else. There have been periods when one could buy a house more cheaply by the month than he could rent a much smaller apartment. The Southern California style of life was made into a costly Procrustean bed. As the freeway net was extended into the rural landscape, it was filled in by the endless suburban smear.

Prospects for Southern California's Future

The intent of this inverted Jeremiad is basically to suggest that, in this instance, perfectly reasonable and healthful individual decisions about the kind of environment one creates to live in, when multiplied by a sufficient number, become disastrously damaging. If the situation is not now so grim as described, present trends will make it so shortly. What options does society have in future modification of the Southern California environment? This is not a cry for planning as opposed to free enterprise development. Residents of Southern California have been planning enthusiastically and expensively for decades, but always for growth, intensification, and extension of present patterns.

*   *   *

1. If population growth, supported by the inertia of an established migratory stream, continues at present rates, Southern California will ultimately be recognized as a less comfortable place to live and a less efficient place in which to produce than the rest of the United States. Appalachia on the Pacific will experience depression and emigration. One might expect also some internal developments which will afford a small compensation, for example, the creation of higher urban densities and the urban amenities associated with them. The distinctive Southern California landscape and living pattern will fade, and the landscape will come to be like that of other American urban areas.

2. Any more optimistic prospect must postulate a notable slowing of the population growth rate, perhaps to one not exceeding the natural increase. With this change, a set of opportunities for preserving and even enhancing the attractions of the natural environment present themselves: for a considerable time a moratorium should be called on the conversion of good agricultural land to subdivisions. The most recently voted change in taxation policy makes this legally feasible. One can regret that the question was not called sooner. A considerable population growth can be accommodated by raising urban density in certain locations, and this pattern is developing. It should not be necessary for a person to reside in a separate house and care for a yard unless he wishes to and doing so should be recognized as a socially more costly option and made more expensive to the chooser.

3. The remaining nonsuburbanized landscape must be recognized as an amenity for the whole society that merits concern and support. Except for tiny areas to be kept accessible only on foot, wilderness is no longer available. The environmental amenities must come from used land. Farmed and otherwise exploited land can be accessible and attractive to the public, particularly the nonmotorized public, with only minor costs to the operation. These costs could readily be compensated by adjustments in taxation. Peculiarly attractive sites, notably the shoreline and the mountain lakes and the most scenic vistas, must be declared off-limits to residential subdivision. Something so limited and vital to all cannot be allotted simply on the basis of individual wealth.

In terms of a century's development the private automobile may have to go, but for a few decades it may still contribute to the good life if it is used reasonably for what only it can supply. It should provide flexible rather than routine mobility, access to seldom-visited localities rather than daily journeys to work or mass entertainment facilities.

The noble experiment on freeways has thoroughly and completely demonstrated its bankruptcy, but we go on tearing up established residential neighborhoods and extending new freeways into rural locales so that further ruining of open land by subdivision is encouraged. Contact with scenery and the environment on a moderately crowded freeway is about equal to that in a tunnel or subway. Only danger and tension are added. The traffic density that would in any way justify the enormous costs in land and for construction of a freeway demonstrates that the route could readily support a public rapid transit system. The so-called cost barriers are a myth, sustained by a peculiar accounting that permits the enormous gasoline tax revenues to be used only for highway construction. The bleakness of long, broad concrete rights-of-way and asphalt deserts of parking lots where cars remain throughout the working day has already encompassed too much of our precious physical environment. They are not reducing the time or cost of the journey to work, and their contributions to personal tension and distress as well as to atmospheric pollution grow steadily.

*   *   *

An optimistic outlook requires assurance that the above structural modifications in the Southern California way of life and the landscape designed for it will ultimately occur. Both points justify agitation. Let these changes begin as soon as possible in order to minimize further irreversible damage. Let other parts of the country learn from, rather than repeat, the Southern California experience.

A Late Recounting of the Vizcaíno Expedition and Plans for the Settlement of California

[T]he cosmographer Enrico Martínez, who had not been with Vizcaíno's expedition but who prepared its maps from field sketches, stated that California was nearly worthless and settling it would be a poor investment.

The entire coastline of continuous land from Cape San Lucas at the tip of California to Cape Mendozino is heavily populated with Indians.... They ... gave us, with love and good will, that which they had.... It was understood from them that the land was good and that it was full of people[.]

—Fray Antonio de la Ascención, March 22, 1632

He who wishes merely to look at a landscape in which the works of man are not apparent will find it ever harder to satisfy his desire.

—"People, Recreation, Wild Lands, and Wilderness"

THE CONSIDERABLE royal investment in the Expedition of 1602–3 up the west coast of the Californias led by Sebastián Vizcaíno resulted in a mass of reports and recommendations, some effective sketch maps by the cosmographer Gerónimo Martín Palacios, redrawn by Enrico Martínez,[1] but very little action.[2] For more than a century what interest there was in California was to be concentrated on the pearl-oyster beds of the eastern coast and later on the evangelization of the Indians. Mathes attributes the loss of momentum to the replacement of the Viceroy Zúñiga,[3] Conde de Monterrey, by Juan de Mendoza y Luna, Marqués de Montesclaros. The latter seems to have opposed consistently both the plans and important official appointments for Vizcaíno. Furthermore, as the return route of the Manila Galleon became better known the costly establishment of a base in California to relieve its crews was regarded as less essential, and mortality from scurvy took regular tolls.

From the first, the most enthusiastic protagonist for the settlement of California, other than perhaps Vizcaíno himself, was the Carmelite friar Antonio de la Ascención, who accompanied the voyage as second chaplain and as assistant cosmographer. He is the primary source of the best-known narrative of the expedition, that of Torquemada.[4] As to actual plans for settlement, however, Fr. Ascención differed sharply from Vizcaíno. The latter planned an isolated rest stop at Monterey Bay, the first landfall made by the Manila Galleon. Ascención's strongest interest was the evangelization of all the natives of Western North America. He proposed to begin at the southern tip of Baja California and work steadily northward, completing a full spiritual conquest of the entire region. Curiously, it was approximately this plan that was begun by the Jesuits nearly a century later and accomplished in 1770 under government auspices.

In 1608, Ascención wrote directly to the king proposing a settlement at San Bernabé, as the embayment behind the line of granitic peaks at Cape San Lucas was called.[5] Writing from the Carmelite convent in Mexico City in 1620, presumably in response to an inquiry from the Audiencia, he prepared an extensive memorial giving both a narrative of the 1602–3 expedition and detailed proposal for the conquest of the Californias beginning with a settlement at San Bernabé.[6] A royal *cédula* of August 2, 1628, specifically requested that the Audiencia obtain Ascención's opinion, along with that of other unnamed survivors of the Vizcaíno expedition, in order that more informed plans for the occupation of California could by drawn up.[7] The document presented here is a further response made nearly three years later. For reasons that are unclear, Fr. Ascención did not choose to present himself personally to the *Oidores* of the Audiencia, but the elaborate notarization at the end of this document suggests that his testimony was regarded as important. This document has been published in Spanish by Mathes.[8] . . .

Since no part of the proposals was put into effect and since other accounts offer a more detailed narrative of the Vizcaíno voyage as well as more detailed and probably more accurate geographic and ethnographic information, the document translated here is at best a footnote on the history of the exploration and ultimate occupation of California. It does represent what may be the last direct impact of Vizcaíno's voyage on Spain's expansion policies northwestward from New Spain, but nearly a century and a half would pass before the Visitor-General José de Gálvez would give substance to Vizcaíno's original scheme. It can also be noted that the king continued to receive quite different advice from other sources. In response to the same 1628 *cédula* on July 30, 1629, the cosmographer Enrico Martínez, who had not been with Vizcaíno's expedition but who prepared its maps from field sketches, stated that California was nearly worthless and settling it would be a poor investment. While not denying the insularity of California or the existence of the Strait of Anian he noted that only 200 leagues of the Gulf of California had ever been explored.[9] Finally, Fr. Ascención wrote another letter to the king on March 4, 1633. He still urged the evangelization of the Indians of California, but the pearl fisheries of the Gulf Coast were to provide economic justification for the operation.[10]

Fr. Ascención must have had notes and copies of his earlier memorials to use in preparing this final one. The dates and places are too consistent to come from a 30-year memory. But a refinement of his perceptions is indicated. The Indians, with the exception of those on Cedros Island, were all friendly, though the narrative in Torquemada tells a more complex story.[11] All harbors, especially those in Baja California, are surrounded by fertile lands, an impression that could be developed only after long absence. The Strait of Anian and the insularity of California, along with other geographic fancies, had become almost established fact in Ascención's mind. Finally, the king of Spain as the world's primate ruler, responsible for the conversion of the world and entitled to its rulership, shows clear and strong. In the mind of this friar in New Spain the *siglo de oro* had not ended.

The body of my translation attempts to represent clearly in English the writer's ideas, departing frequently from his style and syntax. Only in the elaborate notarization have I attempted to preserve the stylistic flavor. Spellings of personal names and place-names have been kept as they are in the Navarrete transcript, with a note if it seemed needed for identification. Accents have been omitted as in the manuscript. [And Aschmann's system has been retained here.] Where Vizcaíno's place-name is not current but the place can be identified with some security the modern name is mentioned in the end notes. It is clear that Navarrete had modernized the 1632 usage and spelling to what was current in 1794. Citations of other documents are to published works, if possible, rather than to the original archival material.

The Year 1632

The opinion given by Father Fr. Antonio de la Ascención, a member of the Order of Barefoot Carmelites in his convent in the City of Los Angeles in New Spain, concerning the location and characteristics of the Californias and the advantages to the Royal Service of His Majesty that would result from their exploration and conquest, including the form and manner in which this might be done as well as the settlement of the harbor of San Bernavé at Cape San Lucas.

In reference to that which the lords of the Royal Council of the Indies wish to know, on the order of His Majesty and in obedience to his royal *cédula* and command, concerning the location and characteristics of the Kingdom of the Californias, I state: That in the year 1602, I journeyed in company with two other members of the Barefoot Carmelites, and of my own religion, named Father Fray Andres de la Asumpcion, who went as our superior, and Father Fray Thomas de Aquino; they are now deceased and in heaven. All three of us were priests and confessors in an armada of three small vessels whose captain and commander was General Sebastian Vizcayno, and as admiral Captain Torivio Gomez de Corban. The armada was sent out by the Count of Monterrey, Don Gaspar de Zuñiga y Acebedo, who was then Viceroy here in New Spain. He had provided, with much prudence and piety, all the things which seemed necessary for the expedition, an expedition which he had been ordered to undertake by command of the king Our Lord Don Felipe III in the first year of his reign. The viceroy ordered and charged me, in the name of His Majesty, that I carry out the office of second cosmographer in the voyage and explorations that we were going to make with the armada. This charge was in spite of the fact that His Majesty had sent as cosmographer Captain Geronimo Martín Palacios, who was a very skilled pilot and experienced in making nautical charts. The Viceroy ordered the latter to sail on the flagship named San Diego, and I sailed on the ship Almiranta; each of us received specific orders. Each of us was to map the land and the islands, and sound the ports and embayments, estuaries, and bays which we were to discover from the Port of Acapulco to Cape San Lucas, which is at the tip of California and is in twenty-three and one-half degrees, under the Tropic of Cancer, and from there to Cape Mendozino, which would be at a latitude of a little less than forty-two degrees.[12] He charged us to carry out the task with which we were entrusted with absolute care in order that with clarity and distinction he might give a complete report to His Majesty about what exists on these coasts and seas and in the lands that surround them.

We departed from the Port of Acapulco on the fifth of May in the year 1602 to carry out these discoveries, activities that were carried out with complete care and diligence, and with a great deal of effort because our voyage proceeded against the northwest wind, which ordinarily prevails along these seacoasts, which themselves extend from the southeast to the northwest. This wind blows with such great force and violence and cold that many times we saw ourselves in danger of being lost and inundated because this fleet in its journey regularly made its tacks from one side to the other close-hauled. It is unbearably hard work, and it was necessary to navigate in this manner in order not to lose sight of land and to make these explorations with care and precision as His Majesty had commanded they be made.

Following the coast of the Kingdoms of New Spain and New Galicia we arrived at Puerto de la Navidad, and from there we went on to the Isles of Mazatlan,[13] which are beyond the Port of Compostela and in the latitude of twenty-three and two-thirds degrees, arriving on the second day of June of that year. These islands are close to the mainland and between them a very good port is formed, and nearby is the Villa of San Sebastian. From these Isles of Mazatlan in order to reach Cape San Lucas, which is at the tip of the Californias, one crosses an arm of the sea, which at that place has a width of fifty leagues. Some call it the Sea of Cortes, others the Red Sea because its waters appear to be reddish, others the Mediterranean Sea of the Californias. This arm of the sea enters toward the north between the coast and land that is formed in part by the Isles of Mazatlan, Culiacan, and the lands and provinces of Nuevo Mexico, and that of the Kingdoms of the Californias. It goes on to give forth into the Strait of Anian, whence it communicates and unites with the Oceanic Sea of the North. And by means of it, according to what I perceive and understand, one is able to make a voyage to our own Spain. And in the passage across this sea at the locality there are four small islands, a single isolated one, which is called San Andres, and the three grouped together, which are called Las Marias. In the mountains which are found in these islands are located the rich mines of Ostoticpac as well as others with rich ores in the same ranges. These are reduced with quicksilver, and they are under the jurisdiction of the Audiencia of Guadalaxara.

Crossing the arm of the Sea of the Californias just mentioned, from these Isles we went on to Cape San Lucas, and in a bay which is close to the point we made a landing. It is a good harbor although it is not sheltered from winds from all directions. In it there is a sufficiency of notable advantages, and it is an appropriate port for the establishment of the first settlement, which, as I will point out farther on, the Spaniards ought to make in this kingdom. It is so because of the richness of the pearl oyster beds, of the abundance of many and varied kinds of fish, and because it is visited by many docile Indians, of a good nature and friends of peace. There is much firewood, good water, good and fertile lands, and a climate that is good and healthy.

From this port, which is called San Bernavé,[14] we traveled onward to explore the entire coast until we arrived at Cape San Sebastian, which is beyond Cape Mendozino at the latitude of forty-three degrees. We continued to give names to the harbors which we were finding and to the islands and embayments in accordance with the [saints'] days on which we reached them. The first harbor that we found after having left the Port of San Bernavé was called that of Magdalena. It is good and has the capacity for many vessels, and at it there are many Indians who received us with peace and love, which gave evidence that they were of a good nature. By way of this inlet an arm of the sea enters far into the land, and it may be that one would therein come upon some major river. From there we continued onward, following the coast and discovering some bays and small islands until we arrived to discover the Bahia de Ballenas. There were many peaceful and docile Indians who gave us a very friendly reception. The land appeared to be good and fertile.[15]

Pursuing the journey we went on to encounter two islands, one of which is called Las Nieves and the other San Roque. In them are very good ports, a great quantity of various kinds of fish, which are good and nutritious, and on the mainland there are many Indians who are peaceful and docile. They received us with affection and were astounded to see people wearing clothes and bearded and at the ships in which we traveled. Near here, on pursuing our journey along the seacoast, we discovered a long mountain range, barren and without vegetation and without a single tree. The entire range is crisscrossed by veins of different colored ores, which appeared to be rich. In the range there must be a very great treasure of silver and gold and of other ores.[16] This sight raised everyone's eyes and even their hearts. We were unable to land here since the mountains cut right into the sea, and the coast is wild and the seas were heavy, so we went onward. Farther along the coast the harbor of San Bartholome was encountered,[17] and along its beaches there were many pieces of ambergris, as we were advised by those who were familiar with the substance when we had described what it was like. Since we did not know what it was we had not paid any attention to it; if it was amber it was a great resource.[18]

We further pursued our journey until we reached Cedros Island, which was recognized by everyone. On it there were some wild and warlike Indians who did not wish us to be where we were, and who threatened us and by signs gave us to know that we should leave the locality. Having reconnoitered the island we returned to the mainland coast, and following it we encountered several good embayments, and the lands inland gave evidences of being fertile, and that the entire area is heavily populated with Indians because all the trails that go inland are heavily traveled and broad. We reached and recognized Cabo del Engaño; farther on is the Isla de Cenizas and the little island of San Geronimo, near which is the famous Bahia de las Virgines,[19] at which a good harbor is formed. Here there are many Indians who are good and peaceful; the land is fertile and there are good trees like those of our own Spain. This port might very well be settled with Spaniards, because the land is good and well inhabited and should not fail to be rich.

From here we proceeded, following the coast and discovering along it several embayments filled with Indians and many islands, small and middle sized, and on all of them there were peaceful and friendly Indians, until we arrived at the harbor of San Diego, which is the greatest one, broader and more capacious than any of the others along this coast. There are many docile and peaceful Indians, and on a sandbar like an island inside the bay there were many pieces of amber as we were informed by those who know the material after we had described its characteristics. If that is what the stuff was, there was a quantity sufficient to load a good-sized ship. No one recognized it, however, and so no attention was paid to it. In the uplands which surround the harbor, it is believed that there is a great treasure because in the little embayments and spits that the water forms in the beach sands there was a great quantity of golden pyrites in the form of small leaves.[20] There is much hunting of wildfowl, lots of fish, and the locality has a good climate.

The Indians here informed us that not far from where we were there had been people who were dressed and bearded like the Spaniards who traveled with us, and that they wore fine clothing. It is assumed that these people populated the area toward the north along the coast of the Mediterranean Sea of California at this latitude. These Indians are acquainted with silver and said that these people had some and that they took it out of these mountains. Here we careened and cleaned our ships and caulked them with wax and pitch in order to be able to sail better in high latitudes.

Having taken on water and wood for fuel we left the harbor of San Diego, and continuing the voyage the islands of Santa Catalina, Santa Barbara, and San Clemente were discovered. All of them are densely populated by Indian fishermen, peaceful and friendly. From the mainland a king of that region came out to view our ships. He invited our General to come to his land and asked his people to come with him, offering gifts for all. It was already nearly nighttime, and we had previously agreed that we would depart the next morning, but we went on in order to see what the gift we were to receive was. While en route a favorable wind came up, and it was necessary to take advantage of it, deferring seeing the gifts until the return trip. With the favorable wind we sailed well during that night, and afterward, traveling through calms as well as it was possible to do, we arrived to recognize the harbor of Monterrey and in it we celebrated Christmas. It is a good harbor and has desirable features and many docile Indians. It is in the latitude of thirty-seven degrees, the same parallel as Sevilla. It is a harbor that is well fitted for Spaniards to occupy because the Philippine Ships that come to New Spain, when they reach it, can find there relief from the labors which caused them torments and deprivations until they arrived here, having suffered those in four or five months of sailing. This is a harbor with abundant trees for ship building, of abundant hunting, of many and excellent fish, and of Indians of good nature. I believe that the surrounding mountains contain great treasures, and there are many animals like those of Spain. The entire area has a climate like that of Castile.

From (this place) we sent back the ship Almiranta with accounts for the Viceroy the Count of Monterrey of all that we had seen and discovered. In his name and in his honor we gave the place the name, Port of Monterrey. Having directed the dispatch of the returning ship we continued our exploration and arrived at Cape Mendozino, and continuing onward we identified another cape, which was named Cape Blanco. Near the latter there is a river to which we gave the name Santa Ynes; it is at the latitude of forty-three degrees.[21] There are more than eight hundred leagues of seacoast between this locality and Cape San Lucas, which is at the [southern] tip of [Baja] California. From here onward the shoreline turns toward the northeast, and here we turned around to return to New Spain. We did this because many of our soldiers and sailors had died, and all of those who remained were very sick, so much so that they could scarcely handle the sails. Also our remaining food supplies were few and without substance. The day on which we began our return voyage to New Spain was the twentieth of January, 1603.

The entire coastline of continuous land from Cape San Lucas at the tip of California to Cape Mendozino is heavily populated with Indians. Wherever we arrived they came to see us as though we were something that they had never seen before. They received with simplicity the things which we gave them, and they gave us, with love and good will, that which they had. By means of signs they invited us into their lands and their houses. It was understood from them that the land was good and that it was full of people, and that there were many mines of silver and gold and a great treasure store of pearls and of amber. This accords, as will be seen in reference, with all that occurred to us on this exploration and that which was seen and discovered. I wrote a brief account, which was sent on to His Majesty. That which I feel, because of what we saw and discovered, is that it would be appropriate in the service of our Two Majesties, divine and human, to act so that some of the harbors referred to above come to be peopled with Spaniards. This is because from them the Holy Gospel could be preached to these peoples and thus salvation could be obtained for as many should as there are in this entire land. And the resources that are in this land would be discovered that they might be enjoyed by His Majesty as the supreme Lord and Emperor of all the Indies that he is. With this his dominions would be expanded to almost double that which he possesses today that is populated by Spaniards. It would be possible to establish settlements at the port of Monterrey, at the port of San Diego, at the harbor of Las Virgines, at the harbor of Magdalena, and at the harbor of San Bernavé at the southern tip of the Californias. The latter should be the first place to be settled, and it could serve as a gateway and starting place for the settlement of the other localities. For that end it should be established with a *plaza de armas* for the conquest and pacification of this great Kingdom of the Californias, and it would serve as the starting place for the preaching of the Holy Gospel.

The first settlement should be at the already mentioned harbor of San Bernavé, which is located at Cape San Lucas because here there is a little lake of very good water and another lake of salt. And there are very good lands for planting crops, cultivating fields, and making gardens, an abundance of firewood, and many docile and peaceful Indians. There are many and excellent fish, pearl-oyster beds, and near this place are mountains with ores of silver and gold.[22] And this is a place with the loveliest of climates. On the point there is a high peak and on it a fortress could be constructed for the defense of the harbor and for support of the settlers in the event that the Indians wished to create some sort of uprising or mutiny against the Spaniards.[23]

Note this: In order to settle this harbor, and from it to pacify this entire Kingdom, and to teach in it the Holy Gospel, I have found a means and method that is easy and cheap. It is necessary to utilize three large frigates or *tartanas* which draw little water. These could be constructed in the Port of Realexo in his Majesty's account. Or they might be purchased since such vessels are not lacking along this coast, along with their crews,[24] providing them all the necessary equipment, sails, rigging, arms, and food supplies.

All should be brought to the Port of Compostela or to that of the Isles of Mazatlan, where there are available all the things that the settlers would need: horses, cows, sheep, goats, and pigs. This livestock could then be embarked on the ships and taken to the port of San Bernavé: Since at that point the Mar Roxo has a width of no more than fifty leagues, the crossing might be made to the other side in two days or less. Thus, it would be possible to stock that locality with these animals with ease, and then they would multiply after their fashion and with them the entire kingdom could be stocked. Thus, the Spaniards would have an abundance of all these resources, both for their sustenance and to provide for their other requirements. With two hundred men, who should be both good soldiers and sailors at the same time, as I indicated in another place, it would be possible to settle and pacify all of this great kingdom. In a brief report that I prepared concerning how His Majesty should have things done in the conquest of recently discovered kingdoms, it is maintained that in them the

Holy Gospel should be preached. Thus, with a good and secure conscience he would come to be the just possessor and lord of the lands that were pacified and subjected to the yoke of the Catholic Church with peace, love, and gentleness; it would not be necessary to wage war against the gentile Indians except in very rare and unusual cases.

The Spaniards who would have landed and settled at the indicated site of Cape San Lucas could form there a fine community. Some could establish fisheries for pearls and for other kinds of fish. Others could prepare gardens and fields for planting wheat, maize, and barley and other seed crops and vegetables, for the land is fertile and has a good climate so that all crops would yield well. With them within a short time the settlement would come to have its necessary supplies, and it would not be necessary to forward further foodstuffs to it. Others could work the mines, signs of the existence of which are present in the vicinity. From the *quintas*,[25] the king Our Lord could recover his expenses, and there would even be a surplus of thousands that might be used to meet other expenses.

In the meantime, while the above-mentioned things were being accomplished, the captain or commander of the Spaniards could send an expedition to explore the entire Mediterranean Sea of California, using the two frigates with orders that each of them carry out a separate mission. One would travel up the coast from the Isles of Mazatlan and the other up the California side, and they would proceed until they reached the latitude of forty-two degrees. There they would change their courses to meet and communicate with each other, giving an account of that which they had seen directly and that which they had learned from the Indians concerning these coasts. Each ship could be manned by thirty men who should be both good soldiers and good sailors, and they should carry the order that they mark carefully every feature along the coast and that they explore all the harbors, embayments, rivers, and islands that were there. They should treat with kindness the Indians whom they encountered, and they should manage to learn from them about the lands in the interior, including data on their people as well as on their cities and on their natural resources. Concerning all this they should prepare a clear and specific report and narrative.

In this fashion, one would come to know if this sea extends to communicate with the Oceanic Sea of the North by the Strait of Anian, and whether by means of it one could make passage and sail on to our own Spain, as I believe to be possible. One could learn in what place and locality the great city of Quivira is located, and of the Rey Coronado (crowned king),[26] in the sector that is on the New Mexican side of the gulf. To learn in what place the Rio del Tizon[27] is located would be of very great advantage and importance, since it might be convenient to His Majesty's service to ship by sea and by river the ordinary supplies that are being sent to the Province of New Mexico.

On the California side, one could come to know in what regions the population of which I spoke in number six are settled,[28] and what valuable resources are present in the mountains there, and what harbors and rivers there are where settlements might be established to use as bases for preaching the Holy Gospel to the natives. Concerning all this the pilots should maintain a complete account so that they can give advice and a full accounting of everything to His Majesty and to his Royal Council of the Indies in order that they can direct the most appropriate plan of action in the service of the Two Majesties.

The third frigate could be used to transport people for the settlement and everything else that would seem to be necessary such as foodstuffs, and animals: mares and horses, young male and female cattle, ewes and rams, goats, and male and female pigs to be used both for breeding and for food. There is an abundance of these animals at Compostela and in the surrounding provinces, so that they could afford easily and with little effort.

Thus, this settlement would come to be populous, rich, and supplied with everything, and there might well develop in it increasing commerce and contracts. The ships from Piru could make port here, since they come and make landfall at this place before they go on to Acapulco.[29] And those who come from the Philippines, who ordinarily arrive with death between their teeth, could there find refreshment and alleviation from the ills that afflict them. And people of the neighboring lands could have dealings and communication with the settlers, for I believe there are peoples who, if they found a gracious reception, would be able to enter into faithful and Christian relations with the Spaniards. I hold it for certain that by way of this sea it is possible to go to the Kingdom of New Mexico, to that of the Rey Coronado, and to Quivira, and to the Kingdom of Anian. From there one could travel to Great China and to Great Tartary and to all the other kingdoms that bound them, and also to sail freely and securely on to our own Spain. In all those places, one might travel preaching the Holy Gospel, with which the entire world would come to be converted to our Holy Faith, *et fiet-unum-ovile, et-unus-pastor*.[30] And our king Felipe IV (whom may God guard and prosper for many and happy years) would become the universal lord of the entire world, a situation which, until today, no other earthly king has been able to enjoy since God created the world. He would be alone a greater lord than all his elders or his predecessors, and this would be easy if that which I am recommending in the treatise which I prepared and am referring to is undertaken and put into execution.

The people who would be chosen to go on this conquest from Mexico, and to populate this new kingdom, could make their journey by land, passing by way of the Province of Mechoaca (Michoacan) and the Province of Guadalaxara on to the Port of Mazatlan. These lands are well peopled with Spaniards, and with abundant and good food supplies, and from Mazatlan one could embark in order to pass over to the Californias. Or to do it another way, in the lands of Compostela and of Culiacan and of Topia there are many habitations of Spaniards who would like to see California settled that they might cross over to live there with their estates and cattle. If only they were given free passage they would go there to serve the king as they did who bore the favor which His Majesty granted to the conquerors and settlers of New Spain. And this would be the easiest and least costly conquest of the many that have been made in these kingdoms until today. From that which has been said in this report it will be evident how poorly informed have been His Majesty and the members of the Council of the Indies since it says in his Royal Cedula, of which I was informed, that this conquest and discovery had been held by the Council to be of little significance. Indeed one sees clearly that the contrary is true by that which I have stated and declared in this brief account and response that I was asked to make by the above-mentioned cedula, which was promulgated on August 2, 1628 in Madrid.[31]

In sum, I say that His Majesty does not have in his kingdoms an opportunity that could be more advantageous nor of greater importance for the expansion of his kingdoms and for the increase of his royal possessions. Further in good conscience, he is obliged to act so that throughout all these kingdoms the Holy Gospel is preached and to bring about, by all means possible, the conversion of their native peoples to Our Holy Faith even though doing so afforded no other advantage. Indeed, one soul is worth more than a thousand worlds together. Feeling this way I sign my name in this city of Los Angeles[32] of New Spain on March 22, 1632. *Fray Antonio de la Acension.*[33]

Proprietary Rights to Fruit on Trees Growing on Residential Property

Memory of taking a neighbor's fruit seems to make a noteworthy contribution to one's sense of belonging and attachment to a home neighborhood, an important asset in the generally rootless land of Southern California.

THE PLANTING of fruit trees on urban and suburban residential lots in Southern California is a common practice. In many instances the justification for this effort must be aesthetic or philanthropic. The owner will pick little or no ripe fruit regardless of climatic or soil conditions. Small to middle-sized boys will strip the trees long before the fruit ripens.

If, however, the planter is judicious in his selection of the species to be planted he may fare better as to harvest. Climatic requirements and optima for the various trees are essentially irrelevant except where conditions are completely beyond the tolerance of a desired species. The child culture will allow rights of private ownership for some kinds of fruit. Others are public property. The owner of the loquat tree (*Eriobotrya japonica*) need not worry about his crop. The neighborhood boys recognize it as public property and never permit the fruit to ripen. An orange or an avocado tree is much safer. Only individuals so hardened as to be willing to undertake actual theft will molest his fruit. Other species enjoy intermediate security.

The following notes are based on personal experience, and afford an insight into the fugitive pattern of a child culture which may persist almost indefinitely but never reach the adult level except in the form of vague memories. For comparative purposes it may be well to identify the source in space and time.

My observations were made just within the eastern city limits of Los Angeles and in the adjacent Belvedere Township during the early 1930s. Most of the residents of the district were relatively poor but not impoverished. A high proportion of the modest houses were and are owner occupied. Almost all were built on separate lots. In ethnic composition there was a mixture of small neighborhoods, a few blocks in extent. Some were occupied by a mixed White American group. Others were definitely Mexican. Perhaps the establishments in the latter areas were consistently more humble. The practice of planting fruit trees in both front and back yards was nearly universal throughout the district. In the early 1930s most of the houses were between 10 and 20 years old, so the many fruit trees planted shortly after construction were close to the optimum bearing age.

Though during the Second World War this district may have produced a little more than its share of juvenile gang warfare, it could not have been considered a particular trouble spot in the '30s. The youths were scarcely pampered and were almost without money, but, with individual exceptions, they were energetic kids who knew their way around but were happy to act within socially prescribed bounds. Stripping certain of your neighbors' trees of fruit was within these bounds, though it should be done surreptitiously. The owner could chase you away if you were observed, but had no right to claim damages for the fruit taken. Most curiously, the owners' rights varied with the kind of fruit.

The loquat was the most peculiarly public tree. Children of the most respectable families would pick the fruit of any tree of this species they could get their hands on, even those of an immediate neighbor. Since they competed with one another, early picking was essential and the fruit seldom matured beyond a faint tinge of yellow. A fully ripe fruit could be found only in the store. The deciduous fruits, plums, apricots, peaches, and apples, were also subject to raiding, and generally while in a fairly green state. If a peach were ever to ripen it would be fairly safe, since it was obviously valuable property, the taking of which would be theft. Figs might be taken, but only in the brief time between ripening and picking by the owner. Green, they are protected by being completely inedible. Oranges and avocados were safely the owner's property.

People being what they are, not all owners approved of this appropriation of their property. There was little that could be done about it. One couldn't watch his trees constantly, and for the public species moral suasion was ineffective. Attempts at vigorous punitive action would arouse the whole neighborhood of boys, who could carry out far more damaging operations against the property of a disliked neighbor. It should be emphasized that the fruit was not taken from commercial plantings. If neighborhood boys had not taken it, it would have been eaten by the owner's family or given away, not sold.

Generally boys between the ages of 8 and 14 picked the fruit. Girls in the same age bracket sometimes participated in the activity as well, though they generally only took from trees in front yards. The toughest kid in the neighborhood might for a brief time protect his own family's fruit trees by physical threats to other boys. But this capacity was soon lost by disinterest and then the inability to undertake the demeaning enterprise of beating up much smaller children.

Because of the risk of running into someone your own age who might protect his own trees or those of his neighbors as his private poaching preserve, there were some physical risks in picking fruit outside your immediate home district, where you knew everyone. Hence when a neighborhood was without boys in the critical age bracket, the owners might enjoy their own fruit. Of course, this would not apply to an extreme case such as a loquat tree in the front yard. As the fruit started to turn color any boy passing by would know that it was not appreciated and appropriate it. One is led to inquire as to the factors which determined the degree to which individual property rights were socially abrogated for particular kinds of fruit. Most of them are apparent in the contrast between the loquat, the most public fruit, and the orange, the most securely private one.

While the prices of the two fruits per pound in a store are roughly the same, commercially grown oranges constitute a major item in the economy of Southern California, and loquats make an almost infinitesimal contribution to the agricultural wealth of the region. Everyone who has driven through the countryside has seen signs offering rewards for information leading to the arrest and conviction of anyone stealing oranges. They are universally recognized as a valuable and salable commodity. Because of its poor shipping characteristics and its unfamiliarity in the market place, the loquat is not ordinarily thought of as an article of commerce;[1] it would be difficult to establish a bill against a child's parents for what he had taken. Furthermore loquats could be picked and consumed before they were ripe enough to have value to anyone other than the small boys who did it. A green orange could not be handled even by those boys, and once it had ripened enough for them it had real value.

While the deciduous fruits, plums of several varieties, apricots, and peaches (apple trees seldom bear fruit in Southern California) are of importance to the economy of parts of California, this is not true in the immediate vicinity of Los Angeles. Again there is the aspect of taking something without recognized economic value, supported by the fact that these fruits could be eaten green. Avocados were protected as a fruit of economic worth in the region, by their large size and consequent unit value, and also by the fact that they were not suitable for eating on the run. Taking fruit was for sport not nutrition, and taking home something of value would have been morally reprehensible.

The placement of the trees in front yards or fenced backyards has a slight effect on their security. The loquat's selection as an attractive front yard ornamental undoubtedly contributed to its vulnerability, particularly to boys wandering outside their immediate neighborhoods. The owner's decision to plant there could be interpreted by any boy who worried about such things as a deliberate invitation. On the other hand the sport of climbing fences into the backyards where figs and deciduous trees were likely to be found was often an end in itself and justified attempts to consume the greenest and sourest fruit.

The preceding remarks are strictly historical, referring to a specific place and time. Such contemporary observations as I can now make, and conversations with adults in various age brackets, suggest that a very similar pattern of fruit appropriation has been widespread in Southern California and of considerable duration. A sample reference from a published source is the following remarks by Mary Frances Kennedy Fisher in a footnote to her translation of Brillat-Savarin's *Physiology of Taste:*[2]

Medlars were called loquats, from the Japanese,[3] when I was a child in Southern California, and they were the only things I ever stole. They always seemed to grow outside the tight-lipped houses of very cross old women who would peek at us marauders and shrill at us. There are very few of the tall, dark green trees left, and most people have never tasted the beautiful voluptuous bruised fruits, nor seen the satin brown seeds, so fine to hold. The last time I saw loquats was in 1947, in the lobby of the Palace Hotel in San Francisco, many of them almost dead ripe, on a long branch which was part of a decoration in the flower shop there. My early experience as a thieving gourmand warned me that they would be at their peak of decay in about six hours.

Her recollection of loquat stealing seems to date from about 1920 and also comes from Los Angeles; it differs from mine in that she was able to find ripe fruits to appropriate. With variations a similar pattern of fruit taking can be recognized elsewhere in the United States.

The activities reported obviously are and were illegal, but at the same time had full social acceptance. A person who attempted to gain the protection that the law afforded him could scarcely continue to live in the neighborhood. A crucial but often ignored question for any society, and particularly our own, is whether formal laws and their enforcement constitute an aid or a hindrance to the maintenance of satisfactory relations within social groups. These notes would seem to support the thesis that such laws are a hindrance. Memory of taking a neighbor's fruit seems to make a noteworthy contribution to one's sense of belonging and attachment to a home neighborhood, an important asset in the generally rootless land of Southern California. This is a remarkable return in comparison with the value of the fruit invested.

Unfortunately, effective fieldwork on the problem is possible only at one, and a very early, stage of an individual's career. I would be interested in specific recollections on the topic from other periods and localities. In particular, the development of this pattern in the vast and otherwise nondescript residential subdivisions created in the last decade and a half might have some interesting sociological implications.

BAJA

CALIFORNIA

Introduction

Conrad J. Bahre

Homer aschmann appreciated and understood the landscape of Baja California with a thoroughness achievable only after long and careful study by a skilled and determined observer; indeed, his writings on Baja are likely to be his most lasting legacy. This claim originates in the depth of his knowledge about the historical antecedents of modern conditions in Baja, as well as in his repeated visits and his several publications on the subject, some reproduced here.

Aschmann first went to Baja California with two fellow students, Thomas Pagenhart and Brigham Arnold, and with their mutual mentor at Berkeley, Carl Sauer, in March of 1949. Recalling those days in a 1981 interview, he said that he had been "concerned about the large Indian population that had been maintained in this very difficult area, which then was nearly empty."[1]

Eventually, he focused his interest on the Central Desert of Baja California, and his classic book *The Central Desert of Baja California: Demography and Ecology* is the product of tenacious library research and arduous and sometimes dangerous fieldwork.[2] The isolated, arid, and harsh landscape of the Central Desert, one of the most remote places in America, held a fascination for Aschmann. His book has come to serve as a manual on how to reconstruct the cultural landscape of an extinct hunting and gathering people and is widely quoted by scholars.

Aschmann was intrigued with the idea that the Central Desert had supported a remarkably dense Amerindian population at the time of first European contact. The native population, largely Cochimí, were among the most primitive inhabitants of the Americas in the 17th century, yet they had adapted to the difficult land with a resourcefulness and ingenuity unmatched since. Only in modern times has the population density in Baja started to rise again, and then only on the extreme northern and southern borders. But what had happened to the earlier population that had occupied the still-empty interior?

Aschmann's objective was to reconstruct the demography and ecology of the extinct Cochimí. He wanted to know who these people were and what their actual numbers were under aboriginal conditions: "What was the nature of the economy and social organization that permitted so many of them to wrest a living from this dry land where now so few people can support themselves? What happened to the Indians and why did they disappear?"[3]

He suspected from the beginning that introduced Old World diseases were responsible for the depopulation, and this proved to be the case, but with no Cochimí living past 1865, he had to reconstruct their lives from clues in the landscape, archaeological remains, archival records and accounts of early explorers and missionaries, and a handful of field reports on surviving Indians, such as the linguistically unrelated Kiliwa and Paipai who live north of the Central Desert. Countless hours translating manuscripts in the collections at the Huntington and Bancroft libraries and many months of fieldwork in the Central Desert provided him with the basis for his conclusions.

One of the things that impressed Aschmann about these Indians was their cultural persistence. So segregated were they from the rest of the North American aboriginal peoples that some of their practices did not change appreciably for thousands of years.

He estimated the aboriginal population of the Central Desert to be about 21,000 persons when the first Jesuit missionaries arrived in 1709. The Cochimí never recovered from the changes introduced by the Europeans; even before the first baptismal census was completed, the natives were being decimated by introduced Old World diseases. By the end of the 18th century most of the Cochimí were dead. In one 20-year period alone, 75 percent of the Cochimí perished in epidemics. The unaccustomed and overcrowded conditions of mission life helped spread disease and contributed to the Cochimí's demise. With their neophytes gone, the missions failed, leaving the Central Desert in a cultural and economic isolation that continues today despite construction of the peninsular highway and several resorts. In the Central Desert, modern civilization has never matched the Cochimí in ecological adjustment.

Reconstruction of the type Aschmann demonstrated in his Baja work requires integrating history, ethnography, archaeology, climatology, botany, and linguistics in a way that only cultural geographers seem to be able to do; specialists in any one of these disciplines rarely see or understand the total landscape. Such wide-ranging familiarity was a hallmark of Aschmann's research and teaching for 40 years.

His disregard for creature comforts and even personal safety in the field was legendary. Hot summer temperatures and scarce natural water supplies were always taxing, and the lack of accessible medical attention nearly cost him his life when he was struck by appendicitis while working near San Ignacio in 1949. After his lucky recovery, Aschmann made numerous other trips to the Central Desert, completing his fieldwork by hitchhiking on the few trucks that periodically transported supplies to mission settlements and ranches up and down the peninsula or between mines and fishing camps.

One of Aschmann's greatest attributes was his ability to generate new ideas, especially in the field. My early understanding of landscape evolution came from the cultural/historical view of landscape championed by Sauer, but my understanding of its relevance to geographic problems was not well developed or clearly defined until I worked with Aschmann. My perceptions of geography, especially as it applied to human impacts on vegetation and vegetation change, came not so much from formal classroom work, but from fieldwork with Aschmann in Baja California and the Norte Chico of Chile. He emphasized that the evolution of a landscape is the result of both physical environmental and biotic interactions through time, especially affected by the long history of human use. In other words, individual cultures differ in how they perceive the physical environment and how it influences them. These perceptions, as conditioned by culture, result in land uses that are evident in the landscape. The cultural/historical method he incorporated in his Baja studies and elsewhere helps to discern the dynamics of landscape change.

Excursions to Baja were always good opportunities to witness Aschmann's abilities in the field. One of his outstanding traits was his facility for interviewing local inhabitants, from the poorest campesino to the highest-ranking government official. He preferred the former, usually considering the campesinos better informed about what interested him because they lived closer to the land.

Aschmann returned to Baja often, over a period of more than four decades. Even before *The Central Desert* was published, he had produced five articles based on his fieldwork there, two of which are included in this volume. "The Introduction of Date Palms into Baja California" (see the section on Flora and Fauna) identifies the year date palms were first introduced into the peninsula and examines their significance as the principal agricultural crop of the oasis settlements of the Central Desert. "Desert Genocide," which preceded his monograph, deals with the effect of introduced Old World diseases on Baja's Amerindian population. It should be noted "that Homer was absolutely outraged when an editor, without consulting him, tacked the title, 'Desert Genocide,' onto his article. The reason he was disturbed was that the Catholic clergy, who controlled some of the vital records Homer was using in Baja California, might take exception to this interpretation of their predecessors' activities and refuse him further access."[4]

Aschmann was particularly successful at collecting, translating, and field checking obscure source material from many disparate collections. In "Historical Sources for a Contact Ethnography of Baja California," he assembled an outstanding compendium of sources, including references and discussions of documents, diaries, and mission records that have come to light since publication of his book. This paper illustrates his mastery over the wide range of diverse documents on the ethnography and geography of Baja California. Especially useful are his comments about the authors and how their characters and backgrounds colored their interpretations of the Indians. Aschmann is at his best here; his analyses show the incisive perceptions of a person intimately familiar with the subject. Because of the fugitive nature of most of such material, many consider this publication an important companion to his larger book. Another significant contribution to the contact ethnography and missionary history of Baja California is *The Natural and Human History of Baja California,* Aschmann's edited translation of the manuscripts of two unknown missionaries.[5]

When Aschmann first entered Baja, the road system was primitive at best, and it did not improve very much for 30 years, except at the northern and southern extremes. By the 1970s, the Mexican government, intent on developing the peninsula, completed a highway running the length of the peninsula. Aschmann's "The Baja California Highway," an expanded version of an earlier article published in *Calafia* in 1974, is a testament to the Baja California he once knew, and writing it was his way of protesting the end of the isolation that had preserved the qualities he loved since first visiting the peninsula. Aschmann deplored the invasion of homogenized Mexican culture into Baja and the onslaught of American tourists speeding to resorts in the south with little appreciation for the Central Desert or its people.

Aschmann took some solace in the fact that the paved highway, by happenstance rather than planning, avoided most of the settlements, water holes, and archaeological sites of the Central Desert. This should afford these places some protection and enable others to study and enjoy, as he did, "a collective place for the oldest and most primitive aboriginal cultures in the New World."

Desert Genocide

The mission population was stable at 350. This gives the high birth rate of 50 per thousand; the child mortality rate was well over 600 per thousand. There was no reserve with which to survive the epidemics that were to come. . . . I kept finding references to the dark side of mission history that have been concealed in other mission literature: the execution of nine Indians at one mission for killing livestock; free use of beatings and chainings for theft and improper sexual behavior.

—Review of *The Letters of Jacob Baegert, 1749–62: Jesuit Missionary in Baja California,* 1982

THE CENTRAL DESERT of Baja California, while not without a certain bizarre beauty, is today one of the most desolate spots in North America. The region extends . . . south of Rosario . . . [about] 250 miles [to just] beyond San Ignacio, and occupies the entire breadth of the peninsula. A single track with occasional side roads runs the length of the peninsula. Over it wheeled vehicles can be driven only with great effort and at some risk to the vehicle. One may travel as much as 40 miles along this road without passing a single inhabitant. Off the road the population is even sparser.

The basic cause of this desolation is drought. The only weather record from within the Central Desert is from San Ignacio, and short record there shows an average annual rainfall of less than four inches. Years entirely without rain have been reported. Only where springs well from the ground, as at the large oasis at San Ignacio, is any sort of agriculture conceivable. In wet years a few scrubby cattle find a fair pasturage in favorable spots, but in dry years many die of hunger and thirst after being reduced to eating the spiny cholla cactus. Of our domesticated animals, only the tough, clever burro can support itself in the Central Desert.

When one reads of the Jesuit mission to the Indians and hears of missions within the Central Desert with Indian populations in the thousands, he can scarcely believe so poor a land could support so large a population. Seeing the great stone mission buildings at San Borja (Figures 1 and 2), Santa Gertrudis, and San Ignacio, the latter far more pretentious than any mission edifice in Upper California, we have concrete and unequivocal evidence of the presence of a large Indian labor force in mission times. These great buildings of hand-hewn stone could only have been built by a larger population than at present resides in their territories.

A series of questions immediately arises. Who were these Indians? How did they manage to live in such large numbers in a country so barren? What has happened to them?

One fact can be stated definitely: the aboriginal population of the Central Desert is extinct. Mission records follow the population decline to the vanishing point be-

Figure 1. Mission San Borja, looking south

Figure 2. Mission compound and adobe ruins from top of stone Mission San Borja, looking northeast

fore the missions were closed. Diaries of gold seekers hiking up the peninsula to the mines in Upper California in the middle of the last century describe mission sites inhabited by a few aged Indians and no children. The present small population is entirely immigrant. The Indians one now sees are generally Yaquis from Sonora or their descendants, who have come to work in the mines.

Our knowledge of the original inhabitants must come from the following three sources: contemporary accounts by early explorers and missionaries, archaeology, and a study of surviving Indian groups to the north such as the Kiliwa and Diegueño, who lived in comparable though less-severe environments and possessed similar technologies and skills. Of these sources, the accounts by the missionaries who actually saw and lived with the Indians are the most valuable. The compendia of Fathers Venegas, Clavigero, and Palou, and the diaries and letters of such men as Fathers Piccolo, Consag, and Ugarte afford most of our information.

The Indians of the Central Desert, generally called the Cochimí by the missionary chroniclers, were members of the Yuman linguistic family, and as such are related to the Diegueño of San Diego County and the tribes of the Colorado River Valley. Like the former and unlike the latter they knew nothing of agriculture. They can be classed as a simple hunting and gathering society. Their material culture was more meager than that of the Diegueño in that they lacked pottery, the domesticated dog, and other items. Their social organization appears to have been simpler, though this may be an environmental response because the struggle for subsistence did not permit many people to live together for more than a few days at a time.

Many students have pointed out that the peninsula of Baja California is an unattractive cul de sac, in which groups have been isolated from contacts with more advanced and progressive Indian cultures elsewhere in North America. The farther south in the peninsula you go, the more ancient and primitive the groups you would expect to find. The Cochimí of the Central Desert were not at the end of this cultural line, but they did live in what was almost certainly the poorest part of the peninsula.

The resources from which the Indians of the Central Desert obtained their food can be grouped in three classes: land animals, marine animals, and vegetable stuffs. All these had to be and could be procured with relatively simple equipment.

The land animals formed the least-important element in the native diet. The country is so barren that the small population of deer and rabbits it supported could contribute only a small fraction of the food needed by the human population. Of greater, but still minor, importance were animals which we would not consider edible. These included rats, snakes, lizards, locusts, and various kinds of grubs, all of which were acceptable food to the Indians.

The marine resources, both fish and shellfish, were much more important. There are repeated references in the mission chronicles to the fact that the coastal bands, or *rancherías,* had a more comfortable existence than did those of the interior. The long coastline on both sides of the Central Desert was certainly an environmental advantage. Shellfish, which are abundant on these coasts, and require little equipment in their procurement, seem to have been of more significance than fish, though at a few coastal localities, such as Cedros Island, the Indians possessed fairly advanced navigational and fishing skills.

The rocky drought-stricken land of the Central Desert supported a surprisingly useful plant assemblage. The fruit of the sweet pitahaya cactus is perhaps the best known of the plant foods, and in its season, the late summer, provided the Indians with their only period of abundance and food security. This plant only grows in the southern half of the Central Desert, however, and because of the seasonal and perishable nature of the fruit, it could not serve as the staple basis of living for the Indians. It was a luxury available for a short time, and at that time travel and ceremonial activities could take place as the people were temporarily freed from their otherwise never-ending quest for a bare existence. The fruits of other kinds of cactus were also eaten. So were various wild seeds and roots. Collecting the seeds was particularly laborious and, without pottery for boiling, they could only be toasted and eaten dry.

The critical element in their whole subsistence pattern appears to have been the roasting of the hearts of the agave, or century plant. An abundance of several edible species of this plant grow in the several parts of the Central Desert. While the slow-growing plant had to be sought over large areas, and required a good deal of labor to make it edible, it constituted the permanently available staple on which the Indian population depended. Though their technology was otherwise simple, the Indians had one critical skill: the ability to take a plant filled with bitter juice, roast it for days in a pit by means of coals and hot rocks to convert it into a sweet, savory, nourishing dish.

Beyond their need for food the Indians of the Central Desert placed few demands on their environment. They smoked a little wild tobacco in tubular stone pipes for ceremonial purposes. Their houses were at best small brush shelters. Clothing for men might consist of a necklace of shells; for women, a grass or skin apron. For ceremonial dress the shamans or medicine men wore long capes made of human hair. The mild climate of Baja California obviated the need for blankets and better dwelling places for protection against cold.

Several methods for calculating the size of primitive populations have been developed by modern anthropologists. . . . I should like to make a special attack on the area-density method. This method consists of applying population densities calculated for other areas where the people were on about the same technological level, and where the physical environment appears to be similar to that of the area in question. The difficulty with this approach is that it is almost exactly backward. You start with the answer and adjust the data to it. The purpose of studying the population density of primitive peoples is to discover what a particular environment afforded in the way of subsistence to a human group with a certain social organization and technology. If you assume that two environments and two cultures are the same in two different places, then there is no problem worth studying. In addition, such an assumption is probably wrong. In one of two apparently identical areas, for example, an obscure plant may yield just enough food to tide a population over the hungriest season, while it is lacking in the other. The area-density figures given below have simply been computed after the populations of various territories were determined by other means. . . .

. . . The surprisingly large populations recorded by the missionaries have induced certain observers to question their veracity. Here the fortunes of history of the peninsula provided a check on the accuracy of the missionaries' accounts. . . . The missions were successively the charges of three Catholic missionary orders, the Jesuits to 1768, the Franciscans to 1773, and finally the Dominicans. . . . The data from the successive orders are so consistent internally that it could scarcely have been falsified by different groups of men with such success.

Working in the northwestern corner of Baja California, Peveril Meigs devised a means of estimating aboriginal population from mission records. He called it the "maximum census method." Basically it assumes two conditions, both of which obtained in all parts of Baja California. Prior to the establishment of the missions the Indians had been isolated from contact with Europeans, and each mission had a definite territory from which its Indian converts were obtained.

As soon as the Indians came into intimate contact with Europeans, and especially when they were settled around the missions, their death rate rose sharply. This was due primarily to the exposure to new diseases for which they had no hereditary immunity, intensified by the unaccustomed crowded conditions around the mission settlement. Consequently, before all the Indians of a mission's territory were brought under mission control, a large fraction of them, both inside and outside the mission, would have died. The aboriginal population would always be larger than any mission population. The amount of this discrepancy is the critical point, and checking the census figures against deaths and conversions where the data permit, I conclude that the aboriginal population in the Central Desert was approximately twice that of the maximum census in each mission's territory. It is interesting to note that Meigs's factor was 2.5. This greater discrepancy appears to be due to a greater resistance to Christianity on the part of the Indians he studied.

One other source of data on aboriginal population, that of contemporary estimates by the missionaries, is not of great utility, for none of the missionaries knew the entire Central Desert in its pristine state. The first missionization within the region occurred as early as 1712, and the area was not fully explored until after 1768. A missionary who saw the northerly part of the region could only have seen the southerly part in a decadent state. The figures which are included below refer to different times for the different mission territories. Though it dropped immediately following the establishment of the missions, it is assumed that in aboriginal times the population did not fluctuate greatly.

The following population figures were calculated with some care from all the mission data I could find (Table 1). They are subject to revision if more data become available, but I believe the changes will be relatively minor.

TABLE 1. Baja Mission Population Figures

Mission Territory	Aboriginal Population	Area in Square Miles	Density per Square Mile
Guadalupe	2,250	1,730	1.30
San Ignacio	5,500	7,450	0.74
Santa Gertrudis	2,800	4,450	0.63
San Borja	3,300	3,600	0.92
Santa María and		2,275	
San Fernando Velicatá	3,500	1,870	0.91
Total	17,350	21,375	0.81

A few notes on the differential densities are pertinent. The most southerly mission, Guadalupe, has the greatest population density. This fact corresponds with the fact that its territory is mountainous and received more precipitation than the areas to the north. This territory possesses the least seacoast, however, thus discrediting the often-expressed opinion that the Indians of Baja California were almost entirely dependent on marine resources. The lower population density of the territories of the Missions San Ignacio and Santa Gertrudis is in large measure a result of the fact that these territories contain the vast salt flats of the Vizcaíno Desert, an almost if not completely uninhabitable region. Otherwise the driest part of the Central Desert would have had an aboriginal population density of about 0.9 persons per square mile.

In comparison with other comparably dry parts of the world which had an aboriginal population which did not practice agriculture, these population figures are phenomenally high. In the Great Basin of the Western United States and in the Australian Desert, aboriginal population densities have been calculated at less than 1 person for 10 square miles. Either or both of two explanations seem applicable: (1) The Central Desert, despite its drought, was richer in foods which would support human life or (2) the Indians of Baja California, despite their apparent primitiveness, had been singularly successful in learning to utilize the resources which their poor land afforded.

Although the missionaries who converted or reduced these people to Christianity were motivated by entirely charitable and humanitarian ideals, probably even deliberate genocide could not have been more effective in exterminating the Indians. To achieve effective indoctrination in Christianity, vigorous efforts were made to keep the Indians in large permanent settlements where their spiritual progress could be watched carefully. Such localities with crowded living quarters and normally possessing a single contaminated water source, were horribly efficient in spreading contagious diseases. Furthermore, the few oases in the Central Desert were so small that only a little agriculture could be developed, and bringing the large Indian population into a few spots made it more difficult for them to gather a living from the wild plants and animals. It is probable that despite the introduction of agriculture and domestic animals, the level of nutrition for the Indians was actually lower in mission than in aboriginal times. The annual reports of the missionaries often speak of hunger and misery prevailing among the Indians. It is unlikely that any actually starved, but dietary deficiencies rendered them more susceptible to disease.

The following table shows the decline of the Indian population during mission times (Table 2).

TABLE 2. Decline of the Indian Population during Mission Times

	1770	1782	1794	1808
Guadalupe	176	105	74	0
San Ignacio	572	241	168	81
Santa Gertrudis	1,244	317	234	137
San Borja	1,538	657	539	192
Santa María	411	0	0	0
San Fernando Velicatá	349	642	525	155
Total	4,290	1,962	1,540	565

In 1770 Guadalupe and San Ignacio were far past their peak, having declined to less than one-fifth of their maximum population. Santa Maria and San Fernando had not yet Christianized all the Indians in their districts and still had a chance to grow from new conversions. In 1775 Santa María was closed and its population attached to San Fernando Velicatá. In 1795 the few Indians left at Guadalupe were moved to other missions. In 1853 Francisco Negrete, a Mexican official, found seven non-Indian families at San Ignacio, and a total of 46 Indians living at the three mission sites to the north. He noted that this population contained no children at all. In another generation the Indians of the Central Desert were extinct.

The factor which completed the elimination of the Indian population was the *mal gálico*, or syphilis. The missionaries endeavored to keep it out, but it entered with the soldiers of their garrisons. The Indians showed no physical resistance to the disease and once it was established their social organization did not prevent almost universal contagion. The disease resulted in raising infant mortality [close] to . . . 100 percent. They were doomed as a population.

With the Indians gone, the Central Desert stayed vacant except at the oasis of San

Ignacio until about 1880, when something of a mining boom developed. This mining boom came and went, and may recur, but today only a few very small properties are being operated. The present population, aside from San Ignacio, consists of a scattering of incredibly poor ranchers and part-time miners, and, along both coasts in some seasons, there are good-sized camps of transient fishermen.

It may be that this history of part of Baja California contains a lesson for us. We tend to think of our way of life as infinitely superior and more efficient than that of simple, nomadic Indians. We can learn that large numbers of these Indians had worked out a means of making a permanent living in a land that appears almost uninhabitable to people brought up in modern civilization. Perhaps we should hesitate before proclaiming our way of life is so good that all the world's peoples should abandon their own traditions and copy it.

Historical Sources for a Contact Ethnography of Baja California

Clothing, architecture, church furnishings, national variations in
Catholic rites, food and the general landscape are described to make
this an excellent eighteenth century travelogue.
—Review of *The Letters of Jacob Baegert, 1749–62:
Jesuit Missionary in Baja California*, 1982

The great merit in Baegert's work will always lie in the naive objectivity
of his observations. . . . and constitute an irreplaceable source in the
reconstruction of that fraction of the culture history of mankind.
—Review of *Observations in Lower California*, 1952

T HE INDIANS of all but northernmost Baja California are extinct; except for
a little help from archaeology and the ethnography of neighboring regions we are
completely dependent on historical documents for our understanding of their culture—a culture that seems to be of critical importance to the whole framework of
ethnologic theory.

Baja California is a long peninsula tied to the North American mainland at the
United States border and the state of California and, more tenuously, across a considerable stretch of extraordinarily sterile salt flats, with the delta and lower valley
of the Colorado River. The Indians of the state of California can be characterized by
extreme stability of residence and a neat and efficient ecologic adjustment to subsistence needs which did not provoke technologic or economic development. The
wild biota was exploited for food and supported a large population. Though domesticated plants were known from contact with tribes to the east, the California Indians did not accept crop cultivation as worth the effort, and archaeology shows that
for millennia large populations sustained themselves without much technological
progress or change in social and political organization.

Along the Lower Colorado the Yuman tribes, who are linguistic congeners of the
Indians of the northern three-fourths of Baja California, as well as the Diegueño of
San Diego and Imperial counties, did plant and harvest crops. Their techniques were
specialized, exploiting seasonal flooding of the Colorado River, and could have no
applicability in the almost riverless peninsula to the south. There is considerable indirect evidence that the flood farming of the Colorado tribes (Yuman and Mojave)
was a recent development, as was the violent war pattern that in recent centuries
chased several other Yuman tribes up the Gila River into physically less attractive
country.[1]

As one proceeds southward in Baja California the productivity of the terrain declines rapidly due to increasing drought. There are fewer plants and animals (although some new and useful desert plant species do occur), and vast areas are uninhabitable most of the time because of the absence of permanent sources of potable water. The shoreline was rich in food resources, but lack of water sources made much of it inaccessible.

The Indians of California may have known periods of food scarcity, even famine, from time to time, but their lands were normally rich, productive, and attractive. The central half of Baja California is desperately poor; man there was hungry if not starving through most of the year. In the far south fairly regular summer rains support a richer flora and fauna, and the isolated southern tip of the peninsula is a not unattractive land for men possessing simple cultures.

Entering from the north, then, groups of Indians must have pushed farther down the cul de sac of the peninsula into progressively more inhospitable terrain. The groups behind them were in better country, and would be more numerous and stronger as well as in contract with other parts of the continent whence progressive technology could diffuse. There was no road back. The initial entrants finally reached a better though limited land. Later comers, even if they had small technical and organizational superiority, would be too few in number, and too hard-pressed to gain a living, to displace them. The little known and apparently isolated Guaicura and Pericú linguistic groups of the far south are separated from the rest of North America by the more recently intrusive Yumans.[2]

Thus Baja California could serve as a collective place for the oldest and most primitive aboriginal cultures in the New World, stratigraphically arranged with the oldest at the bottom. Items of material culture such as the atlatl (spear thrower) could be preserved in use into the ethnographic present; so could elements of religious or ceremonial significance such as the capes of human hair worn by shamans on important occasions. Only Patagonia and Tierra del Fuego present a comparable opportunity for preservation of primitive cultures through geographic isolation.

The data of archaeology add some complexity to this argument. The recent investigations of William C. Massey and especially Brigham A. Arnold have demonstrated that the period of human occupation of the peninsula is very long indeed, probably to be measured in tens of thousands of years.[3] In the Laguna Seca de Chapala area Arnold's careful physiographic analysis, based on the distribution of three distinct assemblages of stone artifacts, shows that man was present during times of much heavier rainfall. The Pleistocene period when cyclonic storm tracks followed more southerly routes must have been involved. Whether the Pericú, the most southerly group in the peninsula, go back to this distant time or whether they are more recent, though certainly very old, immigrants is not yet known.

In dry caves in the far south a number of secondary human burials, the bones wrapped with matting and colored with red ochre, were reported by Ten Kate during the late 19th century. Massey has excavated others and brought them to Berkeley for study.[4] The bones do not appear to be especially old, but they represent a very unusual population. The skulls are consistently and distinctively shaped and are perhaps the narrowest of any human population known, though individual freaks may equal them. Were these physically distinctive humans the historic Pericú, Pleistocene man surviving in isolation down to historic times? We cannot yet say. If so, how did they gain their living, and how did they organize their society? In any event the peculiar significance of the Indians of Baja California to an understanding of the culture history, even the human biological history, of the New World is apparent. Would that they could still be observed, but we must rely primarily on documentary records concerning them.

A parallel, theoretical as well as intrinsic, interest attaches to the question of how men with only simple tools could gain a livelihood in the barren central portion of the peninsula. I believe that I have demonstrated, largely from documentary sources, that they did so in numbers about five times as great as those of the present population.[5] Tools were an important but not the only requisite in this subsistence pattern; extraordinarily detailed knowledge of the detailed distribution of the fauna and flora, and how parts of the latter could be rendered edible, were needed and present; finally a pattern of social organization existed that let small groups of appropriate sizes exploit the sparse and widely distributed food resources in times of scarcity and also brought larger groups together for social and ceremonial interaction on the brief occasions of food abundance.

The Indians in the northern part of Baja California who have survived to the present and can be interviewed by professional anthropologists afford only a little help in understanding the cultures and economies of the lost groups farther south. The Diegueño, Paipai, and even the Kiliwa lived in relatively rich country, and their economies and nonmaterial culture patterns have close affinities with those of California. The Cucupa in the Colorado Delta relate to the Yuma and Mojave, practicing the flood farming and recreational fighting that their distinctive environment permitted. The Seri, across the Gulf of California in Sonora, may be more relevant. They subsist in a barren environment by hunting, gathering, and fishing as did the Indians of Baja California. Even when they were visited by McGee at the end of the last century,[6] however, their culture was fugitive and impoverished. It is in worse shape now. Seeking to avoid cultural extinction under the impact of the Sonora Missions they withdrew to an especially dry and desolate stretch of the Gulf Coast, where a small number still survive.[7] But if we would learn of the Seri, again we must go back to the mission documents of the 18th century and before to get any appreciation of their undistorted aboriginal culture.

When an ethnographer studies a surviving primitive tribe, he visits it, lives within the community if at all feasible, eating and hungering with the natives, and participating in all social and ceremonial activities to which he can gain entry. He also pays native informants to teach him the language and answer questions about social relations, religious beliefs, and other aspects of culture that are not readily accessible to direct observation. Most important he possesses a body of theory about the structure and function as well as the content of a human culture that has been built up by about a century of anthropological study. Without entering into a discussion of the epistemology of ethnographic knowledge, it may be legitimate to say that the modern student knows, within limits, what to look for and what questions to ask to get a reasonably coherent understanding of the culture he is studying. Known regularities in the interrelations of various aspects of social and cultural behavior will afford leads for investigation. The rather consistent relations between kinship terminology and lines of inheritance of property may serve as an example. The competent modern student will not work from a questionnaire and will always be looking for a new and unsuspected culture pattern or element in the group he is studying, but his frame of reference is a refined product of a scientific tradition. The possibility that the ethnographer's frame of reference may narrow his interests to a point where he sees only a small part of the culture he investigates does exist. That danger is probably growing, and it sets a definite limitation to the worth of the work of comparative sociologists. But universalism in the study of culture has long been a vigorously held goal among anthropologists, perhaps more than anything else providing intellectual coherence to their discipline.

Historical documents can be and are asked the same questions about peoples and cultures that are no longer accessible to observation and interrogation. By careful analysis they can be made to provide answers of surprising detail and notable validity. Some of their limitations are obvious, and inescapable; others are more subtle and, though with some danger, may be subject to effective correction. Finally, historical documents can reveal facets of a culture, particularly its dynamic character, that even the most skilled modern ethnographer could not uncover.

The obvious limitation, of course, is that only what was recorded is preserved. The absence of a statement about some culture element or pattern does not mean that it did not exist or even that it was unimportant. The discovery of a new diary or letter concerning a particular people is always exciting, and rarely does such a document fail to add new information or clarify a relationship.

The more subtle problem is that the observer and recorder was almost always from outside the culture he was describing. (The Inca Garcilasso de la Vega is, to some degree, one of the rare exceptions.) He may not have had any comprehension of the significance of a ritual or some bit of social behavior that he observed. Was a grotesque dance step at a religious ceremony a free individual improvisation or a patterned action that could not be varied without destroying the ceremony? He could also interpret social behavior and religious or supernatural belief only in terms of his own cultural norms and, until the development of the anthropological disciplines, his view was likely to be narrow. The puritanical and strongly patriarchal attitude of upper-class Spaniards (and most Spanish Jesuit missionaries were from that class) toward social and sexual relations shows clearly in their discussions of those aspects of the cultures of the Indians of Baja California. When they observed that sexual activity on certain occasions was not restricted to spouses, they could interpret the pattern only as one of a culture too primitive to maintain effective restrictions on the actions of sexually aggressive males. The German, Baegert, coming from a less patriarchal tradition, could see and report that the females commonly made the advances,[8] and, though he deplored it, they were not regarded by other Indians as in any sense "fallen women."

The answer to this limitation is to do an ethnography of the culture of the individual observer. What were the value system and the norms of social behavior in the particular European cultural milieu in which he was reared? The Jesuit missionaries to Baja California had a fairly consistent, religiously derived value system. But they came from many nations and had seen different societies in their boyhoods. Apart from their varying personal acuity we get different sorts of reports from the Spanish-Italian aristocrat Salvatierra, Baegert, the stolid Alsatian peasant, and Consag, a cosmopolitan Hungarian-Croatian nobleman.

The two major compendia on Baja California, Venegas and Clavigero, accentuate the problems. They were written by people who had not seen the peninsula, from reports and answers to questionnaires from missionaries in the field. Both were professional historians and sought to give a coherent account from often contradictory reports. Thus a double distortion may be introduced. But the completeness of these works makes them too important to avoid.

Conversely, the documents on Baja California have their virtues. The Jesuits maintained strong intellectual curiosity concerning the mores of all parts of mankind, even if they were eager to alter or destroy some of them, and there may have been some survival of the extraordinary historical tradition of Sahagún. The very primitiveness of the cultures of Baja California made them an item of special theoretical and theological interest. The missionaries learned the native language and might remain, almost alone, at a mission for a decade or longer. Few modern ethnographers ever get so full and intimate, let alone extended, an exposure to the group they are studying. Also there are the opportunities for diachronic observations over a period from the visits of Cortés and Ulloa in the early 16th century to past the middle of the 18th century. Only after 1700 were the native cultures seriously disturbed by the missions. Thus some appreciation of the internal dynamism of the native cultures, both without and with serious impacts from abroad, may be gleaned. Even if the Indians still survived, we would find the documents of great ethnographic as well as historical interest for what they could tell us of internal modification of the natives cultures over a period of 400 years.

The Four Major Documents

The great compendium of Father Miguel Venegas contains the most comprehensive account of the Indians of southern and central Baja California. Its Jesuit author was assigned to the task of writing a history of the Jesuit mission to California. He utilized official and published historical sources dealing with the peninsula prior to Jesuit entry in the late 17th century. He apparently was given access to all reports from the missionaries and also was allowed to send questionnaires to individual missionaries on the detailed history and state of their particular missions. A copy of one of his follow-up letters was preserved and is published in Ocaranza's *Crónicas y relaciones del Occidente de México*.[9] It interrogates, almost cross-examines, Father Luyando of Mission San Ignacio on some apparent inconsistencies or inaccuracies in the latter's response to an unknown initial questionnaire. It may be pertinent to add that shortly before Father Luyando had been retired from the field as unsuited for the hard life at that frontier mission, Venegas completed his manuscript entitled *Empressas Apostólicas de los PP. Missioneros de la Compañía de Jesús . . . en la Conquista de Californias . . .* in November 1739. A beautiful and apparently perfect copy is in the Bancroft Library and an incomplete one in the Huntington Library.

Historical and biographical detail on the mission activities forms about half of the 2,000 numbered paragraphs. A quarter of the manuscript deals with pre-Jesuit history, and the remainder is a description of the region and its inhabitants, though relevant additional information on those subjects appears in the historical accounts of individual missions. The work is not without redundancy. The assembled descriptions of the Indians suffer from the fact that they are not only interpretations but secondary interpretations by one who had never seen the place or its peoples. Further, the accounts are generalized as to place, and only when similar data appear in the historical section can a report be associated with a particular tribe or group.

The work was not published until 1757, when it was completely worked over by Father Andrés Marcos Burriel and published in three volumes in Madrid.[10] A number of separate documents are included in the third volume, only one of which, Consag's account of his voyage up the Gulf of California in 1746, adds useful material that Venegas did not have access to. There have been many subsequent editions, including an early poor one in English.[11] The recent Mexican edition is good.[12]

Burriel reduced the prolixity of the original and made it more readable. By putting the material concerning the Indians into a single section, and selecting only one of apparently contradictory statements about them, however, he added a tertiary set of interpretations rendering the work far less useful for ethnographic analysis. Often reference to the original manuscript shows that the contradictory statements referred to different tribal groups or times, but frequently they cannot be sorted out.

Father Francisco Javier Clavigero was a professional Jesuit historian, and he completed and published his *Storia della California* in Venice in 1789, after the Jesuits had been expelled from the Spanish domains in 1768. Again he collected his material on the Indians into a single section without reference to the source of the report. He, of course, had access to the reports accumulated between 1739 and 1768, and these, so far as the Indians were concerned, dealt almost exclusively with the newly mission-ized areas between San Ignacio and San Fernando Velicatá. Wherever Clavigero makes statements about Indians that cannot be matched in Venegas or which contradict the latter, they may with some security be referred to the Cochimí who occupied this region. A good English translation was published by the Stanford University Press in 1937.[13]

Nachrichten von der Amerikanischen Halbinsel Californien: mit einem zweyfachen Anhang falscher Nachrichten by Father Johann Jakob Baegert, published in Mann-heim in 1772, is a fundamentally different sort of work. Baegert was a Jesuit mis-sionary who served in Baja California from 1751 to 1768, almost all the time at Mission San Luis Gonzaga in Guaicura territory. He wrote the book shortly after his expul-sion and apparently had little opportunity to consult historical references. His gen-eral historical section is full of minor specific errors. Baegert was, for a Jesuit, an unusually unintellectual man, and his long isolation at a lonely mission did not help to keep him in touch with the intellectual currents of his age. He was forced to write largely on what he had seen and experienced, and was not even very interested in the way in which his observations fitted into a general theory of history and society ex-cept in the most general reference to Catholic dogma.[14] Therein lies the extraordi-nary merit of his work.

His descriptions can be referred to a particular group of Indians, and he reported what he saw whether it made sense to him or not. Many topics of interest, of course, he did not discuss; here is a mass of raw but also objective and unstructured data. Unfortunately, the Guaicura had been under strong mission influence for more than a decade before Baegert's arrival, so he never saw the functioning aboriginal society. One may only hope that a pioneer missionary with a similar naive objectivity has written a comparably extensive manuscript and that it will some day come to light.

A good English translation was published by the University of California Press in 1952.[15] The Mexican edition of 1942 is excellent and contains as a bonus Paul Kirch-hoff's brilliant introductory essay "Las tribus de la Baja California y el libro de P. Bae-gert," which relates Baegert's data to the culture history of Western North America.[16]

The three letters that compromise the *Noticias de la provincia de Californias* by Father Luis Sales are a much smaller work, but they similarly endeavor to give a com-prehensive account of the native culture. Sales, a Dominican missionary, came to Baja California in 1772 and stayed until 1790. His letters were written about 1789 and pub-lished in Valencia in 1794. By the time Sales and the Dominicans reached the penin-sula the Indian cultures south of San Fernando Velicatá were completely disorganized under mission influence. What Sales described must refer to the northwest corner of the peninsula, the Dominican Frontier, where he spent most of his career and where the Indian cultures were functioning during his time. This approaches the Upper Californian culture area and includes some groups that survive, though with reduced native cultural inventories. Although Sales introduced as many or more interpretations of doubtful validity into his descriptions of the Indians as did the earlier Jesuit writers, they are easier to discount because the partial survival of the described cultures permits modern checking and comparison.

Sales's original work is extremely rare, but in 1956 Glen Dawson published an English translation that makes its ethnographic information as well as its valuable historical data on the Dominican period available.[17]

Minor Documentary Sources

Two briefer but important manuscripts contain relatively general descriptions of the Indians of Baja California. In an unpublished volume by Pedro Alonso O'Crouley on the Kingdom of New Spain there is a description of Baja California by the Jesuit Father Norbert Ducrue, dated 1765.[18] Although he only occasionally refers to a specific locality, his discussions of the Indians seem to apply primarily to those living north of San Ignacio and by their localization are thus more valuable. Another original document, now in the Huntington Library, is entitled "Addiciones á las noticias contenidas en la Descripción compendiosa de lo descuvierto, y conocido de la California."[19] It appears to be material collected for Venegas's great work, but it arrived too late to be utilized. Clavigero did utilize both of the above manuscripts, but he mixed their data in with information from other parts of the peninsula, so without the originals the valuable specificity as to place is lost. The "Addiciones" refers very specifically to the region north of San Ignacio and, though, as the title indicates, there is no effort to give a comprehensive description, the specific references to Indian customs are clear and objective accounts.

The authorship is not stated, but it is my belief that these careful notes were prepared by Father Consag in the early 1740s when he was evangelizing wild or gentile Indians on the mission frontier north of San Ignacio. The writing style resembles that which appears in the two diaries of his explorations up the Gulf Coast in 1746 and along the central West Coast of the peninsula in 1751. In all these writings there is a lucidness that exceeds Baegert's. They also share the latter's concreteness in description that inspires ethnographic confidence of a high order. Further, he was dealing with Indians who were just beginning to come under mission influence.

Two published accounts of the natural history of the peninsula have substantial value for their descriptions of the ways the Indians utilized the native flora and fauna of their environment. A compilation by the "Padre Collector" from letters by Salvatierra and other missionaries, apparently written about 1762, was recopied in 1792 and finally published in 1857.[20] It not only gives long lists of animals and plants that were eaten, but discusses their capture and preparation. A description of the use of parts of two different kinds of plants for fish poisoning is a particularly interesting item.

José Longinos Martínez, an official and scientist, made an expedition to both Californias in 1792 to study the natural history, and his report was edited and published by the Huntington Library in 1938.[21] By this time the Indian culture was functioning only in the Dominican Frontier, and Longinos made trips into the interior to visit areas and Indian *rancherías* that were under only light mission influence. Although he was a professional natural historian, his observations were made on horseback. Few of his botanical identifications have validity, and his statements of what the Indians did with the wild plants and animals have the ring of hearsay. They do not inspire the same confidence as do the accounts of the professionally unself-conscious missionaries who lived among and observed the Indians for years.

Two other sorts of documentary materials are likely to contain fugitive bits of ethnographic information of the greatest worth. These are accounts and diaries of explorations, and letters from pioneer missionaries either to their superiors or to friends in Europe. In both types of record the source of any observation can be pinpointed as to both time and place. The information is likely to be sporadic and in any document to refer only to a few particular items of culture. It is also likely to be specific and accurate and not elaborated into a rational but not necessarily true description of a culture complex. The number of such diaries and letters in existence is indeterminate. Only a few have been published, and many may exist unidentified in archives and private collections. A few that are known from secondary sources to have been written cannot now be located.

I shall not attempt a comprehensive listing of either exploration diaries or letters (and sometimes a narrative letter takes on the chronological character of a diary) that contain relevant ethnographic data on the Baja California Indians. Rather I will mention a few valuable items that illustrate their range in time and space and show some of their content. Surviving reports on the early Becerra and Cortés expeditions, which discovered the Cape Region of Baja California in 1534 and thereafter, contain little of interest other than that Indians lived there who were uncivilized and unfriendly.

The Ulloa expedition of 1539–40, however, provides two parallel and rewarding accounts. That of Francisco Preciado was published by Hakluyt. H. R. Wagner's *California Voyages 1539–41* reprints a variant of it.[22] There are reports of expert swimmers along the middle Gulf Coast, clearly a shellfish-gathering economy; and for Cedros Island off the West Coast, and the approximate terminus of the voyage, there is a great deal of information on a numerous population with a highly developed technology for exploiting marine resources. Nearly two centuries later in 1731 (as reported in Venegas),[23] Father Taraval, the next European to visit the island, removed a small, culturally decadent population of Mission San Ignacio. Evidently European epidemics had preceded the missionaries, reducing the native population to a point that its culture had virtually collapsed. Cabrillo in 1541, conversely, reports little of interest on the peninsula.[24] Vizcaíno's voyage up the West Coast in 1603 put in at different places,[25] and the account of a marine-oriented economy at Magdalena Bay that used some sort of fish traps is of interest as is the report of a large ceremonial structure, perhaps a variant of the sweat house, or *temescal,* so well known from Upper California, somewhat north of Cedros Island.

From the middle of the 16th century until the peninsula was turned over to the Jesuits in 1697, and to a limited extent thereafter, the southern part of the Gulf Coast was visited sporadically by pearling expeditions, some legally chartered and some engaged in poaching. These were not literary visitors nor had they any particular interest in the Indians. From their brief reports, however, Massey was able to establish that the atlatl, or dart-thrower, was in use in the Cape Region, along with the bow and arrow, between 1615 and 1640.[26] By the time missionaries reached the area in 1720 that primitive weapon, common in North America at the beginning of the Christian era, had been completely displaced even in this backward end of the earth. Of particular interest is the fact that the historic descriptions of the dart-thrower match the archaeologically recovered objects Massey found in dry caves in 1947 and subsequently.[27]

During their first years at any mission the letters of the missionaries were filled with specific statements concerning the customs of the local Indian population. Similar references occur whenever a missionary reported on an exploration, or *entrada,* in some district more than a day's journey from his headquarters. Once a given mission was well established, however, perhaps five years after its founding, references to native customs disappear from the reports, which then focus on the economic and spiritual progress of the mission. A missionary might, as did Baegert and Sales, after a long stay in the field, put together a general description of the culture of Indians he dealt with. But the maintenance of his mission shortly came to dominate his regular correspondence.

Thus we find letters from Salvatierra describing the Indians around Loreto, Ugarte concerning those at San Xavier and also at Guadalupe, where he cut the logs to build his famous boat, Bravo and Guillén at La Paz, and Piccolo at Mulegé, Tamaral at La Purísima, Hostell at San Luis Gonzaga, and Linck at San Borja, each from a date shortly after the founding of the particular mission. Some are still in manuscript form, commonly in the Archivo General de la Nación in Mexico City or the Archivo General de Indias in Seville but also appearing in smaller collections as at the University of Texas, the British Museum, or the Huntington Library. The archives in Rome have not been examined thoroughly from this standpoint and may well contain a wealth of material. Items have been published individually in Mexico and in California. There are two collections of letters that merit special mention. *Documentos para la Historia de México* has already been mentioned. Series 2, vol. 1, and series 4, vol. 5, contain a number of valuable early letters by Salvatierra and others. *Das Neuen Welt Botts* contains a surprising wealth of material. This multifolio work consists of letters from Jesuit missionaries all over the world written between 1642 and 1760.[28] If they were not originally in German they were translated. Some were to relatives in Germany; others seem to have been obtained, after some delay, from Jesuit headquarters. The heavy attention to Baja California in this worldwide coverage shows how spiritually important to the Jesuits this backward and impoverished peninsula was.

Most of the mission explorations were relatively informal affairs. From a base at the frontier mission a priest and an escort of one to three soldiers, accompanied by a number of converts from his mission, would set forth for a few days or weeks into gentile territory. The shorter expeditions were reported in narrative letters. A diary or journal was often kept on the longer ones. Items likely to be reported were hunting and fishing techniques and new plant foods or new ways of exploiting them. Since the expedition was dated, very specific information as to the seasonality of various subsistence activities can be gleaned. For example, Father Piccolo's letter of June 24, 1709, reporting on a trip to the West Coast from Mulegé,[29] makes it clear that the grass and *bledo* seeds being gathered for food were from annual plants that had grown following winter rains. In November 1716, Piccolo explored the future site of Mission San Ignacio. A major interranchería ceremony associated with the pitahaya harvest was taking place and representatives of 50 rancherías, or bands, were reported to be in attendance. The most thorough and accurate report is in manuscript,[30] although it or some variant is referred to by both Venegas and Clavigero. This is the most circumstantial description of native religious practice that we have from the peninsula, and as such is of enormous worth. The longer formal discussions of religion by the chroniclers are full of interpretative comment that makes them far less reliable.

Exploring in the area north of San Ignacio, Father Consag undertook three formal expeditions. The diary of the first up the Gulf Coast to the mouth of the Colorado River in 1746 was included in the published version of Venegas.[31] The diary of his land journey of 1751 from near Santa Gertrudis up the West Coast of the peninsula has also been published.[32] The diary of his 1753 expedition up the center of the peninsula from the same place cannot be located. Clavigero, however, refers to all three, and the data recorded by this keen observer on the 1751 expedition are of such value, not only on the native subsistence economy but also concerning religious activities and the associated artifacts, that the discovery of the 1753 document is eagerly hoped for. From San Borja in 1766, Father Wenceslao Linck crossed the desert and penetrated the Sierra San Pedro Mártir. His manuscript diary was used by Clavigero, and a copy apparently exists in the Bibliothèque Nationale in Paris,[33] but I have only seen notes taken from it. It also appears to be rich in specific ethnographic detail.

Since some of the Indians from the Dominican Frontier survive, we are not so dependent on historic accounts for their ethnography, but if the historic data were good they would provide an excellent benchmark for studying cultural dynamics in an attenuated contact situation. Except for Sales's generalized account they are not.

The Franciscans and their soldier escorts (Serra, Crespi, Portolá, and Palóu) were just passing through from 1768 to 1773, and show only the most modest interest in the local Indians. The Dominican Father Mora, on his inspection visit of November, 1773, which led to the founding of Mission Rosario, had more to say,[34] but he can scarcely qualify as an objective observer of native customs. He hated them. Arrillaga's explorations in the interior of the frontier in 1796 contain useful comments, though his interest was in physical geography and its relation to mission logistics.[35]

The *Libros de Misiones*, containing records of Indian baptisms, burials, and marriages, constitute a final source of ethnographic data though, of course, an indirect one. Where baptisms were recorded with a Christian first name and a native last name, along with a ranchería identification, a mass of linguistic material is preserved. The fact that missionaries of varying nationalities, in sequence, had the task of . . . [giving written form to] the spoken sounds, and did so differently, actually offers an unusually good chance for a modern linguist to precisely identify the phonemes of the native tongues and sound shifts from one mission area to another. Unfortunately, only rarely can a named ranchería be located, but the number of baptisms from it affords an estimate of the size of that fundamental social entity. Birth and survival rate data can also be extracted from the earliest baptismal records.

The native personal naming tradition was specific—each name had two parts. A fairly limited number of first parts were always associated with one or the other sex. A similarly limited number of suffixes were not sex-linked. Did they refer to a clan affiliation? Libros from Mulegé, Santa Gertrudis, and San Borja all show variants of the same personal names, indicating a cultural as well as a linguistic unity over a distance of some 200 airline miles.[36]

I still have hopes of finding a competent modern linguist with historical interest who might be induced to work over with me this mass of linguistic material for what else it may yield.

Ethnographic Interpretation

Since the 18th-century ethnographies of Venegas and Clavigero cannot be regarded as fully adequate for modern purposes, what additional understanding has been and may be gained by this laborious examination of original sources? This question may best be answered by considering separately some examples of what we now have learned about the Indians of Baja California in each of several cultural categories, and the problems that call for further investigation.

Linguistics

From statements referent to language by the missionaries, surviving place-names, the short vocabularies and translations in Clavigero, the word list obtained by Gabb from the last survivors at Santa Gertrudis in the 1850s,[37] and from the personal names in the surviving mission registers it is demonstrable that from south of Loreto to beyond the United States border all the Indians belonged to the Yuman linguistic family. Further, in the Central Desert, from north of Rosario to south of Mulegé, the dialects were so close together that they can be placed in a single language, Cochimí. To the north and south there were considerable numbers of separate though related languages. As Massey pointed out, this distribution suggests a very long Yuman occupation of the northern three-fourths of the peninsula with the languages diversifying in situ.[38] Of theoretical interest is the fact that in the north and south, where the country was moderately productive, and fairly sedentary residence was feasible, the languages of these small social groups diversified widely just as did the enormously varied Indian languages of Upper California. The fundamentally peaceful nature of these societies may also be relevant. In the Central Desert, where the country was poor and food and water sources often failed locally, forcing migration on the Indians, enough intergroup contact developed to sustain linguistic unity over a very large territory.

The data are poorer for the Guaicura, Huchití, and Pericú languages to the south. The missionaries noted that they were distinct languages and not related to the Yuman. Baegert provides some material on the Guaicura,[39] but little indeed is known concerning the other groups—a few place-names and random words are all that have been found. Massey has postulated a Guaicurian linguistic family including all the southern groups, but the documentation is thin.[40] Since the geographic position of the Guaicurian family suggests very early arrival and long isolation, relating these southerly Indians to other groups would provide an exciting clue to very ancient migrations of peoples and cultures in North America. The eagerness with which the discovery of any valid linguistic material from these southern groups is awaited can readily be appreciated. It might come from personal names in a mission register, or, even better, from a manuscript letter or report from some missionary in the area.

Subsistence Economy

I have endeavored to exploit the documentary sources as completely as possible to get a precise understanding of how so many Indians managed to support themselves in the barren land of the Central Desert.[41] The story is an inspiring example of human ingenuity, in terms of observation rather than technology, in working out a fine ecological adjustment to the most difficult of environmental conditions. I believe that the main story is clear, though new bits of data may refine it. Peveril Meigs has worked out a similar account for the northern part of Baja California.[42] The story for the far south has not yet been extracted from the documents, and, since mission influence and the decadence of native cultures came there earlier, it may be harder to develop. The existence of additional and distinctive biotic resources in this region, and the postulated longer isolation and even greater technological primitiveness on the part of the people, create a special interest in the nature of their economy.

Social Organization

My own efforts to learn the nature of social organization among the Cochimí of the Central Desert were directed to an analysis of how the organization of individuals into family groups, families into rancherías or bands, and a number of rancherías into larger societal entities that cooperated, at least temporarily, for some purpose, was related to the ecological adjustment to the environment. The results were not surprising, confirming conclusions arrived at ethnographically by Julian Steward in the Great Basin[43] and by others in comparably severe environments. The simple family was the basic unit, but it was always related to larger social entities. These relationships were flexible so that their maintenance would not threaten survival in times of drought and famine. Then individual families would disperse and seek sustenance over the widest range possible. On the other hand seasonal or occasional conditions of easy food availability were exploited vigorously to maximize social contact among the greatest possible number of families within the same linguistic realm. Commonly these major gatherings were associated with religious or ceremonial activities. It seems ever more probable that such a pattern of social relations has characterized the human species for most of its history.

The data from the documents might also be interpreted in terms of social structure and function. Were marriage partners chosen endogamously or exogamously? Did clans exist in some form, with or without totemism, and did they encourage regular marriage exchange relations through cross, or parallel, cousin marriage? How did such social structuring relate to intra- and interband solidarity and cooperation? Occasional specific references, as to mother-in-law avoidance by Baegert, or to foods that were taboo to certain individuals or classes of individuals by Hostell or Consag,[44] can provide clues; so may the marriage registers among the Libros de Misiones, with the suffix patterns in personal names and the identification of the specific home rancherías of spouses. The ethnographic ore in the documents is far from being mined out for data concerning native social structure.

Religious Practices

No subject interested the missionaries more than the religious views of the Indians with whom they worked, and on no subject are their accounts and interpretations more suspect. It is hard enough for the most skilled, sensitive, and open-minded modern ethnographer to come to understand what an alien people really believes in a religious and metaphysical sense, and missionaries professionally are not open-minded. Folk tales and other texts in a well-understood native language would help, but neither native texts nor an appropriate knowledge of any native language is available from Baja California. We do have a few circumstantial accounts, largely from diaries of explorations, of particular ceremonies and the paraphernalia associated with them. Accounts of the actual behavior of shamans, both at curing rites and in public functions, also can be identified. Some understanding can be gained of how the shamans as a class of specialists, and the ceremonies themselves, functioned in sustaining the social organization at the ranchería and interranchería levels. This organization has its ecological and subsistence implications as well.

Culture Element Distribution and Survival

The age-area or geographic method is still perhaps unsurpassed as a means of reconstructing the history of nonliterate peoples. Basically it postulates that there have been centers of cultural innovation. These may have shifted through time but were likely to be in rather favorable localities at crossroads where intercultural contacts were likely. A new culture trait or pattern would spread outward, by migration or diffusion, from its place of origin. Often it would be followed by a new pattern which functionally had to displace the first. The outlying areas then would tend to retain the oldest patterns. Some aspects of culture tend to diffuse and displace each other more readily than others, so an isolated cul de sac such as the Baja California Peninsula may have kept certain practices that had been abandoned in most of the New World thousands of years before—some may have come with the first migrants from Asia. Conversely, of course, there is the possibility of cultural elaborations being lost by a people living in an impoverished environment. I shall merely mention a few examples of such ancient survivals in Baja California which have been identified from the documents, in some cases supported by archaeology.

The story on the atlatl, mentioned above, is clear. It survived barely into historic times in the far south of the peninsula. There, as it had been earlier in regions to the north, it was being displaced by the more efficient (though perhaps only slightly so) bow and arrow during the early 17th century.

The use of capes made of human hair by the shamans on ceremonial occasions is widely reported from the central part of the peninsula, and ethnographically even among the Kiliwa,[45] though in no more northerly tribe. These were highly sacred and important objects, which the missionaries went out of their way to destroy. Though not fully documented, we can speculate that such capes were once widespread but that they had to be given up because of the spread of the idea that if a shaman got hold of some bit of your body he could work you great harm by magic. This idea was only beginning to reach Baja California.

From Consag's 1746 trip up the gulf we have clear reference to the domesticated dog first appearing at Bahía San Luis Gonzaga.[46] This oldest of man's domesticated animals was evidently just beginning to spread down the peninsula.

The use of the humpbacked rock scrapers, which are extremely common archaeologically in the center and north of the peninsula, is clearly described in the documents. They were used to plane the flat boards or tablets that were of great import in religious ceremonies.[47] Such humpbacked scrapers are widely distributed in old and very old archaeological sites in the Western United States, for example at the Tank Site in Topanga Canyon.[48] Here is an explanation of the specific use for this mysterious but typologically clearly identifiable artifact.

The list is not complete, and problems of culture complexes that seem peculiar to Baja California remain. Where archaeological interpretation can be definite we may hope for explanations from that source. In other instances a documented report of how an artifact was actually used is needed. Finally, some customs of great interest leave no physical record. I shall close by alluding to the mysterious *maroma*, reported solely from Baja California in the area north of San Ignacio.[49] This is the practice of tying a bit of meat to a string, swallowing it, then extracting it and passing it on to your neighbor to treat in the same fashion until the meat is digested enough to fall apart. The custom clearly had strong implications in terms of social solidarity. I doubt that it is an invention of the Northern Cochimí. More likely it is a survival of an ancient and widespread social practice which with only moderate semantic stretching can be called companionate.

The Baja California Highway

You can get lost on roads which are travelled less than once a month. . . .
Beyond [Ensenada] the peninsula of Baja California stretches some
seven hundred miles, with roads which turn into unmarked trails
across one of the most extreme deserts in North America . . .
—Review of *Lower California Guidebook*, 1958

[This is] a region which has so far been screened by its bad roads
from the typical American tourist. . . . The advent of more visitors . . .
cannot help but change the character of the poor, proud, and to me
very attractive inhabitants.
—Review of *Lower California Guidebook*, 1958

Euʀᴏᴘᴇᴀɴ sᴇᴛᴛʟᴇᴍᴇɴᴛ of Baja California began in 1697 with the founding
of a Jesuit mission in Loreto. Until their expulsion in 1768 the Jesuits extended a chain
of missions over the southern two-thirds of the peninsula to Santa María, their last
one, founded in 1766. Their Franciscan successors, with far greater governmental support, given for geopolitical reasons, founded a mission at San Fernando Velicatá and
pushed on overland to San Diego whence the California mission system was extended. Baja California thus served as a strategic corridor to the frontier province up
which personnel, livestock, plant-propagating materials, tools, and church furni-
ture were carried. It was regarded as a more secure route than the one by sea against
strong northwest winds and a south-setting current. Briefly, from 1775 to 1781, another overland route from Sonora was used, but that was cut by the successful Yuma
Indian revolt.

In 1773 Baja California was transferred to the Dominican order, which mission-
ized the gentile Indians of the Frontier between San Fernando Velicatá and San Diego
and tended the declining older Jesuit establishments through the end of Spanish colonial times and into the period of Mexican independence. Records are less abundant
in the first half of the 19th century than in earlier mission times, but until after the
middle of the latter century there is no report of wheeled vehicles or roads for them
anywhere in the peninsula.[1] Transport was exclusively along mule trails, a network
of which came to connect widely spaced missions and other oasis settlements and
ranches. Less affected by accidental topography than roads, these trails run fairly directly between points of interest. In rugged, subsequently abandoned regions, as
around Mission Santa María, they can still be followed.

A backwash from the California gold rush brought a wave of prospectors into
Baja California, and by 1870 a number of successful gold, silver, and copper-mining

Figure 3. The northeasternmost shaft of Mina de San Fernando

properties had been located as well as a myriad of unsuccessful ones. For a time even high-grade copper ores were hauled as much as 50 kilometers to coastal landings on muleback, as from Mina de San Fernando near San Fernando Velicatá, to the coast at San Carlos (Figure 3).[2] The need for heavy equipment such as boilers and stamp mills, however, was an inducement to construct wagon roads to coastal points, and once they had been established other mines would tie into them. By 1910 the peninsula had a broken net of mine roads, especially in the Northern Territory.[3]

The development of irrigated agriculture in the Mexicali Valley, the accession of the powerfully independent and locally interested Governor Esteban Cantú (1915–20), and the advent of Prohibition in the United States combined to accelerate economic development in the northern part of Baja California. Cantú constructed engineered roads across difficult terrain from Mexicali to Tijuana and from Tijuana to Ensenada. Trucks and cars were available duty-free from across the border. Ranchers and farmers in the valleys and uplands north of San Quintín found or constructed tracks that were passable, at least in dry weather, in a widespread net.

In 1920 the geologist Carl H. Beal made an extensive reconnaissance of the peninsula for Marland Oil Company of Mexico seeking promising sites for petroleum drilling. The results of his work were not published until 1946,[4] with a map which includes his amazingly extensive itinerary, most of it followed by pack train. In January 1922, apparently at the request of the U.S. military district in San Diego, still interested in Baja California as a hangover from World War I, he prepared a 27-page single-spaced typescript entitled "Baja California–Route Studies."[5] In it he identifies all the sections a wheeled vehicle might traverse, noting some wagon roads that an automobile should not attempt. He concludes that an automobile might be able to travel from Tijuana to the onyx mine at El Mármol, though evidently wagons were used to transfer the onyx at that time. The road from Tijuana to Mexicali was established, and from it a number of passable tracks connected many of the ranches and mines on the relatively level plateau of the Sierra Juárez. The track south from Mexicali to San Felipe was passable at some times but carried so little traffic that someone stuck in the sand might die of thirst.

Farther south some disconnected roads from mine to coastal embarcation were noted. The most extensive set had been built by the El Boleo copper mine radiating out of Santa Rosalía. Only the one connecting that town with Mulegé, however, was passable, others having been washed out and not repaired. Finally, two passable roads led south from La Paz to Todos Santos on the Pacific Coast and to San José del Cabo at the tip of the peninsula. For both roads and trails he is meticulous in noting where water can always or only sometimes be obtained, commenting further on its quality. The uncertainty, even danger, involved in traversing the peninsula is implicit.

General and ex-President Abelardo Rodríguez, who became governor of the Northern Territory in 1923, constructed the first paved road, from Tijuana to Ensenada. Even earlier road construction began in the Southern Territory of Baja California with a road pushed to Magdalena Bay in 1921 and others southward to Todos Santos and San José del Cabo. With its widely scattered intensively cultivated oases, the Southern Territory's road building followed the classic pattern. If the terrain obstacles were not too severe, roads would be built to tie together the settlements, following the topographically easiest course, but accepting detours if minor settlements could be brought into the system. Though his economic resources were far smaller, the governor of the Southern Territory was able to tie Comondú to Mulegé in 1927, connecting with the system of the Boleo Copper Company which had independently laid roads south from Santa Rosalía to Mulegé and westward over the divide to San Ignacio.

The Automobile Club of Southern California and Governor Rodríguez, cooperating almost like sovereign powers, undertook to drive wheeled vehicles south from San Quintín to connect with the road system of the Southern Territory. In late 1926 an Auto Club group made it to Rosario,[6] and in 1927 a combined expedition of the Mexican military, including the governor, and the Auto Club drove to San Ignacio, then over the Boleo Company's roads to Santa Rosalía and Mulegé. Mining roads were followed where they existed, routing the track back and forth across the peninsula.[7] In 1928 the Auto Club installed its distinctive signs as far as Mulegé, noting mileages obtained in the previous years.[8] Random roadside vandalism, intensified by

Figure 4. Cabo San Lucas

Mexican nationalism that resents the foreign signs, has obliterated or removed all the signs where the road is still followed. A few survive in spots infrequently visited.

For trucks or well-equipped field vehicles the road was negotiable from Tijuana to Cabo San Lucas, but few tourists attempted it until after World War II (Figure 4). Onyx was hauled north from El Mármol and Cerro Blanco,[9] and in the decade of the 1940s shark liver buyers sought all coves where fishermen might put in. In the late 1940s out of season tomatoes were trucked from the Cape Region to the U.S. border; on a weekly schedule during the 1940s a 1932 Cadillac limousine carried mail and an amazing number of passengers from Tijuana to Santa Rosalía: some used passenger cars were driven from the duty-free border zone for sale in La Paz, and modest but growing numbers of adventurous American tourists pushed southward, many to write books about their experiences.[10]

In 1943 Ulises Irigoyen published in Mexico City a massive two-volume work on Baja California.[11] While it discussed the geography and history of the region in not too accurate detail, as its title suggests the book was primarily a strong appeal to the Mexican national government to build a paved highway the length of the peninsula. Such an enterprise would lead to economic development and strengthen the region's ties to Mexico. The effects of the work were slow in emerging, but when in 1972 the national government did build the highway the expenditures were justified on the same grounds (Figure 5).

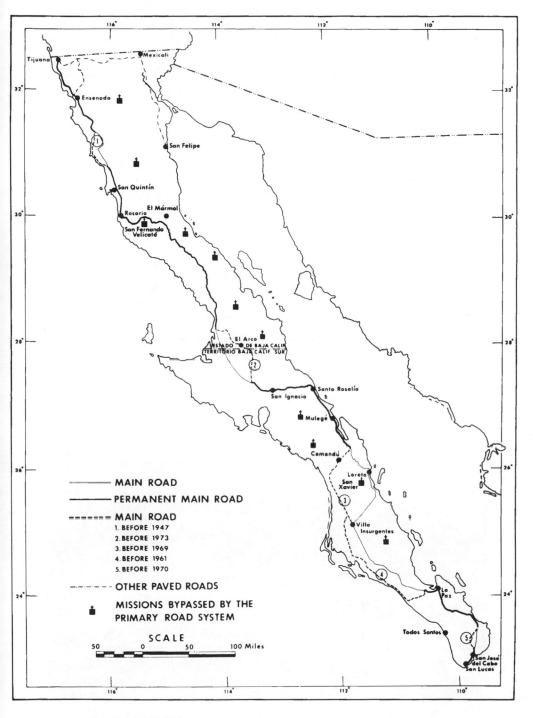

Figure 5. Baja California highway

During World War II the road had been paved south from Ensenada to Santo Tomás. In 1947 and 1948 a major project undertook to extend the paving to San Quintín. Grading was accomplished that far, but funds for asphalt pavement were exhausted at San Telmo, some 75 kilometers short. For 20 years the graded surface, becoming ever more washboarded and rutted, carried heavy truck traffic from the irrigation developments at Colonia Guerrero and San Quintín.

In 1956 a remarkable individual road-making achievement was carried out. Arturo Gross, a part-time miner, prospector, and mine promotor, and long a resident of the Laguna Chapala and Calamajué district was offered 10,000 pesos ($800) by the State government if he could drive his truck up the East Coast from Calamajué to San Felipe. Carrying a pick, shovel, and some blasting material he did it. Within weeks tourists followed with four-wheel drive vehicles. The northern part of the road has been improved, and now there are tourist fishing camps on the formerly completely uninhabited coast.

Curiously, it was the Southern Territory, with far smaller economic resources than the Northern State, that sustained the impetus of road building and improvement, both north and south of La Paz. Soon after 1950 a road was pushed south-westward from Loreto, until then accessible by road only from the north, to join the main peninsular road at Santo Domingo. This road made Mission San Xavier, the outstanding example of Jesuit mission architecture, accessible to tourists. A road was graded

Figure 6. Old highway southeast of Agua Dulce

northward from La Paz to Villa Insurgentes by 1954, and paving proceeded steadily to that point by 1961. For the next few years, repairing washouts caused by severe storms seems to have occupied the road-building resources of the Territory, but in 1968 a major program paved the road south to San José del Cabo. At the same time a project was instituted to complete a paved road north from Villa Insurgentes to San Ignacio, the most northerly oasis in the Southern Territory. A completely new alignment was chosen, crossing the uplands in an east-northeasterly direction to reach the Gulf Coast south of Loreto. Grading preceded paving, often by a year or more, but work progressed steadily and reached San Ignacio in 1972 (Figure 6).[12]

Extending the northern part of the paved road south from San Telmo did not begin until 1968 and in two years progressed only 20 kilometers, and in a year and a half more, to early 1972, made only a like distance, though surveying and grading for a modern road had begun beyond San Quintín. Suddenly the operation was accelerated; federal money became available, and two major contracts were let to grade and pave the entire 600-kilometer intervening stretch to San Ignacio, working from each end. Hundreds of trucks and graders and thousands of laborers were employed. Various stages of construction, from bulldozing a *brecha* to final hardening of roadside gutters in cuts, were carried on simultaneously over 100 kilometer stretches to hasten essential completion of the highway by the end of 1973 (Figure 7).

The heavy investment in the new highway is being justified by its attraction of

Figure 7. New highway under construction in 1973, looking east from east side of Aguajito grade. Note the steep embankments on either side of the highway.

vastly increased numbers of American tourists and the employment that will be created in providing them with services. The American visitors prior to the paving of the highway have been of two classes, the drivers who traveled slowly, enjoying the scenery and the nearly empty country, camping out and spending relatively little money; another group flew to luxury resort hotels, particularly for fishing. The Mexican government's planning assumes that with a paved highway the additional drivers will seek luxury hotel accommodations, and several rather luxurious hotel-restaurants have been established at formerly unpopulated sites as well as new hotels at established resorts such as Cabo San Lucas and Loreto.

The "Baja 1000 Rough Road Race" has attracted annually a further set of tourists, concerned to tear up the countryside rather than look at it. The hope that the paved highway would end this desecration of the landscape was vain. In 1973 the race was run cross-country on a newly staked out track. It has been continued with completely new lineation, but the course has been shortened to 500 kilometers.

Though it is only two lanes wide, less than 10 meters in the least-traveled middle of the route, the new highway was designed and built by modern engineers given free rein. Curves are broad and gentle, grades are moderate, and visibility is generally good. Since water for construction was always scarce and sometimes had to be hauled scores of miles, an ingenious, water-sparing roadbed construction scheme was devised. Crushed gravel, sand, and cement were mixed dry, spread and graded into place, sprinkled with water, and then rolled. The resulting surface is smooth and hard, though how it will hold up will be determined in years ahead. The final surface is oiled and covered with fine gravel.

Except where the highway is actually cut into a hillside, it runs on top of an artificial ridge more than a meter high and only slightly wider than the roadbed. To build this ridge, earth was scraped from as much as a hundred yards on both sides, destroying the vegetation, much of it unusual endemic plants, and leaving a scar that will remain for decades it not centuries. Protection against washouts rather than maintaining the wildly beautiful desert environment clearly had precedence in the engineer's plans.

There are almost no places that a car can be stopped safely, and getting off the ridge on which the road rests is difficult and even dangerous. Clearly the Baja California Highway will funnel tourists directly to the resort centers. Pausing to examine the extraordinary flora and the attractive desert terrain, the features that attracted the driving tourist of the past, is discouraged and often made impossible. One could drive to La Paz without being conscious of more than a long dull highway interrupted by a few settlements.

The alignment of overland transport routes in Baja California has changed in one rather consistent pattern from earliest historic times. The earliest mule trails and probably their Indian trail predecessors went rather directly from water source to water source. These streams and tanks were settlement sites, and in general are con-

Figure 8. Resort near Cabo San Lucas

centrated in the rugged uplands of the center and eastern edge of the peninsula. The mines which gave rise to the first wagon roads tended also to be in the rougher country, but they sought the shortest and easiest route to the coast, either Pacific or Gulf. The pattern of swinging back and forth across the peninsula that marks the original road for wheeled vehicles derives from two tendencies, the effort to utilize the mining roads whenever feasible and seeking lower and leveler land. Water sources and settlements were still connected if possible, but a number of oases that had held missions—San Borja, Santa Gertrudis, Guadalupe, and San Xavier—either long did without any road connection or were tied to the main road by long, poorly maintained side tracks.

The new highway continues this trend. The biggest shifts in alignment involve staying far out on the flats of the Vizcaíno desert almost to the latitude of San Ignacio before heading east to that point, thus bypassing the former mining and trading centers of Calmalli and El Arco, and following the Gulf Coast well south of Loreto before crossing the drainage divide into the Magdalena Plains. The mission oases of La Purísima, Comondú, and San Xavier are bypassed.

In its most recently completed sector, from Rosario to San Ignacio, the highway has been consistently displaced one to three kilometers west of the old road except west and north of San Ignacio, where there is a completely new alignment. All the tiny settlements along the old road that eked out a precarious existence serving tourists have been bypassed as have some larger ones. In some instances, their residents have been able to move to a new site on the highway, but this requires more capital than many possess. Further the new alignment, in contrast to the old, is not focused on hitting the infrequent spots where water can be obtained.

Finally, the long-term residents who have depended on tourists geared their services to the minimal requirements of the rough-road camper. The tourist whom the new highway is designed to attract will be served by new entrepreneurs from Mexico City, who will provide, at high prices, what might be found in an American resort (Figure 8). Profits are going to the investors and managers imported from the mainland. Mexico's problems of underemployment and her need to develop lucrative economic activities cannot be ignored. One can only hope that the benefits gained by the crassest touristic development of the wildlands and shores of Baja California will be worth it.

LATIN

AMERICA

Introduction

James J. Parsons

Soon after his initial and exhaustive examination of the historical geography and human ecology of central Baja California, Homer Aschmann directed some of his attention further south toward similarly margined arid landscapes and their occupants' adaptation over time to the limitations these landscapes imposed. His earliest stop was the first true desert encountered in the Americas: the Colombian Guajira peninsula, a New Jersey–sized land jutting into the Caribbean immediately east of the towering Sierra Nevada de Santa Marta, its tip, the northernmost point on the South American continent. The Guajira—one of those blank spots on the map of which little seemed to be known, at least in the English-speaking world—supported a tropical thorn-scrub vegetation and had sufficient rainfall to sustain only a minimal agriculture along its interior southern margins.

At Berkeley Carl Sauer, Aschmann's mentor, had negotiated an open-ended contract with the Geography Branch of the Office of Naval Research (ONR) that was modestly to fund field studies by students and faculty in a loosely defined "Caribbean" area. Although emphasis was to be on coastal studies, including beach and shore conditions, climate, and vegetation, the contract also left room for studies of human occupance. For Aschmann this offered a golden opportunity. Sauer had been greatly impressed by the quality of the observations and the sound and original thinking in Aschmann's Baja California work and, significantly, support was available.[1]

The parched Guajira peninsula, homeland of the Guajiro, the largest indigenous group in Colombia (which also included substantial additional numbers in adjacent Venezuela), fit appropriately into the ONR contract provisions. Despite its strategic location, the peninsula was one of the least-developed parts of Latin America. The Navy was interested in training students who might operate comfortably and effectively under difficult environmental conditions and without the accustomed conveniences of modern civilization. As he had demonstrated in German POW camps and while working in Baja, Aschmann was a "tough hombre" and thus fit the Navy's profile. The ONR was to support him for two summers of fieldwork, chiefly in Colombia but with a brief foray into the Guajiro barrio in Maracaibo, Venezuela.

Life in the Guajira centered on the wells and water holes, the numbers of which were being actively increased by the government's Provisión de Aguas division, on whose trucks Aschmann was to depend for transportation. Landforms and vegetation reminded him of the Vizcaíno Desert of central Baja, although the rainfall, highly variable, was somewhat greater in the Guajira, ranging from less than 5 inches in the Alta Guajira in the extreme north to 15–20 inches in the Baja Guajira around

Río Hacha. Regarding the difficulty of living on the very modest ONR subsistence allowance, Aschmann once commented to his mentor, "You don't understand, Mr. Sauer, how many *cervezas* it takes to get through the day in that heat." His initial reconnaissance report, made available in mimeographed form to a limited ONR distribution, was soon published in Spanish translation in the *Boletín* of the Instituto Colombiano de Geografía in Bogotá. It was a kind of baseline study in the growing literature on the area, and it formed the foundation for several subsequent articles he wrote on the Guajira.[2]

These articles elaborated on the themes of the cultural vitality and close identification with the uncompromisingly barren land of the Guajiro, who, in the manner of the Navajo Indians of the American Southwest, had adopted a herding economy based chiefly on goats and sheep sometime after the Conquest. He noted among these people such distinguishing institutions as bride purchase and matriarchal descent and inheritance, onto which had been grafted patriarchal property management and authority, presumably as an adaptation to stock raising. As in pastoral societies elsewhere, where wealth tended to be measured in stock numbers and where the rich were getting richer and the poor poorer, significant leveling mechanisms among individuals of different economic strata included the prestige associated with elaborate and costly funerals and animal loss from periodic drought and the drying up of water holes. Clothing, generally imported, was the principal external evidence of wealth.

Organized into some 30 clans, the Guajiro (Guayú, or Wayú, as they called themselves) had maintained their independence in part because the landscape, as in Baja, was unattractive to others and because of their resolute insistence on speaking their own language. The Colombian and Syrian storekeepers hoping to do business with them were forced to learn their language. These traders, Aschmann noted, commonly married wealthy young Indian women. Their mestizo children, reared in a matrilocal society, retained the language and culture of their mothers and formed a natural bridge to "outside" Colombian society, which facilitated the reciprocal process of acculturation. With the exception of Maracaibo's influence, the powerful pull of city life so important in our day was still not much in evidence. Proud, and with a highly developed sense of personal dignity, the Guajiro lacked the submissiveness and sense of inferiority of many American Indian groups, effectively defending themselves against the assimilative power of "white" society and even commanding its respect. Aschmann describes how in Valledupar, a mestizo town on the road to Santa Marta, stately Indian women in colorful mantas could be seen "walking down the street as if they owned it."[3] Here too comparison with the Navajo was invited.

Although the government's campaign to increase the number of wells and water holes was designed to improve economic conditions, Aschmann saw that the principal consequence of such efforts would quite certainly be increased pressure on an already degraded range, especially in the vicinity of facilities. The impingement of Colombian *colonos* on the better-watered southern and western margins was already beginning to disrupt the traditional transhumance that provided a safety valve for pastoralists in dry years. Today the majority of what has become the Department of Guajira, with its estimated 200,000 population, is no longer "Indian." Virtually all authority is in "Colombian" hands.

In the years since Aschmann's sensitively nuanced reports, outside influences have enormously impacted Guajiro society and the Guajira landscape. In 1977 the Exxon Corporation, in a joint venture with the Colombian government coal agency (Carbocol), initiated a feasibility study for the development of the vast Eocene bituminous coal deposits at the base of the peninsula. Since 1980, when the El Cerrejón project was declared economically viable, a fully integrated, $3.5 billion mining and shipping facility has been constructed that has made Colombia one of the world's leading exporters of coal. The low-sulfur, high-calorie product moves chiefly to European markets under long-term purchase contracts. Surface manifestations of this largest resource development project ever initiated in Colombia include a gargantuan open-pit mine, huge storage silos, a 90-mile railroad, dockside handling and storage equipment, a deep-water export terminal, company housing for many of the 4,000 employees, and a major desalination plant at Bahía Portete (Puerto Bolívar), where a harbor has been dredged to handle ships of up to 150,000 tons deadweight capacity.

The high-tech railroad, on which two 100-car unit trains roll night and day, had to be built to allow for flash floods caused by infrequent but very heavy rains in this normally arid region. Some 15 million tons of coal will be exported annually during the projected 23-year life of the partnership. Reserves are expected to exceed 3 million tons.

The potential and reality of the inevitable transformation of Guajira land and life resulting from this monstrous transnational enterprise has been partially documented in a 1984 report by Cultural Survival, by recent collections of essays published by the Instituto Geográfico Agustín Codazzi, and by the National University.[4] Although the better-watered, southwestern part of the department, including the vicinity of the mine and the capital city of Río Hacha, is already well "Colombianized," north of the Río Hacha-Maicao-Maracaibo road, except along the railroad and in the vicinity of the new port, the omnipresent herds of goats, the women in traditional clothing, and the typically Guajiran dispersed homesteads (rancherías) make it apparent that this is still Guajiro territory.

The notorious Colombian drug trade has also significantly affected Guajira life. The peninsula's long and unprotected coastline and open Venezuelan border (long contested for its petroleum prospects) have traditionally invited small-time smuggling of American cigarettes, liquors, fine lingerie, appliances, and other high-tariff items. Green coffee has moved to the offshore island of Aruba to escape the restrictions of the International Coffee Agreement. Aschmann observed nearly 40 years ago that the flowing dresses and ample mantas of Guajiro women could not have been better designed for carrying valuables across the international border, the Hispanic tradition of sexual propriety protecting them from search. But the drug trade, initially locally grown marijuana ("Colombian gold") from the Sierra Nevada and more recently cocaine originating from beyond the peninsula, has since much intensified the traffic, with dozens of impromptu airports reportedly scattered across the desert. The remoteness of the Guajira and its coastal location have made it an attractive base for this illicit trade, but the native people have been only marginally advantaged by it.

The contrasting fate of the desert dwellers of Baja California and those of the Guajira whetted Aschmann's curiosity about the differential adaptations and adjustments to aridity of these and other traditional societies. For example, in 1960 he was in the Canary Islands, as well as Madeira and the Azores, with his eyes wide open as always. His observations are summarized in a particularly thoughtful paper, "Divergent Trends in Agricultural Productivity on Two Dry Islands," published for the first time elsewhere in this volume.

He was particularly struck by the differences between the ingenious adaptations to extreme dryness by farmers on the neighboring desert islands of Lanzarote and Fuerteventura in the eastern Canaries. The moisture-saving gavia system of trapping occasional flood waters, the labor-intensive use of volcanic lapilli mulch, and the planting of vines in deep pits especially attracted his attention. The last two practices supported a thriving export agriculture on Lanzarote, whereas upland grazing associated with flood farming on Fuerteventura had left that island an eroded skeleton. In the Canaries the economic hardship generated by desertification has been alleviated by emigration and, more recently, *gran turismo.*

Aschmann's continuing preoccupation with traditional societies and their impacts on landscape is reflected in the more general papers here reproduced on Indian groups and subsistence patterns in Latin America. The perspective is always historical, with a comparative ecologic base, an approach characteristic of virtually every contribution included in this book.

Local conditions and experience are shown to have been diverse and critical. In the West Indies as in Baja California the population was exterminated, whereas in highland Mesoamerica and in the Andes at least some groups thrived. The regional variation in the toll of introduced disease was great. So was the assemblage of crop plants available to local populations and the extent of the adoption by native peoples of European livestock and ways of farming. Differing rates of acculturation and mestizization reflected different attitudes and approaches of both conquerors and conquered. Spanish administrative policies varied significantly from place to place. So did local environments, high or low, wet or dry, with soils youthful or mature and slopes steep or moderate.

En route to the Guajira in 1954, Aschmann stopped off in Honduras to visit the United Fruit Company's agricultural school at Zamorano. He noticed there that small farmers favored slopes over flatter valley lands and wondered if this might be explained by the greater ease in clearing the sloped land by controlled burning under digging stick or hoe cultivation systems. His "Hillside Farms, Valley Ranches," published in *Landscape* in 1956, is perhaps the most eloquent and convincing argument in this direction yet published.

In the late 1960s a *convenio* between the University of California and the University of Chile encouraging faculty exchanges gave Aschmann the opportunity to visit the littoral drylands of the Chilean Norte Chico and the Atacama. Here aboriginal survivals were insignificant. He was to focus his attention on constructing a historical model of the rise and decline of the individual mine, a kind of theoretical "think piece," its factual underpinnings deriving in part from Leland Pederson's *The Mining Industry of the Norte Chico, Chile*.[5] This project was an outgrowth of the conspicuous presence of abandoned mines and their appurtenances—nitrate, silver, and especially copper—in the northern Chile desert.

Individual mines, he observed in "The Natural History of a Mine," go through an established series of stages, each presenting a specific set of problems and opportunities to mine operators, the labor force, and society at large. The fundamental universality of the histories of mines and the transient character of this form of resource exploitation is underscored. The model, he suggested, might have relevance for national policy (e.g., taxation, labor relations, price controls, subsidies), particularly in a country like Chile in which regional disequilibrium is a chronic problem.

Later papers, such as "The Immortality of Latin American States," ask Why? Despite endemic political instability and enduring local boundary disputes, the nations of Spanish America have been fixed and immutable entities since the exit of Spain as a colonial power some 170 years ago. When compared to the shifting fortunes of the political entities of Africa or Europe, this seemed to Aschmann quite remarkable. Their durability, he argues, reflects complex political, economic, and ideological forces, including an intense nationalistic ardor typically inculcated by the school system, the sense that each has the "right" to exist, and the opportunity that each represents for politicians and bureaucrats.

In Paraguay, where he spent a more recent sabbatical, Aschmann returned to a continuing interest in language as a cultural force.[6] There he found bilingualism (Spanish and Guaraní) not only enduring but increasing in strength, and in a population that he judged to be no more Indian than that of Chile and less so than that of Mexico. Tracing the complexities of the history of Guaraní, a language made official in Paraguay in 1937, he emphasized the role of a 19th-century creole elite and the leaders of the dominant Colorado political party, with its base in the mestizo peasantry. For Paraguayans bilingualism was a valued identifying marker when facing the outside world, even a badge of honor to be worn with pride.

Whether in Paraguay or elsewhere, one can only admire Aschmann's way of wondering, of using his own eyes and ears to establish "ground truth," of making connections. His reasoning is consistently sound, critical, and original, and he is never afraid to pass judgment. "It seems to me" and "as I understand it" are phrases frequently encountered in his writings. There is no trace of sophistry in his work, only the honest ring of one who wants to know. His bias was that of the historical-cultural geographer, of anthropogeography, and he was familiar as few geographers have been with the anthropological literature and anthropological ideas. He thrived at the trough of inquiry, never hesitant to take a position on controversial matters. He was a credit to his profession, to all who call themselves geographers—and Latin Americanists.

The Cultural Vitality of the Guajira Indians of Colombia and Venezuela

When rains have been good, small plots of ground are cleared and planted to a variety of annual crops, but the typical field is likely to be abandoned after a year or two.

The parallels between the Guajira and the similarly vital Navajo culture are so patent that they do not need much elaboration.

THE GUAJIRA PENINSULA forms the northernmost extremity of South America, jutting northeastward into the Caribbean just west of the enormously oil-rich basin of Lake Maracaibo (Figure 9). Formal political control of the peninsula is shared by Colombia and Venezuela, with the larger western and northern sectors belonging to Colombia. Save for a few soldiers and officials, missionaries, and traders, the Guajira Indians constitute the population of the region, and they extend their settlements into the more humid lands to the south and southwest. In 1951 Colombia counted over 50,000 Guajiras; assuming accuracy within 10 percent for this census would be complimentary, but it is as likely to be high as low. There are some 40,000 Guajira Indians in Venezuela, about half of them concentrated in the Barrio Ziruma on the northern edge of the city of Maracaibo.

This Indian population appears to be increasing, and it is clearly maintaining its linguistic and cultural vitality. Isolation in an inaccessible region cannot be used to explain the persistence of this culture; this was one of the first areas in the Americas exposed to European contact. The homeland of the Guajiras, however, is one of unusual unattractiveness. Along a remarkably steep climatic gradient the climate ranges from a dry savanna at the southern base of the peninsula to a full desert at its northern edge. The short record available indicates an average annual rainfall of about 5 inches at Bahía Honda. This increases to about 25 inches at Río Hacha and Carraipía (Figures 10 and 11). As in all drylands the amount of rainfall varies greatly from year to year, and any sort of farming is precarious even in the most favorable localities.

The topography of the peninsula aggravates its agricultural disutility. The wetter southern and western base, the Baja Guajira, is almost perfectly flat; the hill ranges that get up to 3,000 feet in the Alta Guajira catch a regular mist from the incessant trade winds, but do not get enough precipitation to support streams to irrigate more than a few acres in the tillable adjacent lowlands. The most critical element in the geology of the Guajira has no surface expression. A fault line cuts the peninsula off from the better watered Sierra Nevada de Santa Marta and the Montes de Oca. Streams descending from the Montes de Oca sink underground in the zone of the fault line, and

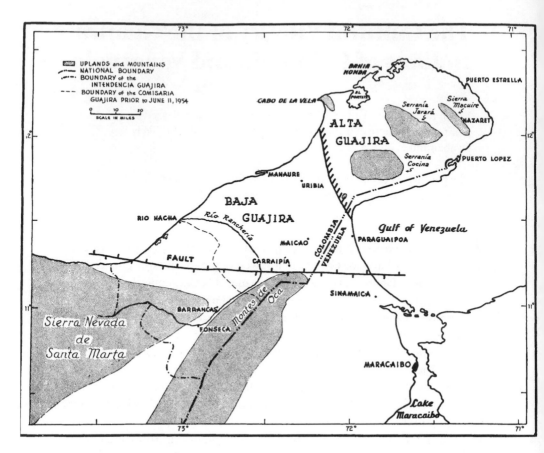

Figure 9. The Guajira Peninsula

the groundwater flows eastward to the Gulf of Venezuela, where it supports extensive coconut plantations; it does not maintain the substantial flow of underground freshwater that might be expected in the Baja Guajira. Although the water table in the Baja Guajira is generally high, all but the top lens of groundwater is highly saline, the salts apparently derived from the marine sediments into which water from the occasional heavy rains has soaked. A well used only to water stock is likely to turn brackish after a decade; one pumped for irrigation purposes will have a much shorter useful life. During the complete droughts that often endure for more than a year over the whole peninsula, the pits, or *jagüeyes,* from which water for livestock is drawn are likely to dry up or become salty, forcing complete abandonment of whole districts.

Most of the people of the Guajira are engaged in stock raising. Mixed flocks of goats and sheep, along with burros, range the entire region. Cattle are grazed around the uplands in the east and in the better-watered districts to the south and west, and

Figure 10. Clearing in Montaña, five kilometers south of Carraipía

Figure 11. Clearing in process two kilometers south of Carraipía

a wealthy Indian will keep horses and mules. There are both large and small stock owners, but, even for big stock owners, the grazing economy requires much labor during the dry season, since water for livestock must be raised from the jagüeyes in earthen jugs or five-gallon oil cans. An owner of great herds thus must maintain a large number of herdsmen for times when they will be needed to draw water. Members of this propertyless, dependent class are variously termed employees, retainers, or slaves.

When rains have been good, small plots of ground are cleared and planted to a variety of annual crops, but the typical field is likely to be abandoned after a year or two, because of the failure of the seasonal rains rather than the exhaustion of the soil. Few Guajiras can or do rely on farming as their primary source of subsistence, though there are a few dozen irrigated *buertas* at the foot of the Sierra Macuire (Figure 12). At the southern and western edges of the peninsula farming is a more reliable basis of livelihood, but it is more likely to be carried on by non-Indians; or the Indian who settles down to farm permanently tends to lose his cultural identity.

There are a number of small Indian fishing villages along the coast; the fishermen

Figure 12. Looking southwest at Sierra Macuire from northeast of Nazaret

use dugout canoes made from the great trees that grow in the humid district at the northern foot of the Sierra Nevada de Santa Marta.

The long coastline of the Guajira Peninsula has invited smuggling activities since the 16th century, and these continue to the present day, enhanced by the presence of the Venezuelan-Colombian border that cuts across the region. The numerous and surprisingly well-stocked stores of the area are only in relatively small part supported by the products of the region. Smuggled goods are an important item, and the traveler from either Colombia or Venezuela is likely to buy heavily of them. While immigrants rather than Indians are likely to administer the bigger mercantile operations, these prosperous storekeepers, often of Levantine origin, commonly marry wealthy Indian girls and assimilate into the Indian community. At a lower social and economic level, the characteristic dress of the Indian women, the flowing *manta*, could not have been designed better for carrying valuables across the international boundary, and doing so in both directions is a regular source of income for many Indian women; they are protected from search by the Hispanic tradition of sexual propriety.

Guajira social organization is a fully functioning system, sufficiently distinct from the typical Latin American pattern to merit comment. Inheritance and descent are reckoned matrilineally, and through his mother an individual belongs to a clan. Blood feuds are perpetuated between clans, and members feel a responsibility to protect fellow members from both violence and starvation. Property rights, however, are definitely not shared by all clan members, and come only from specified inheritance or individual earnings. A large clan, such as Epieyú, will include both wealthy and completely impoverished families.

Juxtaposed on this scheme is a strikingly incongruent system of bride purchase, with animals forming the preferred currency of exchange. Polygyny is perfectly acceptable if the man can pay for a number of wives. Bride price is a matter of pride, and the daughters of wealthy families must be sold high; one may command several hundred cattle. According to custom the husband gets his wife's cost back for the first daughter, but the price of all subsequent daughters goes to the wife's family. In marriage negotiations, however, two values are recognized which have a Colombian or Venezuelan rather than an Indian flavor. Virginity is as highly prized as among upper-class Spaniards, and its absence will force a bride price reduction. Whiteness of skin is so eagerly sought that if the family can afford it a girl will be kept in shaded seclusion from puberty to marriage.

The appearance of the Guajira population strongly suggests that intermarriage with both Europeans and negroes has long been going on. Commonly a man buys an Indian bride, usually the most expensive that he can afford, therefore one from a rich family. Children of such a union have established status in their mother's clan, and for many generations have generally identified themselves with it. Indianness of an individual is a cultural matter, marked by language, dress, following Indian marriage and inheritance practices, and self-identification. In their forbidding land the

Guajiras have been and apparently still are assimilating the more prosperous immigrants. Offspring of a poor Colombian who has purchased a cheap Indian bride are more likely to abandon Indian ways, but such individuals are regarded as lower class by all concerned, and are unlikely to be emulated.

Within the peninsula command of the Indian language, Guayú, is a requisite for all traders, and this tongue appears to be holding its own, acquiring non-Indian bilinguals about as rapidly as the Indians are learning Spanish as a second language. Symptomatic, and perhaps a causal factor in this situation, is the high status enjoyed by Indians even beyond the borders of the Guajira Peninsula. A leading citizen of the small but ancient town of Barrancas who had no readily recognizable Indian physical characteristics had no hesitation in recognizing naked Indians as his relatives, and when they visited his home they were treated as front door guests; if they got drunk and in trouble they were to be given both shelter and protection and avuncular chastisement. Guajira women visiting as far away as Valledupar put on their dress mantas and walk down main street as if they owned it. It should be added that elsewhere in Colombia, Indianness is definitely not something to be proud of; Indians slink into the edges of towns to trade, and persons conscious of their social position are shocked by any intimation of Indian ancestry.

Two rather unusual aspects of the Indian culture may assist it in maintaining a high social status in the Hispanic world that surrounds it. In the first place, the Indian belief in sexual exclusiveness, the premium paid for virginity at marriage, and the notion that illicit sexual activity is a tort that must be paid for in blood or money is an overdrawn caricature of upper- and middle-class Latin American attitudes on the same subject. Secondly, Guajira society is by no means egalitarian economically or socially. As in the rest of Colombia or Venezuela, there is an enormous social distance between the wealthy, most of whom have inherited their property, and the poor; and, rich or poor, there is no difficulty in equating an Indian with his coordinate class in the national society. Curiously, this equation is accepted on both sides.

There may be some partially, if not absolutely, causal connection between an economy based on nomadic herding and a society sharply stratified on an economic basis. Even if we assumed that initially all families held the same number of livestock, in an environment which is periodically subject to droughts and starvation some would preserve a larger fraction of their herd through these times of stress. Those who kept their herds in good shape could build them up faster and select for better animals more effectively in succeeding periods of rain and good pasture. In all subsequent periods of stress the owner of the superior herd would enjoy a cumulative advantage, and the disadvantaged herdsman would eventually have to kill his last goat for food and become completely dependent on his advantaged fellow. We find this pattern repeated again and again among the pastoralists of the Old World. The Guajiras, who adopted a pastoral way of life less than five centuries ago, and probably only in the last 200 years, provide almost experimental confirmation of this thesis about the re-

lation between a pastoral ecology and the development of an economically stratified social structure.

There is no intention to convey the impression that the Guajira culture is one of
idyllic benevolence and happiness, for such would certainly be erroneous. The individual in this society, either poor or rich, is placed under severe stresses. If he inherits important property, he should strive, under highly restricted environmental
conditions, to increase it. If he does not inherit property he is likely to suffer social
abasement, the inability to marry, and even acute physical hunger to be assuaged only
by gnawing on the stems of cacti. The strains induced by these social and economic
stresses show up in an enormously high incidence of drunkenness and both sober
and drunken fights. Among the many poor Guajiras who have emigrated to the city
of Maracaibo to work as unskilled laborers, the incidence of insanity, so severe as to
require institutionalization even by the tolerant local standards, is a concern to the
Venezuelan authorities. When they leave home the Guajiras are not timid and retir-
ing, nor are they adjusted to an acceptance of their lot. They are unhappy individuals on the make.

But in gross they survive. Perhaps such a culture which subjects its bearers to such
severe stresses is the only type that can survive in competition with the similarly
stress-producing and much larger cultures that surround it in this shrinking world.

The parallels between the Guajira and the similarly vital Navajo culture are so
patent that they do not need much elaboration. But one further instance may be
mentioned. A society in which the rich get richer and the poor poorer needs some
means of partial equalization of opportunities for consumption, a means provided
institutionally by the Navajo chant. The funeral rites for a rich man among the Guajira perform a similar function. These affairs, which may last two months or more,
seriously go about consuming a substantial fraction of the deceased's property, which
is mostly livestock. To these ceremonies everyone who has time to spend, even casual
passers-by such as I, is invited to butcher, cook, and eat a large fraction of the departed's legacy. Some animals will also be sold to provide liquor for the guests. The
heirs of a deceased *rico* would dishonor both the departed and themselves should
they fail to put on a funerary feast proportionate to the size of his legacy.

The Subsistence Problem in Mesoamerican History

First by his command of fire and then more intensively with that deliberate control over the biota of limited areas which we call agriculture, man has come to be a major semi-independent variable in the forces modifying the environment.

THOSE OF US who like to consider ourselves culture historians share a basic outlook almost without regard to the respective disciplines in which we were trained. The crucial element in this outlook, as I see it, is an overpowering desire to learn as much as possible about what actually happened in specific societies at particular times and places. Presumably this specified curiosity is reinforced by a desire to know why these events occurred, and, if possible, to generalize in terms of sequences in social development, but it is the emphasis on the initial curiosity which distinguishes the culture historian among the social scientists.

Because of this inexorable focusing on a specific historical problem we almost invariably find ourselves utilizing the data and techniques of many disciplines in an effort to throw light on some phase of human history. Generally, a problem as initially selected and formulated will start from somewhere within the more specialized competence of the investigator, but where it leads him is likely to be limited only by his imagination. As a geographer, I ordinarily begin by worrying about subsistence problems in a given locality. What were the methods used to get a living from the terrain, and how successful was the effort? The food supply obviously is the central and dominant element in this problem of gaining subsistence, though in certain situations, getting clothing, shelter, and fuel, or materials for essential tools may become important. This is not simply a matter of technology and labor applied to a given and static physical situation, since the environment that supplies these needs is subject to drastic modification by human action.

No implication is intended that an understanding of an environment and the means a human society has developed to extract a living from it is adequate to explain the unfolding of that society's history. Nonmaterial value systems, borrowed or developed internally, will color the society's cultural development, and these must be discovered and evaluated in their own terms. Nonetheless, the need to get a living from the land persists, and directly and indirectly governs a significant fraction of cultural behavior.

Mesoamerica and Its Physical Environments

There are few if any areas in the Western Hemisphere more suited to an investigation of the interaction of man and land than Mesoamerica, a region defined by cultural and historical criteria rather than by physical or geographical ones. Within this small part of the earth's land surface are represented numerous examples of a surprisingly large fraction of the physical environments to be found anywhere on earth. Temperatures in these latitudes are governed primarily by altitude rather than latitude, and Mesoamerican topography ranges from sea level to elevations too cold for permanent human occupation. Precipitation values are not well recorded instrumentally, but the natural vegetation would indicate that both in hot lowlands and cool mountain slopes there are rainfall values ranging from superabundant moisture to semidesertic conditions. The great range of lithic materials and complex tectonic histories together with the diverse climates have produced an enormous variety of slope and soil conditions.

It is important to emphasize the complexity and instability of a specific natural environment in terms of those features which are of special importance to man. Man is ultimately dependent on the biota of his environment for his sustenance. The specific components of this biota result from the interaction of biological evolution and the geographical dispersal of living things within the physical limitations imposed by climate, topography, and soil. In addition to climatic changes associated with the general circulation of the atmosphere and possibly due to extraterrestrial causes, modifications in topography can effect significant variation in local climates. The violence of volcanic activity in the immediate past at several localities in Mesoamerica assures us that these are forces to consider even within the span of human history. Above all, the character of the biota and its capacity to sustain a human population is dependent on soil, that thin veneer of inorganic and organic matter in which plants grow. The nature of a soil results from the nature of its parent rock, the climate in which it formed, the plants and animals which have lived and are living in it, and the time during which it has been exposed to the last two influences. This time factor is in part a function of erosion rate, a product of topographic slope, climate, and the protection afforded by the vegetative cover.

First by his command of fire and then more intensively with that deliberate control over the biota of limited areas which we call agriculture, man has come to be a major semi-independent variable in the forces modifying the environment. Ordinarily men select what they deem to be favorable localities and soils for agricultural activities. Then they proceed to replace the natural biota with plants and animals which are more satisfactory in meeting their needs. It is scarcely surprising that this action should often endanger the delicate ecological balance which initially made the area attractive for agriculture. If it is to provide more than transient benefits to human sustenance, an agricultural technology must include some means for preserving in symbiotic balance the environment in which the introduced plants and animals can continue to grow.

Mesoamerica is a region which has known agriculture for some millennia. At several times and places, considerable population densities coupled with complex sociopolitical organization have been achieved. There has been time for the impact of man on the environment, and possibly even that of the environment on man, to have produced observable effects. Since, despite a general cultural similarity throughout the region, aboriginal Mesoamerica was never a single cultural entity, the student enjoys the possibility of following a multiplicity of parallel histories, in each of which a society did its best to solve the problem of extracting a living from its environment.

The Subsistence Technology of Mesoamerica

The core of the subsistence economy throughout the region was horticulture, the cultivation of crop plants with the aid of hand tools. Hunting and the collection of wild vegetable stuffs also was practiced, but except in virtually abandoned or unoccupied areas these activities could provide only a tiny fraction of the sustenance needed for the existing populations. The popularization of [Paul] Kirchoff's term "superior cultivators" to apply to the complex civilizations of Mesoamerica promotes a misconception if it is understood to mean that farming techniques in Mesoamerica were notably superior to those practiced by the horticultural peoples to the north and south. It is in the other aspects of culture that the superiority, if that is the proper term, of Mesoamerica becomes evident.

Within Mesoamerica the apparent distinctions between highland and lowland, and wetland and dryland farming are based almost entirely on differences in the crop plants used in the respective areas. The effect of environmental limitations on the individual species and varieties is simple and direct. Among individual tribes occupying comparable terrain it is difficult to find differences in their horticultural techniques.

Specifically, farming consisted of clearing a plot of ground by fire. In the more humid areas trees were girdled so that dead leaves and branches could be burned. On slopes where a pronounced dry season existed simple ignition was all that was needed. Ground was broken with a digging stick and the various crop plants were seeded individually with cultivation consisting of heaping up a hill, again with the digging stick, around individual plants. Fruit trees might be planted from cuttings near settlements and protected and harvested for many years, but except in localities with unusually fertile soils, field crops occupied a given plot only one or a few seasons before a new clearing was made and planted. Only a tiny fraction of the arable land of Mesoamerica was worked on in anything other than this simple slash-and-burn, or *milpa,* system, a system characteristic of the simplest cultivation to be found anywhere in the New World.

Mesoamerican farming shows up in a notably favorable light only in terms of the extraordinarily rich assemblage of crop plants which it utilized. Nowhere in the world has any plant been developed which will match maize in terms of yield per acre, nutritive value, and ease of cultivation by hand methods. A vast number of varieties or races, adapted to nearly any of the diverse physical environments of the region, existed. Several kinds of cotton, beans, squash, and such vegetables as tomatoes and chiles similarly have proven their worth by spreading to other parts of the world. Agaves and amaranths are important foodstuffs, though they have not enjoyed similar worldwide acceptance. Going downslope into more tropical localities the basic assemblage was further enriched by tree crops of which cacao and avocado, perhaps the best known, are only samples.

The careful nurture of specialized varieties, of maize for example, adapted not only to yield well under various environmental conditions, but also to offer special taste, texture, and nutritional characteristics to the consumer must be the product of a long tradition of careful plant breeding, one dependent on the individual handling of seeds both in planting and harvesting, a system in sharp contrast to the broadcast treatment of grains in the Near Eastern and Mediterranean civilizations.

The exceptionally fine genetic material with which the Mesoamerican planters worked afforded them two advantages. In reasonably fertile soils, even though hand labor was squandered in cultivation, yields were sufficient to support an appreciable fraction of the population working at tasks other than farming, at least for a major fraction of the year. The surviving cultural monuments in many parts of the region provide ample testimony to this effect. Secondly, despite an acute scarcity of animal products in the more densely peopled districts, diets were remarkably well balanced within the vegetable foods locally grown. In general these were populations of great biological vitality, capable of explosive natural increase whenever reasonably favorable sociopolitical conditions prevailed.

In comparison with the horticulturalists of Central and South America, those of Mesoamerica showed a notable disinterest in root crops, though the latter were not entirely lacking. A possible result of this was a general inability to farm effectively in wet lowlands. The domestication of animals likewise was little advanced. Mesoamerica was far from having a true agriculture based on control of both animals and plants in the exploitation of its land resources. The point I wish to make in recounting these characteristics of Mesoamerican farming is that they apply as well to the Aztecs and Mayas as to what are regarded as backward groups within the region, the Otomí or the Mije. Except for their somewhat poorer crop assemblage, a poverty which is clearly associated with the environmental limitations of an extratropical land, they would also apply to the farming peoples of the Eastern United States.

Evidence is appearing that some additional farming techniques were applied at several localities, though to a limited extent. Whether the *chinampas* in the Valley of Mexico should be thought of as drainage or irrigation operations, they seem to have expanded the ability of the inhabitants to use a certain kind of land more or less permanently, but they were strictly localized. During the past decade there has been a notable surge of interest on the part of Mesoamerican culture historians and archaeologists directed toward identifying aboriginal irrigation systems. Both documentary sources and archaeological reconnaissance have been used;[1] the first approach has been singularly more productive, producing evidence of hundreds of irrigated spots in all but the wettest parts of Central Mexico. As far as I can determine from the literature, however, all sites that have been identified positively involved only a few acres and seem to have served to water gardens and fruit trees. At lowland localities it was cacao *huertas* [orchards] that were watered. This control of the water supply of crops certainly constitutes a notable technological advance, but up to the time of the upper class rather than making a significant contribution to the subsistence of a large part of the Mesoamerican population.

The aboriginal use of terracing has been noted at several localities from Central Mexico to Chiapas and Guatemala. Protection from soil erosion and increasing moisture retention are postulated as reasons for two distinctive types of terraces. These developments, though not carried out to such an extent as in the Andes, can be admitted as an improvement in subsistence technology, though it must be added that they did not seem to work, since most of the terrace remnants are to be found in subsequently abandoned areas.[2]

The Mesoamerican high cultures are recognized as such because of characteristics quite apart from subsistence technology. The social structures, religious systems, organized long-distance commerce, at least in the Aztec state, as well as military and imperial organization, were distinctive and impressive. Probably as an outgrowth of religious thought, there were great achievements in astromathematical science and art. In comparison with the other regions of North America where cultivation was practiced there were notable technological superiorities exhibited in Mesoamerica. Various forms of ceramic work, metallurgy, weaving and dyeing of textiles, and monumental stone architecture either existed only in Mesoamerica or had been developed there to a far greater degree of technical perfection. But in all cases, even in metallurgy, this satisfied only an aesthetic sense. I can think of no instance where the highly developed Mesoamerican technology in these fields actually contributed to the ease of life in Veblen's terms; supporting more people in more comfort with less expenditure of effort.

The highest civilizations of the aboriginal North American continent, while exhibiting impressive achievements in other phases of culture, made do with a food production system of a remarkably primitive sort. The complex superstructure of the social and politicoreligious organization of large, closely settled populations was capable of placing an overwhelming burden on their narrow base of subsistence technology. There seems to have been no effort to maintain or build up soil fertility by manuring, admittedly a difficult proposition without domesticated grazing animals.

Cultivating only with a digging stick a man could farm only a very limited amount of land, perhaps not much more than an acre. High per-acre yields were essential if he was to support himself and family.

Only especially favored localities could produce adequate yields on a more or less permanent basis. Some support to the soil was provided by the practice of growing nitrogen-fixing legumes in the same field, or even the same hills, with maize, and turning under the plants. But maize is a notoriously heavy-feeding crop and on most soils declining yields appear after one or a few years.

The Indian response to this situation is to abandon the field to brush, grass, or forest, letting it rest for some years before it is recleared and planted. If the cycle of regeneration is long enough, recovery of the soil occurs, giving good crops again for a short time, but this system demands extensive lands for small populations. If population pressure or land scarcity demands too prompt reclearing and planting there is less than complete soil recovery, and a need for still more rapid abandonment and subsequent reclearing.

Indian cultivation in Mesoamerica was and is concentrated to a remarkable extent in hilly upland terrain. One reason for this, of course, is the relative abundance of such surfaces in that part of the world. Burning, the principal method of land clearing, is also distinctly easier both to accomplish and to control on steep slopes. Finally, the dynamics of tropical soil formation and removal by erosion create a situation in which the soils on slopes are, temporarily at least, likely to be more favorable for farming than are lowland soils.

The mature or zonal soils of the tropics, the characteristics of which are governed primarily by the climate and natural vegetation under which they were formed, are heavily leached and so low in fertility that a single cropping may exhaust them for a generation. The Indians sought immature soils, alluvial or denudational, for their horticulture. Only the upper, steeper alluvial slopes were attractive, since the heavy clays deposited in flatter places are almost impossible to till with a digging stick. Where a denudational surface is cultivated to a moderate extent, the rate of erosion of the surface may be just in balance with the weathering of material from the parent rock underneath so that crops are grown in a continuously renewed and relatively fertile soil. The Indian system of planting in hills, in contrast with plowing, provides fairly good protection against erosion, making this balance easier to maintain. The agricultural worth of these denudational soils is determined primarily by the chemical and mineralogical characteristics of the parent material. Limestone, granite or other acidic rocks, and basic lavas are three widely distributed rock types in the uplands of Mesoamerica, each affording different farming opportunities to the cultivator. The basic lavas are by far the most attractive, and farmed in this fashion they may account for the long-maintained high population density in Western and Central Guatemala and in spots on either side of the east-west volcanic chain across the middle of Mexico.

Each upland soil, however, has a limit to the proportion of time it may be kept in cultivation without danger of excessive sheet wash, followed by gullying and complete destruction. With a highly organized society and a growing population shortening the fallow period, insidious sheet erosion, giving no forewarning by declining yields, may suddenly pass the critical limit, and fields be destroyed by gullies in a single season.

O. F. Cook noted another risk from alternate clearing by burning and then abandoning land in that under certain climatic and edaphic conditions a natural forest may become a grassland, a vegetative formation which digging stick cultivation cannot till.[3] In modern times such grasslands can be used for grazing, but the Indians had to abandon the district, perhaps for centuries, while the slow process of ecological succession replaced the grass by a manageable forest cover.

In certain environmental situations soil destruction, once it has passed the critical limit, is almost irreversible. On the dry hills north of the Valley of Mexico, the Teotlalpan examined by S. F. Cook,[4] soil erosion, apparently following deforestation, has stripped the surface down to the *tepetate*, a completely sterile surface which may persist for millennia. This is a clear case of removal of topsoil, since the tepetate is formed by subsurface lime deposition. The steep slopes of the dissected crystalline rock country in Oaxaca and Guerrero, though covered by some *monte* [brush], now carry practically no soil. In Oaxaca pre-contact aboriginal ruins would appear to refer to greater populations than the limited valley surface can now support.[5] The postulation of extensive farming on steep slopes that now hold no soil is reasonable if not proven.

On the other hand there were localities where Indian land use seems not to have resulted in serious soil depletion. The gentler volcanic surfaces in Western Guatemala are maintained in high productivity, though some of the steeper ones are ravaged with gullies. The basins of Mexico and Puebla retain much of their land capital, and the broad alluvial valleys of the West Coast were depopulated at the time of the Conquest while still productive and prosperous. In other cases, as in the limestone country of both dry Northern and wet Southern Yucatán, it is doubtful that human activity had much to do with the thin soils and relatively low agricultural potential of the area.

A Synoptic Look at Mesoamerican Culture History from the Subsistence Viewpoint

At some time in the distant past, a period we can call formative without regard for archaeological precision in defining the term, a number of societies were experiencing the cultural luxuries made possible by a food security resting on the horticultural control of crop plants. This could scarcely have been simultaneous in all of Mesoamerica, since considerable time was necessary to develop the varieties of maize, for example, so well adapted to each of the diverse climatic environments where

Cortés found it. But probably by the beginning of the Christian era or earlier, this point had been reached almost everywhere but in the humid lowlands near the Gulf Coast.

The crop assemblage of Mesoamerica consists of plants adapted to cultivation where distinct wet and dry seasons exist. This is strikingly true of almost all varieties of maize, beans, and cotton, for example, and suggests that the first steps in plant domestication were taken in the drier parts of the region. We cannot ignore the curious fact that the earliest archaeological records of maize cultivation keep coming from almost desertic regions north of Mesoamerica, though this fact is hard to adjust to botanical concepts of possible wild ancestors. It may of course be merely the result of the chances of preservation and archaeological discovery.

Growing populations; more complex social organization; trade, in ideas as well as goods, with nearby regions possessing distinctive environments and products; and the rapid elaboration of various crafts, arts, and the nonmaterial aspects of culture were characteristic of several local centers. Shortly, however, discordances develop. Subsistence becomes more difficult in certain upland areas, Oaxaca for example. The maintenance of the expensive superstructure of such things as religious temples is difficult for a society which must expend an ever larger fraction of its available labor in just securing its food supply. Conquest or attack on areas which appeared still to retain their prosperity; migration to as yet unexploited areas to skim the cream of their natural resources (perhaps Mayan entry into the northern lowlands can be thus explained); or abandonment of the most expensive aspects of a high culture and getting by in smaller, simpler, poorer social units, are all possible reactions. All are likely to have been represented in Mesoamerica's history. Probably only a tiny fraction of the later ones are recorded in linguistic islands, rapidly diffused pottery types, destroyed cities, and Indian legends and chronicles. The state of flux may have begun well before the beginning of the Christian era.

We can expect that most, though apparently not all, of the attacks of the poor on the rich were beaten off. But in any event the victors found themselves with a tradition of warfare, a tradition which seems to have generated ever increasing internal momentum, perhaps reaching its climax in Aztec and Tarascan imperialism. Probably most migrants found their new homes no better than their old ones, though the Maya found enough to create magnificent monuments in Yucatán before decay set in.

Historical tradition and archaeological discovery has demonstrated a definite retreat of the northern limits of the culturally defined Mesoamerican area in the precontact period. These abandoned northern areas are on the dry margin of the lands suitable for cultivation by aboriginal techniques. Similarly most of the great ruins which were abandoned before the Conquest, those of the humid Gulf Coast lowlands and amid the steep slopes of the Oaxacan mountains, are in areas in which the balance between agricultural adequacy and inadequacy due to depleted soils is particularly delicate.

Destruction of a high culture within a district by conquest is of course possible, but the invasion by barbarians from the north can easily be overestimated as a force in Mesoamerican history. In places where the environment remained attractive, as in the basins of Mexico and Puebla, another high culture soon replaced it. Where the ecological balance had deteriorated, however, failure to reestablish the attributes of complex culture and dense population often endured at least until the importation of a new agricultural system by the Spaniards called for a complete reevaluation of the agricultural worth of areas.

The disorganization accompanying this migratory, aggressive phase in Meso-american history had its culturally progressive as well as destructive aspects. To more intensive and extensive intercultural contact might be attributed much of the highest cultural elaboration by which the region is characterized. The Mexican West Coast from Jalisco to Culiacán affords a kind of evidence for this notion. The large population of this region has left masses of archaeological materials. Their horticultural techniques compare favorably with any in Mesoamerica, and in the broad alluvial valleys no major ecological disbalance seems to have occurred. Isolation beyond the barrancas on the western edge of the Meseta afforded some protection from the more aggressive groups to the east; no pattern of intertribal aggressiveness had developed locally. But the West Coast is now thought to have been a stagnant cultural backwater, lost until the present century after its collapse under the brutal blows of Nuño de Guzmán.[6]

The Aztec empire introduced a new economic complex, which may have anticipated some of the effects of the Spanish Conquest even though time was not granted for its full unfolding. This complex might be termed commercial imperialism. For cacao, dyestuffs, metals, and feathers if not for the cruder goods of commerce and tribute, it is difficult to interpret the collections for Tenochtitlán as other than an attempt on the part of the privileged class to enjoy the products of the diverse environments of an extensive region, a motivation not much different from that which led to the expansion of Europe in the 16th century.

There is reason to suspect that the cacao plantations exploited so promptly by the Spaniards, and so notoriously associated with depopulation and abandonment of the lowlands, were taken over from an already-established Aztec pattern of imperial exploitation. The conquest of Cortés is part of a continuum, not a completely new type of experience for Mesoamerica. The basic subsistence problems which faced the Indians a thousand years ago remain, colored locally by the characteristics of each distinctive physical environment.

Sporadic comments on post-Conquest land exploitation may be appended: introduced and newly virulent endemic diseases had much to do with the sudden depopulation of virtually all the wet lowlands, but a more precise definition of this event, perhaps the most drastic change induced by the Conquest, is appropriate. What part did specialized plantations of commercially sought crops play? The present century is witnessing the first significant reoccupation of the ravished lowlands. Partly this involves new commercial crops such as bananas and sugar cane, but a down slope expansion of subsistence farming, centered on maize and initiated with fire clearing, is of parallel importance. With 400 years of recuperation the wet lowlands may have returned to a physical state like that in which the Maya found them. It will be interesting to watch the way this environment reacts to these new agricultural impacts.

The biotic elements and farming techniques introduced by Europeans opened some new agricultural possibilities in Mesoamerica. Grasslands changed from barriers to cultivation to valued pasture with the coming of domesticated animals and the plow, permitting more extensive tillage and thus the use of drier, less-productive land, made possible the extension of agriculture northward in the gentle alluvial slopes of the Bajío. The use of wheat and barley, Old World crops with lower moisture requirements than maize, facilitated this advance. Though the stimulus of a market in the northern mines had much to do with the settlement of the Bajío, the region has settled into an economy directed toward local subsistence production. Modern irrigation works have created a number of oases of intense and possibly permanent agriculture on the dry fringes and north of the aboriginal frontier of Mesoamerica.

While creating new land use opportunities, the European farming techniques have introduced new damage and danger. The pasturing of goats and sheep on hill slopes has removed the last protection for the soil, extending the exposure of bare rock and tepetate, especially along the dry northern margins of Mesoamerica. Plow cultivation is probably more conducive to soil erosion than is digging-stick tillage, and in addition poorer and often more vulnerable lands are exploited. Finally, a growing urbanization is providing a market, and taxes and modern economic values are forcing the growers of agricultural staples to produce and sell a surplus beyond their own subsistence needs. This, in effect, is a continuation and intensification of the Aztec tribute collection insofar as its effect on land resources is concerned.

The growth of population in Mesoamerica in the past two generations has been enormous, exceeding both in rate and absolute amount anything which happened in the most favorable pre-Columbian times. Destruction of, or damage to, the resource base continues or is newly initiated in various environmental and cultural situations. Past experiences in the same or comparable localities, if they can only be discovered and understood, may warn of future dangers.

Hillside Farms, Valley Ranches: Land-Clearing Costs and Settlement Patterns in South America

[T]he still-forested hills are being cleared and taken into cultivation more rapidly than the undeveloped lowland areas.

EVEN A CASUAL comparison of maps showing population density and relief in tropical America shows a striking concentration of people in areas with considerable elevation and relief. The grosser aspects of this population distribution, such as the contrast between the Andes and the Amazon basin, might be explained in terms of the poor soils of the Amazon lowlands, but in local situations, subject to immediate field observations, such *simpliste* explanations are obviously inadequate.

The northern departments of Colombia—Magdalena, Atlántico, Bolívar, and Norte de Santander—include both broad alluvial plains, some covered with excellent soils, and uplands, culminating in the great bulk of the Sierra Nevada de Santa Marta. One can frequently observe uncleared forest or broad pastures in the lowlands with the adjacent hills cleared and planted in small fields of maize, yucca, and *plátano*, often on slopes greater than 50 percent. Paralleling this land utilization contrast, there are numerous scattered farmhouses or rural villages in the hill country and few if any human habitations in the lowlands. There are, of course, some farms in the lowlands and some uncultivated hillsides, but the impression the traveler gains corresponds perfectly with that presented by small-scale maps of population distribution. Closer observation often accentuates the impression. What appears from a distance to be forested ravine on a steep hillside turns out to be a carefully tilled *finca* with a mixture of several kinds of fruit trees and an understory of coffee or cacao.

Recognition of this concentration of people in the uplands of tropical Latin America is no new discovery. The disadvantages of hill farming, however, must have been apparent to any cultivator. Yields are normally lower, and more physical effort is required for the actual cultivation. The problem of soil erosion may for a long time escape the consciousness of the farmer, particularly if fields are shifted frequently, but its cumulative effect would tend to force him ultimately into regions of gentle slopes.

To the list of factors which may have encouraged the surprisingly intensive agricultural activity in the uplands, factors which range from simple climatic preference to more favorable soil character, I suggest the addition of another, the relative ease of clearing steep slopes. This is based on a physical phenomenon known to every fire fighter: forest and brush fires burn more intensively up steep slopes than on the level. Except in some plantation operations which, through large outside capital investments, control much heavy machinery, burning constitutes a major element in all land-clearing operations in the more humid parts of tropical America. It is substantially the only means to get rid of the great mass of vegetable matter which, whether growing or decaying, impedes any crop cultivation. Burning, however, involves much more effort than merely touching a match to some dry leaves or brush. Even during the dry season, a major fraction of the woody plants contain enough water to protect them from destruction by heat, and they will resprout immediately after the next rains, or as soon as any crop could begin.

According to my observations, cultivation in the tropics is almost completely restricted to humid areas in which much of the perennial natural vegetation is evergreen. O. F. Cook noted that in unusually wet years when burns were poor, root sprouting of the brush so seriously reduced crop yields that some isolated Guatemalan villages were threatened by famine.[1] On level surfaces it is rare that a fire will maintain and spread itself in the natural vegetation, even during the dry season. Burning up a steep slope, however, a fire will often maintain itself and generate enough heat to kill practically all plants.

These generalizations are, of course, subject to considerable variation in terms of the local conditions, especially the floristic composition of the vegetation and the length of the dry season, but the advantage of easier and more intensive burning will remain with the steeper slopes. The reverse advantage also pertains to the slope; that is, fires are more subject to control. Except in country which is probably too dry for farming, a fire will not burn down a slope. Thus, with minimal effort, the area to be burned can be determined as the fire is set, and there is less danger of damage from a runaway blaze than exists on level surfaces.

As I have observed it, clearing of level forested areas involves felling the larger trees and girdling the smaller ones. Branches from the larger trees and underbrush are cut and piled, and a series of bonfires set. After everything which will maintain a fire has been burned, the charred larger tree trunks and stumps still remain. These must be dragged to the side of the clearing before plowing can be undertaken. The accomplishment of these operations requires several months, and four or five men are employed on a 10- or 20-acre plot. A tractor may be needed for a few weeks. Such a clearing operation can only be undertaken by one who possesses considerable capital.

Clearing a three- or four-acre patch of forested hillside also begins with girdling the trees, but then a fire is set at the bottom of the slope and from a single setting generally will burn out the smaller trees and bushes as well as the leaves and smaller branches of the larger trees. When the larger tree trunks are felled they will roll or can be rolled by hand to the bottom of the now rather bare slope; there they are left. On slopes too steep to plow, dead stumps constitute little disadvantage. An individual can clear a plot of this size in a few weeks with no more equipment than an axe and a machete, and be ready to put in his crop for the next rainy season.

Where soil conditions and cultivation practice require abandonment of fields for several years after cropping, the greater ease of burning brush on steep slopes affords an accumulating labor-saving advantage. In dry situations a hill can be made ready for planting by simply setting a fire at its base. Even if the regrowth must be cut before burning, there is no necessity of gathering and piling the brush. Simply slashing it will be adequate, and again the investment of labor and capital is smaller for the steep slopes.

The result of this greater ease of preparing hill lands for planting is that these lands can be utilized by individuals possessing a minimum of capital. The ultimately more productive level lands cannot be exploited agriculturally except by the relatively wealthy: those who can maintain a labor force for several months in addition to the actual growing season. To gain the advantages of plow as opposed to hoe cultivation, further capital for removing stumps and for draft animals or tractors is essential. Thus, in large measure, the flatlands come to be exploited and owned in large units by a limited number of wealthy, often absentee, landlords, while the steep slopes can be cultivated in small plots by a numerous poor peasantry.

This interpretation may be applied in Latin America on the aboriginal as well as the modern level. Without entering deeply into the complex question of the ultimate origins of plant domestication and agriculture in the Americas, we can consider the economic problems facing a social group or tribe that was making the transition from a hunting, gathering, and fishing economy, to one relying on crop cultivation. The group might be either originating the new economy or, more likely, borrowing both the plants and the cultivation system from a neighboring group.

The almost invariable concomitant of a hunting and gathering economy is, if not an insecurity of food supply, at least a need to seek sustenance continuously, generally on a day-to-day basis. Such a society can be thought of as without capital. It could not afford to apply much of its labor to a task the rewards of which would not become available for many months. The fact that a slope could be cleared by fire and made ready for planting with only a few hours of labor would be a great advantage. Groups living in hill or mountain lands would be more amenable to converting their economy from one of gathering to one of crop cultivation, and there would be a tendency for lowland peoples to move into hill country as they developed an agricultural economy.

As a primitive society grew in cultural grace after the adoption of agriculture, the storage of surplus food could provide the capital necessary to maintain a labor force while it undertook the relatively difficult task of clearing more level lands, and to endure the longer wait for the great rewards which these lands would afford. It is now widely accepted, on the basis of considerable botanical and a little archaeological evidence, that irrigation agriculture in both the Old World and the New is a secondary development, possible only after agriculture, originated in the uplands, had made possible the accumulation of a capital surplus and social organization competent to undertake irrigation projects.[2] The same sort of accumulation, although of more modest scope, probably had to precede the clearing of forests in tropical lowland areas. If no disturbance intervened to disrupt the society and economy in the lowlands, their greater productivity would, of course, soon permit them to maintain a much greater population density.

While recognizing the present and past high density of population in the hill lands of the west coast of Mexico, Carl Sauer postulated that the base of agricultural settlement was in the most productive agricultural lands of the valley bottoms. He conceived that pressure of an expanding population formed the inducement to clear by burning and cultivate the neighboring hill and mountain slopes.[3] The thesis presented here amounts to an inversion of this hypothesis.

In these terms, the hilly upland regions appear as natural hearths for the development of incipient civilizations, and the more productive lowlands as colonial areas. A recent archaeological investigation of the Colombian highlands around Bogotá provides some pertinent evidence.[4] The cultural remains in the area of the historically famous Chibchan empire proved, under careful archaeological examination, to be both very scarce and very shallow. This was especially true in the high flat basins, the savannas themselves. The surrounding hill country showed far more evidence of occupation, although there, too, no deep middens were found which would indicate ancient permanent settlement. This, however, is to be expected if a shifting hillside cultivation was practiced. A reasonable interpretation of this archaeological evidence is that at the time of the conquest the Chibcha, having recently established rather highly organized kingdoms, were just beginning to move out of the hill country into the flatlands.

Under pressure of conquest and enforced migration, the lowlands which require capital, at least in the form of stored foodstuffs, to permit their exploitation, would be most vulnerable to social disorganization and abandonment. The hilly lands could serve as refuges. A family could, by burning, clear land and plant crops without great labor expenditure and could maintain itself by wild foods until the first harvest.

It is difficult to document these postulated retreats to the hills on the part of sedentary cultivators as they occurred in pre-Columbian times. The distribution of native languages in northernmost Colombia, however, with the more highly organized Chibchan tribes tending to be restricted to the roughest terrain,[5] can be clarified by such an explanation. There are better records of what occurred in the decades immediately following the Spanish conquest. Here we repeatedly encounter records of the disappearance of a high civilization and a dense population in lowland regions.[6] The territory in some instances has remained nearly vacant down to the present, or has been occupied only by small, deculturated, nomadic groups, such as the Lacandones in eastern Chiapas and northern Guatemala. In other instances, as in the lowlands of Colima, Mexico, or the coastal valleys of Peru, the land was so attractive that it was, after a time, developed by the capital of the politically secure conquerors. On the steep hills of the adjacent highlands the aboriginal population was able to maintain itself, racially if not culturally.

The above discussion is not intended to imply that other factors, such as greater freedom from introduced diseases and greater inaccessibility to slave raiders, did not play a part in protecting the Indian populations of highland as opposed to coastal areas. But the ability to reestablish agriculture on overgrown hill lands, when the capital of the group had been almost completely destroyed by conquest, pillage, and flight, must have played a significant part in the survival of agricultural institutions and considerable Indian populations.[7]

In much of modern Latin America the varying types of land utilization arising from the differential costs of land clearing are conditioned by and condition the economically stratified class system. The contrasts are particularly apparent in localities such as the Department of Magdalena in Colombia, where there are still extensive *baldíos* in the process of being cleared and settled, and much of the land has only been taken under cultivation within the last generation. Title to newly cleared land can be obtained very cheaply from the government by demonstration of occupation and development, though often the occupant of a poor hill farm will not trouble with the legal complexities involved in gaining a title.

The poor man looking for land to farm generally seeks it in the easily cleared hills. If he is almost completely without capital he must seek land within walking distance of his home village, and in the regions of older settlement only the poorest sort of land may be available. To pioneer in a new region some capital is necessary, but the man who intends to work his own farm will seek it on a steep slope where he can hope to clear and plant his plot in a single season.

The level lands are seldom approached by other than persons wealthy enough to employ a labor gang to fell and burn the trees and brush, and a tractor to drag off tree trunks, pull out stumps, and later to plow the soil. Normally an extensive clearing involving twenty to several hundred acres will be made in a single season, and commonly the sponsor will continue to live in town, overseeing the operation by visits. Often the labor force comes from a town several miles away and merely camps near its working place during the week. The land remains without permanent occupants.

In sparsely populated hill country, the poor farmers may abandon their cleared slopes after a few years as the crop yields decline when the first blush of fertility wears off the virgin soils. The ease of clearing a new plot by a similar burning operation makes this possible. As population increases, the cycle of return and reclearing must be shortened, with a progressive decline in soil productivity, but the reburning continues to be relatively cheap in terms of labor expenditure.

The disadvantages of this system are appallingly apparent to a visitor from the United States. The onset of gully erosion may be delayed for a few cycles of abandonment and reburning, but the stripping of the moisture-absorptive and fertile topsoil proceeds rapidly, and the farmers themselves are fully conscious of declining yield after slopes have been cultivated one or two years. In the most densely populated districts production may be so low that the entire population is threatened with hunger. Per-capita production is low, since cultivation and harvesting must be exclusively hand operations. The cost of food produced in this fashion is high indeed in terms of labor. A day's farm work will buy only three or four kilos of maize.

Against these economic disadvantages of hill farming, a social advantage must be weighed. The hill farmers, though poor and often without opportunity for formal education, characteristically constitute a vigorous, reasonably healthy, and self-reliant peasantry. They are conscious of their independence, and not subject to the hopelessness which afflicts the urban proletariat or the landless agricultural laborer of the lowlands, where sharp class distinctions exist. The hill farmer is not likely to lose his land to an absentee landlord, since the land yields practically no economic surplus which could serve as rent, but merely enough to maintain the working forces. In places where a valuable tree crop with high labor demands for its cultivation, such as coffee or cacao, can be produced, the hill farmer may gain a cash income, but in general, a rising standard of living for persons engaged in farming steep hillsides is probably impossible. To gain that result a large fraction of the labor resource would have to be employed where it could produce more.

Where a wealthy landowner has invested sizable sums to clear lowland forests a different cycle of land utilization ensues. The landowner's concern is not to increase total production but to increase profits and rent. Normally he will plant maize, cotton, or some other field crop to garner the high yields afforded by the initial soil fertility, but he will rarely fertilize and attempt to grow a staple food crop on a permanent basis. A choice is made. In the most favored localities, intensive plantation cultivation with heavy labor application and fertilization may be instituted. More commonly, even on good soils, the land is devoted to improved pasture and cattle raising. This accounts for the peculiar appearance of the cultivated landscape where the good flatland is used only extensively, and the steep hillsides are farmed with the greatest possible intensity. It must be admitted that this system has a virtue in the strictest conservational terms. The best soils are protected and perhaps even enriched by their utilization as pasture. It is the effect of the casual utilization of the lowlands on the economy as a whole, and the greater pressure on the upland soils which results, that are questionable.

There are multiple motivations for the practice of grazing cattle on the potentially most productive lands. The social prestige, which in Latin America has long been associated with being a cattle rancher, should not be ignored. But this kind of land use also offers economic advantages to the landowner. Since the labor costs in herding livestock and maintaining a pasture are minimal, and land taxes are very low, a system maintained by the political power of privileged wealth, a regular return to the landowner is almost certain. Fluctuations in international markets are of little

111

concern when the production costs remain much lower than the value of the output. Of course, per-acre productivity is low. Only a person with extensive holdings can net enough to maintain himself. The initial investment, both in the laborious clearing operation and in stocking the range, is so great that only the wealthy can afford to become ranchers.

The establishment of a commercial plantation involves a comparable investment concentrated on a smaller plot of ground. The clearing cost naturally constitutes a smaller fraction of the investment. Since the operating costs of a plantation are relatively high, there is a risk determined by the state of the market. Generally speaking, a person wealthy enough to establish either a plantation or an extensive pasture will elect the former only if there is prospect for tremendous profits. Such profits can come from specialty crops for export or items protected by high tariffs; sugar or tobacco often fall in the latter category.

The unfortunate aspect of this system, in terms of the economic and social values of the United States, is that the potential advantages of cheap and efficient production, resulting from large operating units working the best land, are not realized in cheap foodstuffs and a raised standard of living for the urban population. These large units produce specialty crops, either for export or restricted by price to relatively wealthy natives. It might be added that the wages of agricultural laborers are pitifully low. As the hill lands that a man with little or no capital can clear and farm are filled up and then have their fertility exhausted, or their soil lost through erosion, the surplus hill population has the poor choice of going to the lowlands to seek farm work or to the cities, unfitted for anything other than unskilled labor. With the high cost of staple foodstuffs, life in the cities is scarcely attractive for unskilled laborers.

The circle is extremely difficult to break. The low productivity of the hill farms makes it essentially impossible for the owners of even the best of them to amass enough capital to undertake the clearing of anything but other easily burned hill slopes. Some of their sons do this and the still-forested hills are being cleared and taken into cultivation more rapidly than the undeveloped lowland areas. It is to be expected that even this release will no longer be available in a generation. The low-paid agricultural laborer has even less chance to escape his economic class. At best he might attempt to clear a hill slope. He could never gain enough capital to undertake the more expensive operation of clearing and farming a lowland tract, and, if he tried, might run into expensive legal contests for which by education he is in no way prepared, before he could gain title to his hard-won land. This hard-to-clear land is, so to speak, reserved for the rich, and barring a revolutionary change in national social policies is likely to remain so.

Mexico stands as something of an exception, and from this standpoint it is perhaps easier to appreciate the depth and scope of the great Mexican revolution. In recent decades, with government assistance, there has been rapid settlement in the formerly nearly unoccupied wet lowlands of Mexico.

In conclusion, it must be recognized that differential clearing costs are not the only cause of the inordinately intensive use of the hill slopes of tropical America. On the physical side, the greater amount and regularity of precipitation in the uplands is particularly attractive to the poor farmer, who might literally starve if he lost a crop due to drought. The generally superior health conditions and the more comfortable climate are real advantages possessed by the uplands. On the other hand, the ease of clearing seems to me to be very close to a basic condition determining the areas selected for the first agricultural developments. The other advantages, the forces of vested interests, and the inertia of populations associated with the affection which rural people feel for their native districts, tend to perpetuate the population concentration in the uplands, despite the apparent economic advantage that would attend a reverse distribution of population and agricultural enterprise.

The Natural History of a Mine

The mine is inherently transient, but it may have the option of promoting future development or creating a permanent desert.

The stripping of the Norte Chico of its woody vegetation to sustain widely scattered smelters in the early and middle 19th century is still perceptible in the landscape.

THIS ESSAY MIGHT be classed in the realm of *geographic economics* rather than *economic geography*. Considerations that are eminently geographic, that is, are tied to the peculiar characteristics of particular localities, are examined. However, the models presented are designed primarily to inform economic policy, specifically policies that would exploit local mineral resources to achieve economic development of economically underdeveloped countries or regions.

Gold is where you find it. For thousands of years efforts have been made to develop a theory that would predict the occurrence of concentrations of desired minerals in sufficient intensity to make their extraction economically feasible. Geologic knowledge certainly has increased, and there have been greater improvements in prospecting techniques which can be applied intensively in geologically promising zones. The discovery of a great mine, however, continues to be a largely fortuitous event, although once a major mineralized district has been identified, exploring it to its limits can be a more ordered investigation with generally predictable returns. It would seem that the processes of mineral implantation and secondary enrichment, occurring sporadically over vast periods of geologic time, that occasionally lead to singularly intense concentrations of one or more rare and valuable elements or minerals will long remain mysterious. The isolated prospector with his burro, pick, and pan, and the elaborately equipped geophysical- and chemical-prospecting team perform the same economic function in essentially the same manner. They hope to bump into something good, more or less by accident. Their professional skill consists of a capacity to recognize what they have encountered.

General Characteristics of Mineral Deposits

Once a mine has been discovered, its development and exploitation, leading to eventual depletion and abandonment, seem to follow an inevitable course, though the size and richness of the deposit will determine the duration of its several stages. The determination of this course is significantly conditioned by a number of physical characteristics that apply to essentially all economic mineral deposits.

*   *   *

a. The economically workable mineral concentration is finite. Its establishment frequently took place in the distant past under geologic conditions no longer present, or in the cases where mineral implantation is continuing, it almost always occurs at a rate so slow that it must be measured on a geologic time scale. Exploitation is destruction, and at a foreseeable point in time the mine will be exhausted. Mining engineers speak of exploiting a wasting resource though they have developed a remarkable capacity to avoid thinking too much about the bleak ultimate end of the operation in which they are at any time engaged.

b. Mining itself consists of high-grading a mineral deposit. The richest veins or deposits that can be identified will be worked first, and subsequent operations will inevitably deal with poorer ores, continuing as long as those that remain still can be worked profitably. The relatively small size of the mining claims that traditionally have been allowed by states as diverse as Britain, the Spanish Empire, the United States, and the modern Latin American nations has tended to reinforce this pattern of high-grading. Heavily capitalized mass-mining procedures capable of exploiting vast bodies of low-grade ores, notably copper, tend to weaken it. But it may be suggested that the high-grading tendency is based on a fundamental desire to economize on the effort invested in achieving the identifiable end, the production of a valuable mineral.

c. A mine starts at a point on the surface. As material is removed the pit shaft, or tunnel, deepens and the extraction process becomes progressively more expensive. Transport and lifting costs always increase; in underground workings the protection of shafts and tunnels from cave-ins and increasingly complex ventilation and drainage problems multiply costs.

d. The most superficial part of any mineral deposit is likely to be the richest for two reasons. (1) This is where the mine was found. A portion of the earth's crust was sectioned by erosional forces, and at one or more points of exposure the special mineral concentrations could be located. It is statistically probable that a deeper probe at those points will enter materials more closely resembling the surrounding country rock. There have been famous exceptions. Great pockets of extraordinarily rich silver and copper ores were encountered underground at Chañarcillo, Tamayo, Carrizal Alto, and elsewhere in the Chilean Norte Chico, often associated with the intersection of veins of differing orientation or with veins that cut through different sorts of rocks.[1] (2) Further, the natural processes of ore enrichment seem almost invariably to be most effectively beneficial near the surface. In silver and copper ores, for example, those near the surface may be oxides while the deeper ores are sulfides. The former are commonly of higher tenor and are always easier to process from a technical standpoint. Similarly the mechanical processes that concentrate gold in placer deposits occur only at the surface, although sometimes such deposits have been buried through subsequent geologic activity. Thus, even in the largest and richest of ore bodies or mineralized districts there is almost a certainty that in its later stages a mine will be working progressively poorer ores.

General Characteristics of Mining Operations

In the mining operation itself and in the processing of ores, technology has not been static. Especially during the past century there has been an enormous increase in the capacity to handle lower-grade ores, and with few exceptions, notably aluminum, the chemical character of the native mineral in which a desired element is infixed has ceased to be a barrier to its exploitation except in a limited, cost sense. The trends in technological development have held a consistent direction, all leading to a notably heavier capital investment in each operating mining unit.

The mining operation itself falls into two parts: an extractive or rock-breaking one accomplished with explosives, bars, or in soft ore simply with shovels; and transporting the loosened material to the pithead, concentrator, smelter, or market. The latter phase is in general the more costly. With increasing mechanization and growth in the size of machinery used, it has become possible to handle material at notably lower costs per ton. In the United States, for example, it is economically feasible to extract copper ores with as little as 0.5 percent copper content, and the Gran Minería, or great mines in Chile, are utilizing copper ores with a tenor of less than 1 percent. Mass mining of this sort is feasible only in huge ore bodies, from which the enormous capital investment can be amortized. It is also capable of using up such ore bodies in relatively short time spans. Relatively speaking, narrow, rich veins are no longer essential, and vast rock masses with widely and thinly disseminated values can for the first time constitute a mine. In some instances, although the rich veins have been exhausted, the surrounding rock holds enough value to make it exploitable in mass.

Progress in massive chemical treatment of ores, again involving heavy capital investment reinforced by the declining tenors of the ores being worked, has resulted in a tendency to carry on an increasing amount of concentrating, smelting, and refining as close as possible to the point where the ore is extracted. Again the size of the capital investment demands an enormous ore reserve for its justification. In some instances, notably in the Chilean Norte Chico, it has been possible to afford certain economies of scale by placing a large refinery or concentrating plant in a convenient location amidst a number of small mines.

With the exception of coal mines with great reserves, which have for nearly two centuries demonstrated their ability to attract permanent industrial developments to the immediate vicinity of their pitheads, most major modern mining activities are carried on in otherwise little-settled and relatively isolated districts. An element, of course, is that the ore bodies closer to population centers, such as the famous silver mines of Lavrion in Attica, have long ago been depleted if not worked out. The three major copper mines of Chile are quite typical: one, El Teniente, is in the Andes and the others, Chuquicamata and Potrerillos (El Salvador), are far in the desert north. In such cases major additional investments in living quarters and services for the labor force (in dry regions an exotic water supply often must be obtained) and transport facilities capable of bearing heavy loads must be introduced, often with the sole justification of supporting the mining operation. In some degree these investments too must be amortized by the product of the ore body before it is depleted.

Mineral Values and Prices

It is the price at which a given metal or other mineral can be sold that ultimately determines whether a deposit containing it is a minable ore or just so much rock. Forces acting in opposing directions over the past few centuries have created the modern unstable tableau of mineral prices. Increasing industrial demands for minerals and the rising costs of extraction associated with depletion of mines, lower-grade ores, and increasing inaccessibility of unworked deposits exert an upward pressure on prices. Improved mass-mining techniques along with modern chemical and physical means of treating ores have tended to reduce them as have important new discoveries. A growing technical capability to substitute a number of materials for a particular use is one of the few forces that would tend to stabilize prices.

An examination of the resultant in real prices of the action of these conflicting forces on each of the significant minerals of commerce is beyond the scope of this study, but a few generalizations seem possible. Trends in mineral prices have been highly variable over time with the most heavily used substances, iron and the mineral fuels, showing a degree of price stability as do the precious metals used for monetary rather than industrial purposes. Price instability seems to reach its greatest intensity in those minerals of just less than the maximum intensity of use, and peaks perhaps in copper.[2] As a given mineral, often a metal, increases its industrial importance it is likely to move into an era of increasing price instability, not simply one of rising prices. The development of uranium mining in the past 25 years affords an excellent example. It would seem fairly safe to predict that the same pattern of price instability, at least in many minerals, will continue into the future, occasionally yielding a bonanza, but more characteristically plaguing the mine operator as he tries to plan the exploitation of a deposit. In particular, capital investments will be amortized as fast as possible, reinforcing all other tendencies toward high-grading the deposit.

Graphic Expression of a Mine's History

Despite their almost infinite diversity in terms of kind of material produced, size and richness of the deposit exploited, and accessibility, individual mines go through an established series of stages, and each stage presents a specific set of problems and opportunities to the operator, to the labor force, and to the society at large. In their simplest form these stages are illustrated in Figure 13.1. The scales on this model for both money and time are, of course, highly variable, and the variation is not always parallel in the two elements. A major fraction of the mines of history have passed through the first three stages and entered the fourth; often the abandonment of the mine completes this stage within a single decade. In the past 150 years very few mines have persisted as long as 50 years without passing well into the fourth stage, although in earlier times when man had less inanimate energy at his disposal, the depletion of an ore body could proceed less rapidly and the third stage might endure for cen-

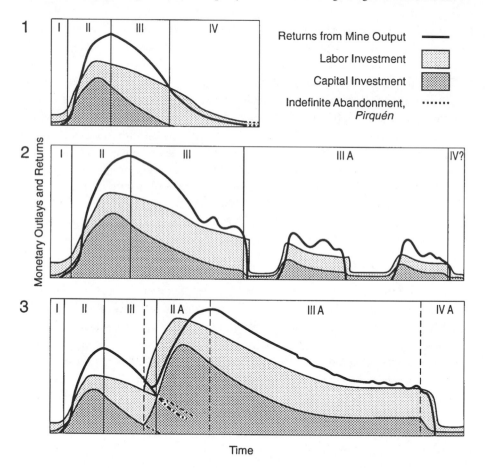

Figure 13(1–3). 1—Stages in mine development: (I) prospecting and exploration; (II) investment and development; (III) stable operation; (IV) decline. 2—Mine development influenced by prices: (IIIA) intermittent operation governed by prices. 3—Mine development with renewed investment: (IIA) renewed investment; (IIIA) secondary operational; (IVA) cessation of operation or abandonment (in stages III and IIA the dashed lines represent prospective developments if new investment had not occurred).

118

turies. The history of modern mines of greater durability would have to be described by more complex models as in Figures 13.2 and 13.3.

For small deposits of a high value, easily processed mineral such as rich pockets of placer gold, curves of different slope might better describe the first two stages. In such cases the amount of capital investment would be relatively less in relation to labor investment, and really profitable returns might be generated from large gold nuggets almost from the instant of discovery. Conversely, when mass mining of low-grade ores is involved, as in modern large copper mines, there may be no returns at all until after substantial investments of both labor and capital have been made, i.e., well into Stage II.

Some characteristics of each of the stages may be noted.

Stage I

Until recently the prospector's wearying efforts were sustained largely by hopes of vast rewards that were seldom realized. His grubstake was painfully amassed through work in other enterprises, or might be supplied in modest form as the most risky of investments by an individual or corporate backer who would receive most of the value of any discovery that might be made. The moderately high labor input shown in the model represents the innumerable prospectors who found nothing, ore samples scratched from the ground that never paid their assay charges, and incipient mines that were never developed beyond this stage. Modern prospecting teams, usually salaried and with corporate backing, with substantial geophysical and geochemical equipment would show a considerably higher level of capital investment. Sample core drilling or comparable activities to determine the extent and quality of the now-located deposit are properly placed in the latter part of Stage I, and are represented by increasing capital investment.

Stage II

At the beginning of this stage someone has decided that the discovery represents a mine with profitable prospects. It may be the discoverer, his backers, or an investor who has bought the rights. Once the decision is made, capital investment and the importation of a labor force is likely to proceed as rapidly as is technically feasible. Not only mine workings such as shafts and tunnels are involved. Rock-crushing, concentrating, and smelting and refining plants are often installed (Figure 14). Except in those ever rare instances in which a new mine is developed in an already settled district, there must also be a substantial investment in an infrastructure not directly related to mining itself. The major elements in this infrastructure are a transportation system, a factor that clearly enhances the mine's productivity, and living quarters and social services for the labor force (Figure 15). The latter may be financed directly by the mine or indirectly through relatively high wages, but in any case they

Figure 14. Ore-sorting operation at base of hill below El Tofo, 20 miles north of La Serena-Coquimbo

Figure 15. Company town of El Tofo

are a proper charge against the mine's output. Commonly this investment comes relatively late, and continues well into Stage III in the mine's history.

With a heavy input of capital and labor, mineral output increases rapidly, and if judgments as to the extent of the deposit are accurate, quickly rises to highly profitable levels.[3] Stage II terminates at the point where the mine reaches its highest level of profitability. Although gross output may remain high, and even rise, factors come into play which cut down on net to the mine's owner or operator.

Stage III

Where a major deposit is being worked this stage of stable operation may endure for some time, but three factors tend to reduce its profitability. Inexorably the mine deepens or extends, increasing the cost of bringing out the mineral, and with only rare interruptions the tenor of the ore tends to decline. The state or society begins to extract a growing toll, either through taxes or in requirements that the mine construct and maintain social service facilities for its labor force and often for the region as a whole. Finally, this is the period when the labor force is in its best position to claim an ever larger share of the return from a mine's output. In antiquity, when miners were truly slaves, this factor presumably did not operate, nor did it under the *encomienda* system in Spain's New World empire. The inability of this labor to improve its returns permitted long-continued production of the gold from basically poor placers in Chile, the West Indies, and elsewhere in America during the 16th and early 17th centuries; the Indian's only defense was to die.

Even before laws favoring labor organizations were enacted during the last few decades, the stage of the stable operation of a mine was the one in which the worker could most effectively express his demands to the mine owner or operator. A fixed investment was in a place that could seldom be moved or oriented to any other productive activity, and using it in the mine was still profitable. Originally of diverse origins and sentiments, the workers had come to feel an identity of condition and purpose that could sustain a struggle for economic power. The history of repression on the part of management and destructive violence on the part of labor is well known. It is suggested here that as long as the individual mine continues to be profitable, i.e., remains within Stage III of its development, the labor force will be able to increase its share of the return. Even a policy of repression of workers has the effect of increasing labor costs through the employment of more guards and related personnel. It is when the individual mines approach the end of Stage III that the struggle reaches its bitterest point. Examples are numerous and come from many places. Coal mines in Harlan County [Kentucky] or Lota [300 miles south of Santiago], or the Chilean nitrate fields in the 1920s show similar patterns. And inexorably the mine becomes more costly to work both because of its depletion and physical extension, and the heavier labor charges.

Stage IV

At some point the mine ceases to be profitable, though commonly it does not immediately cease production. Significant new investment, however, does cease. The later operation of a mine may be financed in part by the destruction of works that were necessary for its continuing operation. In underground mines, pillars of ore that protected the tunnels might be extracted along with shoring for use elsewhere underground; thus extensive areas are made inaccessible because of danger of roof collapse. Machinery of all types is simply worn out rather than maintained and replaced. These activities represent the salvaging of a small fraction of fixed investment that otherwise would only be abandoned. Occasionally a heavily financed corporation with other profitable operations will close a mine suddenly without running it down, hoping to reopen it in good condition at some future time when price conditions are more favorable or a technological development reduces costs. Clearly only a major, long-lived corporate enterprise can justify such actions; ghost company towns such as Jerome, Arizona, or many abandoned nitrate *oficinas* in Northern Chile have had a long quiet period of slow decay and rather poor future prospects. Such ghost towns are singularly characteristic of desert areas. Elsewhere, flooding and the more rapid decay of fixed investments under humid conditions soon erase them from the landscape.

Reaching Stage IV represents a particularly grim development for the labor force. Normally it occurs at just the point when labor organizations have gained their maximum power and effectiveness, and have thereby made the declining mine unprofitable because of a granted wage increase. Suddenly the inevitable decline in productivity of the mine makes them unable to achieve gains for their members or even to sustain their present incomes. At this point agitation for nationalization and continued operation at a loss, or government subvention such as tax rebates and maintenance at government expense of part of the infrastructure that once was the responsibility of the mine, becomes intense. Because the labor force has political power some of these concessions often are granted, but when this occurs the mine ceases to be a contributor to national wealth. Rather it is an economic burden that impoverishes the economy. To a more than comfortable degree the coal mines of Chile, along with those in many other parts of the world, have for some years been in this situation.

Where a good deal of the infrastructure, particularly in the area of housing, is in the private possession of the labor force, that force is vulnerable to pressure to accept lower wages, a singularly unpleasant option which keeps the mine operational for an extended period. Even less attractive is the possible acceptance of deteriorating safety and health provisions as a means of reducing operating costs. The break in the labor investment (Figure 13.1) well into Stage IV represents these developments.

In Northern Chile, in mines that exploit minerals of such value that they can be transported by primitive means, an institution has developed that permits a nearly exhausted mine to continue to operate sporadically over long periods of time. This is the *pirquén* and it was being described and complained about by royal mining inspectors during the 18th century;[4] it still exists. A mine no longer profitable to work in an organized fashion could be abandoned and just taken over by one or more *pirquineros*, or it might be leased to them formally for a nominal sum. The pirquinero worked on his own account, almost without capital investment, picking over old dumps, taking ore from support pillars, and following tiny veins by hand and pick. A mine given over to pirquén soon becomes dangerous even to enter, though this might not deter a later generation from following the same practice. The pirquinero's rewards for his labor are singularly meager. Often in the Norte Chico they must be supplemented by goat herding for a bare survival. His economically miserable situation is sustained by hopes, generally vain, of making a new discovery. Thus the pirquén supports, from its extremely marginal returns, some prospecting costs— Stage I perhaps of another mine.

Figure 13.2 introduces into the model the effects of cyclical price fluctuations on the history of a large mine. It is probably applicable only to mines that command extensive low-grade deposits and which are controlled by powerful corporate entities, either private or national. Such entities must be carrying on important economic activities in addition to operating the mine in question. Finally, for a mine to show such a history, the organization and power of its labor force could never have gained for it a dominant position.

Most critically, the pattern of price changes that might produce such a history is fairly sharply defined. There must be strong cyclical fluctuation on a base of sustained demand and a long-term stable or rising price structure. Such conditions are fulfilled by metals of fundamental industrial importance, which are of sufficient value in relation to bulk that they can be shipped all over the world, and thus have their prices determined by an unstable world market. Copper is the classic example, but other nonferrous metals and the ferro-alloys tend to satisfy these conditions. An interesting development of the last 50 years is that the increasing technical and industrial importance of silver has taken it out of the precious metal class with fairly stable prices, and placed it in the much more volatile industrial material category.

Even fairly violent price changes have little effect on the operation of a mine in Stages I, II, and the early part of III. The richest and most accessible deposits may be and normally are worked at a profit. As ores of lower tenor from deeper shafts that are more expensive to exploit come to be worked, fluctuation in price may make the difference between profit and loss. To the extent that they can control the management of the mine, both the labor force and society through the state are likely to urge continued production and stockpiling, though seldom will they accept wage cuts to return to profitable operation. Even a state-managed mine if it is of major dimensions, however, cannot long be operated at a loss or without continuing income without serious damage to the structure of the economy.

Some major corporations, and possibly some socialist states, can make a costly decision on the basis of confidence that despite short-time fluctuations the long-term price trend is stable and rising. The choices are either to abandon the mine completely, work it at lower costs by not maintaining and renewing the capital investment thus making it hard ever to reactivate, or maintain at their expense the capital investment both in the mine workings and the supporting infrastructure. The latter alternative is illustrated in Figure 13.2. It will be noted in Stage IIIA that while income from the mine ceases completely during the inactive periods, considerable expenses connected with the orderly shutdown of operations and maintenance and protection of works continue. There are also start-up expenses before profitable operation can be renewed.

Because of the unpredictability of price fluctuations the decision described is highly risky. Most disastrous would be a technological innovation or discovery of substitute material that permanently affected the price structure of the product. Synthetic nitrates provide Chile with a painful example. Rarely would any operator invest to keep a mine ready for reactivation for more than a decade if a favorable price structure did not develop. Even that duration would only exist in a developed country with abundant capital and low interest rates. In Chile five years is probably an absolute maximum. On the other hand, the capacity to adjust production to take advantage of the best prices results in great efficiencies in the exploitation of a given mineral deposit toward the end of human welfare. As a comparison of Figures 13.1 and 13.2 shows, the total output of the mine is increased and the period of its effective exploitation lengthened. By investing modestly to preserve the extant capital investment, an enormous waste of resources in reactivating the mine is avoided, so only a moderate price appreciation is needed to reactivate it. Finally, when the mine is operating it is operating profitably and severe downward pressure on wages is avoided.

The most intractable element in the productive forces that might make it impossible for a mine to sustain itself in Stage IIIA, that of intermittent operation, is the labor force. Laying it off is unpleasant at best and is likely to produce social disturbance. Once that force is laid off and has found other employment, reassembling it may be costly, though such costs provoke an upward pressure on wages that may be considered socially desirable. Chilean labor law has for decades endeavored to make layoffs and the intermittent operation of an enterprise difficult, if not impossible, perhaps to the considerable detriment of the entire economy. An optimal situation would be found if the mine were located in a settled region with an economy rich and diversified enough to absorb the mine workers without forcing them to emigrate and lose their personal investment in living quarters and community facilities. Then, to reattract them the mine would have to offer perceptible advantages in wages and working conditions. The above statements may seem to say merely that it is nice to be rich for both the individuals and societies. One may still observe that union rules, particularly in respect to seniority, that deter an individual from changing occupations freely introduce a rigidity that is economically detrimental to the entire community.

In the mining region of Chile, notably in the Norte Grande, the economic barrenness of entire regions has long posed a severe physical limitation. Once a mine closed down, be it a producer of nitrates, copper, or precious metals, it was improbable that any other economic activity in the vicinity could be found. In the Norte Chico, however, the limitation has been less severe. The capacity of Illapel to survive as a mining community for centuries, sustaining the residential investment of the mine workers' families after scores of individual mines were worked out and temporarily or permanently abandoned, is worthy of intensive investigation.

It must be admitted that a pattern, intermediate between those illustrated in Figures 13.1 and 13.2 and a notably less efficient use of all resources than the latter, has been much more common both in Chile and elsewhere. In this case the mine is just abandoned. When price improvement or new technological processes, often ones that permit reworking the old tailings, occur a new enterprise takes up and reactivates the claim. Such developments are singularly costly in capital investment because deterioration of both the investment in infrastructure and the mine workings themselves has intervened. A specially frequent deterrent to reactivation is flooding of shafts and tunnels. Such reactivations may be so degenerate that they are scarcely distinguishable from the pirquén, with minimal wages, dangerous working conditions, and other attributes of social distress. In vacant and extremely arid regions such as the Norte Grande, however, deterioration of the capital investment through both natural decay and human scavenging is minimized, and mine abandonment and reactivation are economically less irrational.

While the pattern of exploitation illustrated in Figure 13.2 may greatly prolong the useful life of a mine and more than double its output the end is still inevitable. Eventually the deposit will be exhausted.

Of the myriad of mines discovered, worked, and depleted only a few ever experience the rejuvenation illustrated in Figure 13.3. Because the sort of rejuvenation indicated is only possible in large-scale operations, however, many once individually operated mines in a district may have to be combined into a single operation to permit the rejuvenation to take place. Its main feature is capital investment on a greatly increased scale. Since the goal of the investment is to reduce costs per ton of material extracted and shipped, if the enterprise is to be successful there cannot be a proportionate increase in labor costs. Mechanization to reduce the amount of labor needed to extract, process, or ship the minerals is an invariable feature.

Rejuvenation seldom occurs during the first bonanza period of a mine's operation, when rich ores are close to the surface. Rather, it is introduced late in Stage III as returns are falling off and costs are rising. It may even take place in the decadent stage or after abandonment. It will be noted in Figure 13.3 that the curves for Stages

I, II, and III are replicas of those in Figure 13.1. The identification of a huge body of ore is a requisite. It may be of too low a grade to be worked profitably under the initial system or be inaccessible because of depth or flooding without heavy new investment. Multiple holdings along a vein or other deposit serve to deter attempts at rejuvenation. Mining claim laws, whether derived from English or Spanish traditions, have long favored the fragmentation of holdings and operations in any major mineral deposit. Only after such operations have ceased to be significantly profitable is it feasible to obtain control of enough reserves to justify a major investment.

A mine may be rejuvenated by increasing the efficiency of any or several of the activities that come between the mineral in the ground and its delivery to a consumer. These activities may be grouped in three sections. The first involves extraction of the material and its transport from within the mine to the pithead or, in Chilean terms, *cancha*. Concentrating, smelting, or other processing is a second element, one which may occur close to the pithead or at some fairly distant point en route to market, often a seaport. Finally, improving transport facilities to get ore out and supplies in may bring about reductions in costs. In addition to results achieved by constructing roads or railroads, a mine can be rejuvenated by improving port facilities hundreds of miles away. In the case of the last two sets of activities, the investment that rejuvenates a mine or group of mines may be made by other than the mine operator himself. Custom smelters in or near mining districts and privately financed and operated railroads have more than a century's history in Chile, and more recently road construction and public smelters at Las Ventanas and Paipote as well as beneficiation plants at Illapel, Domeyko, and elsewhere have been built with public funds with the express intent of rejuvenating as many as possible of the declining small and middle-sized mines in the several regions.

The introduction of new and advanced technology can but need not be a part of the rejuvenation process; often only the investment needed to apply long-known techniques in isolated areas is involved. In Chile's mining history the area of concentrating and processing ores is the one in which technological developments have been able to exert their strongest impact, but even here progress has not been uniform. Rich ores are often given little or no treatment at the mine and only as their tenor declines will the investment needed to concentrate them and reduce shipping costs be considered. Any new mine, of course, takes advantage of the state of the mining art at the time of its inception, particularly in those activities where the new technique requires only moderate capital outlay. Fire and water to break up hard rock gave way to black powder blasting and then essentially everywhere to dynamite, but *apires* still hauled bags of rich ore up ladders until mines were abandoned because the investment in tunnels through sterile rock and rails could not or did not need to be financed.

A few features of the model shown in Figure 13.3 may be noted. There is a considerable time gap between the initiation of the investment that rejuvenates a mine and the return to profitable operation. Because of the heavy investment and the lower-grade ore treated, even though total production may greatly exceed that in the original bonanza period, the level of profits is likely to be lower. If the investment was judicious, however, the mine's life will be extended enough to provide full amortization. The operating phase, Stage IIIA, is again one in which labor costs tend to rise, characteristically through union-demanded wage increases. In considerable measure to counteract such costs there is likely to be a continuing capital investment in labor-saving mechanization as long as the mine is profitable. Finally, in the later stages of the mine's operation, because of high fixed costs, the degree of profitability is likely to be strongly affected by even relatively slight price fluctuations.

At some point, however, ore of a tenor that makes operation profitable will be gone. A second or even a third rejuvenation by new investments that make it possible to handle still lower-grade ores is possible but infrequent. The end of a rejuvenated, and by definition a heavily capitalized mine, tends to be definite, and the long declining phase, Stage IV in Figure 13.1, seldom develops. Intermittent operations related to price fluctuations, as illustrated in Figure 13.2, are somewhat more probable.

The Mine and the Mining District

The natural conditions that give rise to mineral concentrations in the earth's crust are likely to affect fairly extensive areas, frequently with special concentrations along large veins or at nodal points, with lower-grade ores or thinner veins in the intermediate region. In the first rush after discovery of a mineralized district the entire area is likely to be staked out in claims, which often extend into the completely sterile periphery. Many claims never pass Stage I to become mines, but in major mining districts such as Chañarcillo or Andacollo there were more than a hundred mines each.[5] Each of the individual mines of the district would show a history like that illustrated in Figure 13.1, though progress to the stage of decadence for some might take only three or 4 years while others might still be operating profitably after 30. Thus Figure 13.1, with both the monetary and temporal scales expanded, might represent the history of a district as well as a single mine.

In a large mining district the investment in infrastructure and the benefits derived therefrom could be shared among a number of mines. Elements of importance in this infrastructure are the transportation facilities to outside markets and sources of supply, concentrating or smelting plants, and a settlement offering greater social services and facilities than the camp at the mine to which miners could repair on weekends and in which their families might reside. This fixed investment might prolong the life of individual mines letting them work at a small profit lower-grade deposits that would not in themselves pay for the needed infrastructure. In the same way, abandoned mines in the district or places bypassed in the first rush because of relatively poor ores might become workable once the infrastructure was created. In the Norte Chico of Chile still a further level of support was provided by valley oases with urban centers that were central to and served a number of *minerales*. Copiapó, Vallenar, and La Serena-Coquimbo are major examples[6] as Denver would be for the silver mines of the Central Rockies or Sacramento for the northern part of the Mother Lode of California.

The Purpose of a Mine

It is not proposed that the historical models offered in this paper will be of particular use to a mine operator in planning the exploitation of his own property. The particular characteristics of its deposits will remain his overwhelming concern. The fundamental university of mines' histories, however, does have relevance to national policy, particularly for a country such as Chile in which the exploitation of mineral deposits plays such a critical role in the national economy, and where regional disequilibria constitute a chronic problem. Some general notions about the ultimate social purposes served by mining may be reviewed first.

Until the time of Agrícola (mid-16th century) there was serious question as to whether it was morally appropriate for men to engage professionally in extracting from the earth minerals that they had not put there and could not replace.[7] The question still holds philosophical interest, but on the practical level modern industrial society has committed itself; a continuing and increasing flow of a wide variety of minerals into the world's industrial centers is absolutely essential for our very survival. The laws and tradition that govern society's relation to mining operations, however, especially in regions that once were part of the Spanish Empire, stem from a less idealistic source. Mines were easy to tax. The king retained title to mineral deposits and claimed a fifth of the gross output of any mine within his realm. The more mining activity that was carried on the greater his revenue was. All classes were allowed to prospect for and claim mines, even on privately owned lands, and the size of any claim was restricted so that as many mines as possible could go into production. A claim could not be held unless it was worked. Should a conflict over water for agricultural or mining use develop, the mine was likely to be favored.[8] Placing an export tax on certain minerals, notably nitrates, in the late 19th and early 20th centuries, the Chilean government sought to succeed to the king's former position in regard to mineral wealth.

Another view is that the wealth a mine yields during its bonanza phase is deserved by either the prospector who discovered it or the entrepreneur who risked his capital to put the mine into production. This premise also tends to bring about maximum development of mineral resources and a supply of minerals at minimal prices to the world's industrial markets. The premise is distinctly English in origin and has been carried over into the United States as well as into the British Dominions.

A more recent viewpoint stands in sharp contrast, especially in regard to its effects on the productive process. It is that the largest possible fraction of the returns from a mining operation should reward the labor force that invariably performs exhausting, often dangerous work under conditions of isolation and social privation. A modern student gives this viewpoint immediate sympathy, perhaps more intense if he is aware of mining history and knows how long mines just used up slave or forced labor, or has read the poignant descriptions of working conditions in coal mines, be they of British mines of the 19th century, Baldomero Lillo's stories of the Chilean mines at Lota at the turn of the century, or accounts of the mines at Vorkuta in the 1940s. During the past few decades labor organizations in mining enterprises, exploiting the general sympathy noted above to gain enabling legislation, have made enormous progress in capturing a greater fraction of any successful mine's returns for its miners. At the same time, the very nature of mines as described above and the technological developments to which they are subject combine to make a strictly labor-oriented division of returns dangerous and destructive. Except in the initial bonanza period, and then only in certain deposits, the individual workman, though skilled and industrious, can produce little from a mineral deposit. It is the capital invested in applying inanimate energy and engineering a plan for mass extraction, concentration, and shipment of the mineral that makes the mine worker's daily output so substantial. Almost by definition the worker does not control his capital. It must come from capitalist entrepreneurs or from a society that includes many elements other than mine workers. Either investor will want returns on his capital and have alternate placers in which to invest it.

Perhaps a more critical feature is shown in the models presented here. Any mine's maximum yield and profitability occurs very early in its cycle of exploitation. Later, costs will rise as the less accessible and commonly lower-grade portions of a deposit must be worked. A major new infusion of capital may lower costs and rejuvenate the mine, but the cycle of decline soon reasserts itself. After a few years of operation at a given level of investment in mechanization any mine will have a smaller net return to share with its workers, regardless of the management's willingness to do so. Conversely there is no prospect so abhorrent to the labor force, either as individuals or organized in a union, as that of a continuing decline in real wages.

These considerations seem to lead to an inescapable conclusion that early in a good mine's history it will yield profits that rationally cannot be distributed to the labor force. There is no implication here that wages should not be high; they should be at least enough to induce free labor to move to the mine and live under difficult conditions. But the society will not benefit if wages are fixed at a level that cannot be sustained through the operational life of the mine, or which will substantially shorten that life. It does not seem significant whether the financing and management of the mine come from domestic or foreign private capitalists, or from the national society as a whole (through an agency such as CORFO—the Chilean Corporación de Fomento de la Producción). Advantages and disadvantages can be identified for each source. It is assumed that the society as a whole (at this stage in the world's development, the nation-state in which the mine is physically located) will exercise some direction over how the wealth the mine generates will be distributed. A set of considerations that recognizes the inherent prognosis of any mine's development may be presented.

Capital Amortization

There must be provision to return from the mine's profits before the deposit is exhausted the capital that was invested to develop it, plus a reasonable interest that compensates for the risk that the mine might be a failure. At the political extremes this applies equally to a mine financed by foreign capital and to one financed by the government of the country within which it is located. Should this provision be ignored in the former case the foreign capital and management and technical skills will not be available in the future. In the latter case other sectors of the society will be financing mining activity at the cost of their own living standards. Above all it is essential to maintain a body of capital available to activate new mines, rejuvenate old ones by further investment in mechanization and newer, more efficient technology, or to invest in other productive enterprises.

Recapture of Excessive Gains

Several mechanisms have been employed historically to bring to the society as a whole as much as possible of the enormous return that an especially rich mineral deposit may yield in the early phases of its exploitation. Property taxes and export or *ad valorem* taxes on the mineral output fall into one group. They are direct and potentially equitable, but must be applied to mining enterprises with a temporal sensitivity that governments find hard to exhibit. At just the time that a mine proves to be profitable it has its greatest capacity to bear a tax burden. As the mine is depleted or becomes more expensive to operate this capacity declines. Sustaining the tax load, something that governments ever in need of revenue find hard to avoid will force the mine into decline or abandonment before its time, Stage IV in Figure 13.1. Then intense local unemployment and the premature abandonment of capital investment can only make the society poorer. The disastrous condition of the Chilean nitrate fields in the 1920s was at least exacerbated by sustaining an export tax on that mineral for too long a period. Should taxes of this sort be retained it is essential that they be applied in a manner which takes account of each mine's position in its own course of development. It is suggested that progressive individual and corporate income taxes may be a more effective means of making each mine's early wealth yield social returns without paralyzing its operation in the later, leaner period.

Regional Infrastructure

As an essential part of its earliest development a mine or the grouped enterprises of a mining district must create or have created for them a set of facilities that can serve not only the mines but the entire region. It happens with great frequency that the exploitation of mineral resources constitutes the first effective settlement of an extensive region. Even when the mine itself is located high in the mountains or in an area barren for other reasons so that the immediate locality is unlikely to support other economic activity, a transportation facility that can handle heavy cargos must be built either to connect it with an existing settlement or with a newly constructed seaport. Thus the intervening terrain will be made more advantageous for settlement. It does not seem unreasonable for the society to request that this transportation facility be routed in such a way as to support regional economic development in other directions, if such routing carries only a modestly greater cost than the cheapest one to serve the mine's immediate needs. Such encouragement of settlement will later redound to the mine's benefit in providing supplies, a labor market, and a reduction in isolation, and thus a more attractive ambience for the mine's labor force. These features will be important in keeping down mining costs during a later period when the richest ores are gone and profit margins are low.

In desert regions a mine may need to provide its own water supply, or its own drainage operations may incidentally create one. Unfortunately, processing the ore may also contaminate part of the limited extant water supply, as it did for example, at Chañaral. In any case, it is in society's interest that an effort be made to integrate the mine's water requirements and provision with present and potential future agriculture and urban developments in the basin, maximizing opportunities for multiple use of the same water, and minimizing damage from contamination either by purification or segregation. The mine is inherently transient, but it may have the option of promoting future development or creating a permanent desert.

In both transportation and water manipulation the mine is most capable of contributing to a regionally supportive infrastructure during its early life, Stage II, that of investment and development, or even more probably stage IIA in Figure 13.3, when a massive new investment is being undertaken. At this time the margin between its returns and costs is the widest. Demands that a mine rectify earlier errors in creating its own infrastructure after it has passed well into Stage III may raise its costs insufferably and hasten the end of its operations. Conversely, as mines move toward the end of Stage III and the prospect of the end of profitable operations becomes imminent, both their management and their labor forces are often able to unite to put all the political pressure they can on society at large to create for themselves at public expense an improved infrastructure that would reduce costs and prolong the period of profitable operation.[9] It would be well for public authorities to note carefully the place where the mine to be supported falls in its own cycle of development. Unless there are committed funds for a rejuvenation of the mine itself, even a heavy public investment in the infrastructure may keep the mine operating profitably very few years longer, and the worth of the public investment would be little in an area in the process of abandonment.

Another sector of the infrastructure is the community in which the mine's labor force lives. Typically housing during the period of development is either makeshift shacks in the case of small mines or company barracks in the case of larger, better financed ones. Often an open city of the makeshift sort grows up independently, as close to the mine as it can get, to provide dubious recreation to the mine workers and to be financed indirectly by their salaries. Later in the mine's history, during the period of labor gains, the creation of a real community for the miners and their families merits continuing investment. Chile affords a variety of examples of such communities, and indeed there is some trend toward progress from the grim massed tenements in the company town of Lota Alto. It is suggested that a free town that can serve as a regional center for the area is a better solution if there is any physical prospect for it, even if it means transporting the help to their places of work. Providing credit for housing and higher salaries rather than subsidized low rents appears to be the wiser use of a given part of the total cost of labor. A wise infrastructural investment for both the society and the mining enterprise, however, is in the urban amenities such as theaters, well-equipped schools, and parks that will tend to make the place where the miners live a regional economic and cultural center, thus broadening its economic base. As a negative example one might ask whether anyone in the neighboring province of Arauco ever goes to Lota to shop despite the fact that it is the nearest city and far larger than any within the province. Investment in the residential community, if in part supported by individual decisions, can and should continue as long as the mine retains its labor force. It can also support the rejuvenation of the mine or the development of others within the district, as Illapel seems to have done. And except in the most desolate locales, as in the Norte Grande, [it] may be the basis for economic activities of a permanent nature when the mine comes to its inevitable end.

Maintaining the Mine for a Maximum Useful Life

The minerals extracted from a given deposit serve several functions. They provide raw material for national or world industry, or for consumption, and in the latter case afford foreign exchange to the country in which the deposits are located. Further, the mine affords employment and a market which stimulates the regional economy. These benefits are maximized as the deposit can be most completely utilized and the level of exploitation sustained at a fairly uniform rate over the period of operation. The models presented in this essay show two inescapable limits. Any deposit is finite and ultimately will be mined out, and most of the capital investment will have to be abandoned. The labor force will have to move or find other economic activities. A somewhat less absolute limit arises from the condition that only early in its history of exploitation, or shortly after a cycle of reinvestment, is the mine highly profitable. Later there is a period in which only leaner ores, thinner seams in the case of coal, or deposits less accessible because of depth will remain, and it is barely economic to work the mine. It is neither politically feasible nor morally attractive to sustain operations at the cost of the labor force by reducing wages, but there have been abundant instances when a mine was prematurely abandoned because the labor force at the height of its growing organizational strength gained wage increases that immediately or shortly forced the mine into State IV, that of unprofitability.

Wise public policy will encourage, in addition to a satisfactory wage level, the heavy investment of capital in features that will improve living conditions (social services such as schools and hospitals) and working conditions early in the mine's profitable history so that this capital charge will serve for the life of the mine and not come as a burden later when the mine is less able to bear it. As the mine nears its end, however, particularly if the state has come to play a major part in its operation, as it did for example in the coal mines of Arauco, there is strong pressure to develop an elaborate infrastructure with public funds. It may be justified as reducing costs and thus prolonging the mine's life, but is likely to be a means of relieving unemployment. The models presented here would indicate that such late construction is a bad investment and that the money might better be spent in relocating the population or developing other industries for their employment. After a mine enters Stage IV in its history, its continued operation can only be sustained by a continuing waste of funds or by cutting rewards to the labor force below acceptable standards. Only massive investment in the mine itself, new technology, or a substantial change in price structure on the world market, such as silver has experienced in the past two years, can revive a mine when it has reached this point.

The Mine as a Market

A mine consumes goods and services as well as producing a product. Its first impact is specially focused on the labor market. Rarely is a new mine developed in a region sufficiently densely settled to supply its labor force locally, and recruiting labor has a national and often an international impact. A consistently positive effect is that of providing work for miners in near or distant mines that have reached the stage of decline, a group welcomed for its skill but sometimes shunned because of its affinity for strong labor organizations. In 19th-century Chile there were continuing complaints that withdrawal of labor for the mines was impeding the agricultural development of Central Chile. This paper will not attempt to assess the question of whether the mining industry proved beneficial in improving the status of agricultural labor, or unfortunately provided a minimally effective safety value that deferred too long any effective steps toward agrarian reform and more efficient and productive use of agricultural labor. At the other end of its cycle the mine turns this force back into the labor market, often in a region with few other economic resources, as in the case of the Norte Chico of Chile, or even with no other resources, as in much of the Norte Grande. Ideally all would leave to work in other mines. Unfortunately many do not, and in intolerable poverty in the Norte Chico ex-miners scratch a living from a combination of pirquén, the scavenging and often destroying of abandoned mines, and overgrazing the landscape with their flocks of goats.

Mines have traditionally served as a basis for national development, attracting foreign capital that may develop a generally useful infrastructure in addition to developing the mine, providing foreign exchange, and attracting immigrants. Sometimes, as in Chile's case, they temporarily support most government activities through export taxes, provoking serious malaise when they reach a stage in which they cannot bear this burden.

The goods a mine consumes can be put in four categories: (1) supplies for the labor force, (2) capital investment goods from pit props to heavy machinery, (3) combustibles for processing the mineral and transporting it, and (4) water. Clearly, obtaining these goods as close as possible to the mine will be advantageous, but, as needed, a profitable mine can seek its supplies successively in national and international markets. In supporting its labor force a mine constitutes a powerful stimulus to local agriculture, particularly in regard to perishable, hard-to-ship commodities. In dry areas this market may finance irrigation works of permanent value and productivity. Certainly the Elqui and Huasco valleys would not have reached their present level of agricultural development without this stimulus. In the past, but probably not again, the need for transport also stimulated agriculture to supply mules and fodder for them, leaving a sensible gap in the agricultural economies of all the valleys of the Norte Chico when technological change eliminated most of the need for draft animals. The market for capital goods will at least be national and can serve as support for the broader industrialization of a country. A singularly delicate timing of legal pressures is required if using the mining market as a basis for industrialization is to be healthy. In its early, most profitable phase a mine can pay heavily for the capital goods it needs, either through import duties or by buying from new and still inefficient industries that a developing country might be trying to establish. At a later stage, however, being forced to do so will raise costs and hasten the shutdown of the mine. A protected, assured market in a profitable mining industry may allow suppliers, protected by tariff barriers, to avoid increasing their efficiency and seeking broader markets. Forcing mines to serve as a basis for developing the national industries is clearly a legitimate means of recapturing for society some of the wealth a mine may produce during its early stages. But this laudable social effort must be sharply limited temporally or there will be a risk of forcing a premature end to mining activity and leaving the country with inefficient industries that have no other market.

In the areas of combustibles and water, especially in arid or semiarid regions, the market provided by the mine may have effects that are beneficial for the regional and national economy or are disastrous to the region's long-term welfare. The stripping of the Norte Chico of its woody vegetation to sustain widely scattered smelters in the early and middle 19th century is still perceptible in the landscape. Fragile ecologic communities were destroyed that will not regenerate for many centuries, if ever. It is fortunate that remnants of an extraordinary, fog-sustained forest survived at Fray Jorge and Tongoy, if only for scientific study. Stripped hillsides have increased soil erosion and made stream flow more erratic and unmanageable. Great forests of the hardwood *guayacán* were burnt in the great smelter that bore the same name and elsewhere, and now little is left to supply a potential craft industry. Across the Bolivian border the process continues, using the slow-growing *llareta* in industrial quantities to process sulfur from the volcanos of the Altiplano.

The Chilean coal industry and, to some degree as secondary effects, the glass, firebrick, and steel industries, have as a major part of their origins the market offered by mining smelters and the railroads that serve the mines. It is historically unfortunate that Chilean coal reached the Norte Chico so late, after many mining districts were so depleted that they could not afford to create transport facilities to bring in coal and continued instead to destroy the natural vegetation until both it and the ore that could be worked under those conditions were gone. Again the time in a mine's history in which the more desirable provisioning of its market was afforded proved to be critical.

Major mines must develop a water supply, and in their early, profitable phases they can bear heavy costs and construct long pipelines to do so. They can also take more accessible water that is being or might be used for irrigated agriculture and degrade or pollute it so that it is useless or even harmful to the land over which it flows. Public policy concerning a mine's access to limited water supplies will do well to note both that the mine's life is finite though the damage it can do to the landscape may be permanent, and that early, and only early, in its history a mine has the economic capacity to develop new water supplies or to take costly engineering steps to minimize the damage done by its pollution.[10]

Indian Societies and Communities in Latin America: A Historical Perspective

We are about ready for a major study that will attempt to answer ...
[the questions:] Where, to what extent, and why have Indian cultures
and communities survived in Latin America? ... a geographic perspective promises a sound framework for such a study.

[A] satisfying cultural geography of Latin America or any of its countries or major regions will not be produced until we understand not
only the activities of individuals in the diverse communities but also
the motivations and goals that induce them.

A LITTLE LESS than five centuries ago what we now identify as Latin America was occupied by a kaleidoscope of societies, each with a long, complex, and largely
unrecorded history. That all are called Indians no longer obscures the fact that an extraordinarily broad spectrum of economic and technologic capabilities and social
and political complexity is represented, often in immediate juxtaposition. The location of each group and its economic and social organization were the resultant of
migration, cultural growth, and development involving both autochthonous evolution and intergroup borrowings, and successful or unsuccessful competition with
neighboring societies. The pre-Columbian culture history of Latin America has its intrinsic fascination, and a modest number of geographers have contributed to its unraveling, though they have been dependent on archaeologic, linguistic, and contact
ethnographic investigations for most of their basic data. In Mesoamerica and the Central Andes formal native historic traditions, supported in the former area by some documents and inscriptions, permit the use of more characteristic historical methods.

My concern in this paper, however, is focused on the Indian societies in post-
Columbian and modern times. Where and why did they survive both racially and
culturally? Where and why did they survive physically but have their cultures altered or replaced? What sorts of relationships with their environments, physical and
social, do communities that are identifiably Indian exhibit today, and how are their
members reacting to the assimilative and acculturative pressures of the national states
in which they are located? In addition to providing intrinsically interesting information and descriptive insights into the character of a considerable number of the
Latin American states and regions, studies directed to answering the above questions
may afford primary contributions to answers to more general and theoretical ques-

tions. A few examples are: What social and other attributes do those societies have that successfully maintain their integrity although [they are] surrounded by other more numerous and economically more powerful groups? When two distinct societies live in fairly close contact over a long period of time what sorts of culture elements and complexes are likely to flow from one to the other, especially which elements are capable of passing from the richer to the poorer group? What are the effects on individuals and small communities of slow or rapid assimilation into at least the economic system of a more complex society? Can distinctive patterns of land use and exploitation exist side by side in the same sort of terrain or will that system which is in the short run economically more productive or profitable inevitably take over? A singular advantage in studying Indian communities in Latin America is that there are so many of them, typically almost completely independent of each other, that a sample of significant size can be developed. General tendencies as opposed to developments resulting from idiosyncratic local conditions or histories may be identified. Both historical and contemporary studies can contribute to the sample.

Historical Concerns

Destruction

The European contact with the Indies that began with Columbus in 1492 is an essentially unique event. Alien invaders found a large, comfortable, peaceful population in the Greater Antilles that was singularly unprepared to resist conquest and oppression. An improbably unfortunate combination of greed and egoism on the part of each Spaniard who arrived in the New World, almost identical characteristics among their leaders, both Columbus and his contemporaries and successors, and in Spain an ineffective mixture of idealism, need for revenues, infighting among courtiers and ministers, and bureaucratic bungling resulted in a sudden and almost absolute destruction of the native population, its wealth, and most of its cultural heritage. In short succession the same destruction was carried out in the Bahamas and South Florida and almost all the way around the rim of the Caribbean. The results were economically disastrous to the conquerors and hence, we can assume, unintended. But from the beginning of the 16th century there was no question that European presence was fully established in the Indies. There was no one else there.

Las Casas's *Brevissima Relacion de la Destruccion de las Indias*, published in 1552, remains our fundamental record of this event, but Sauer's masterful analysis, based on wide field familiarity, Columbus's letters, the lately published *Historia* of Las Casas,[1] and the decades of Peter Martyr, among other early chroniclers, has placed in geographic perspective the results of an invading society's full destruction of a native one. Some important crops and perhaps their modern cultivation patterns and a few words such as hammock and hurricane, maize and tobacco are all that remain. The Indian absence forms a major basis for Augelli's incisive dichotomization of Middle

America into rimland and mainland sectors, which is continued in West and Augelli's text on Middle America.[2]

In a few places around the Caribbean, however, Indian societies survived considerably longer, even to the present. These include the Caribs of the Lesser Antilles, the Guajiras, the San Blas of Northeastern Panamá, and the Miskitos of coastal Nicaragua and Honduras. Causal explanations for these survivals are not uniform, although a strong sense of ethnic identity, even in poverty, is a necessary condition. The Caribs were warlike and effective fighters. The Guajira Peninsula and the Miskito Coast were notably unattractive to Europeans though for different reasons. I have no satisfying explanation for the survival of the San Blas Indians.

The record of almost complete destruction of Indian societies is not limited to the initial Spanish contacts around the Caribbean. The East Coast of Brazil and the La Plata estuary were exploitatively stripped of their natives for crude economic gain, and in the early 1530s Nuño de Guzmán left desolate the Mexican West Coast from Tepic to Culiacán.[3] Baja California, on the other hand, lost all its natives because dedicated missionaries sought to aid them.[4] In Northern Mexico the presence of important mines is invariably associated with the absence of local Indians, but the pattern does not appear to hold in the Central Andes. A careful examination of documentary records is likely to show that an overwhelming decline in the Indian population of an area has as its immediate cause introduced diseases such as smallpox and measles, toward which the Indians showed almost no resistance. But the problem remains when we ask why these diseases were lethal to societies as well as to individuals in some areas and not in others.

Survival in Densely Settled Areas

In two extensive regions, the Mesoamerican highlands from Guatemala to Central Mexico and the Andean highlands from southernmost Colombia to just south of Bolivia, large Indian communities survive. All have been subjected to a substantial though not uniform degree of acculturation, but it may be accepted that regular household use of an Indian language is an effective indication of Indianness. By this criterion major fractions of the populations of Guatemala, Ecuador, Peru, and Bolivia are Indian. The same was true in Mexico into this century and there are still many hundreds of thousands of monoglot speakers of Indian languages in that country.

All these highland Indians descend from highly organized and complex societies that were conquered promptly by the Spanish invaders, although dangerous revolts occurred in the Andes as late as the 1780s.[5] Having observed the effects of the demographic disaster in the Antilles, Cortés and the viceroys who took over Mexico and Peru were aware that a continuing flow of wealth depended on the maintenance of an Indian labor force. There were abuses in forced removal of labor gangs to distant mines and putting highland Indians to work in lowland plantations, and epidemic diseases could not be controlled. But it was acknowledged that enough people had to be allowed to work their own lands to provide food for themselves, the Spaniards, and the forced labor. Their initial numbers were large and, though modern studies have demonstrated tremendous mortality and population declines in the first decades after conquest,[6] in most highland areas survivors were sufficient in number to maintain a fully viable social and demographic entity. In the present century, and especially in the last generation, there has been an explosively accelerating population growth among the Indians as well as in the several national populations. There may be as many individuals of highland Indian ancestry living today as there ever were; there will be more very soon, though, of course, there has been considerable racial mixing.

Once the civilized highland Indians came to be recognized as human vassals of the Spanish Crown a series of deliberate procedures were applied to them. As much of their labor as could be spared from their own subsistence activities was applied to revenue-producing activities and their support. Native governance was encouraged at the village level, but the Spaniards supplied perquisites for the native leaders and demanded that they deliver a requisitioned labor force. Conversion to Christianity was mandatory. All these measures and a number of others held strong acculturative potency. Forced labor under alien or socially distant rulers was not new to the sedentary Indians of the highlands, and they could endure it better than could freer and more primitive groups, and many men did return to home villages with new experiences and sometimes skills. It was in the higher levels of the Indian social strata that acculturation was most drastic. The upper levels of the native priesthoods were simply cut off, often physically. Only within the more isolated villages could priest or shaman practice his cult clandestinely, and the complex religious structures were simplified to folk cults, which in many places still survive, syncretistically or parallel with the official Catholicism. Civil leadership was co-opted by the Spaniards, and children of the higher Indian nobility, especially in Peru, were often given European educations and placed in honored though carefully watched and powerless positions. Village leaders, *caciques* and *curacas*, were given authority, responsibilities, and special favors throughout the colonial period. Their position was always a delicate one, identifying with their own people of the village but supported by and responsible to their oppressors. Regardless of which side they chose to give their loyalty, their effectiveness and personal success were enhanced by a maximal knowledge of Spanish ways.

The result of these Spanish colonial policies and practices, in addition to a substantial physical survival of the Indians, was a complete elimination of the civilized, centralized, and urbanized content of the native cultures and civilizations, both in terms of material culture and of social organization. They were replaced by European values and practices. It is interesting to note that the great rebellion of Tupac Amaru of the 1780s in Southern Peru sought only low-level administrative reform. Its leader, to his execution, professed loyalty to the Crown and the Catholic Church.[7] Indianness survives in the villages where crafts, agricultural practices, and intravillage social relations, modified to be sure, maintain a continuum with pre-Columbian patterns. They resemble the "Formative" or "Archaic" villages of 3,000 years ago more than they do the Aztec and Inca empires the Spaniards conquered. Because the good or even adequate farmland left to the Indian villages is scarce, outsiders, Indian or otherwise, are unwelcome and village endogamy is closely followed. Even where all villages in a region except the administrative centers are Indian, as in Western Guatemala, Southern Peru, and much of highland Bolivia, each village is socially isolated from the others except for some formalized and traditional market exchanges.[8] Its contacts with the outside are through agencies of the national government as they once were with officers of the Crown.

The retention or loss of Indian languages among the highland Indians presents a number of problems. Two imperial languages, Nahua for the Aztecs and Quechua for the Incas, were adopted by the Spaniards for their own administrative purposes. They were spread farther and used more widely in the first century after the Conquest than they had been before. Their speakers, however, were more subject to acculturative pressures, since these languages were learned by Spanish officials and traders. Important but recessive languages, Otomí in Mexico and Aymara in Southern Peru and Bolivia, for example, although they were losing to Nahua and Quechua respectively both before and immediately after the Conquest, have been more durable during the past two centuries. Their speakers have been more isolated from acculturative pressures. Since Mexican Independence and especially since the Revolution of this century Nahua has been fading rapidly and has few monoglot speakers. The impoverished Otomí in their mountain *rancherías* seem likely to preserve their linguistic identity longer. Quechua, however, seems to be more vital. It is hard to see how its wider, international distribution can be a causal factor, since the Indian villages are isolated polities. The backward social structure of the Andean states, which, as opposed to Mexico, have been content to leave the Indians largely out of the national life also protects them to some degree from linguistic assimilation.

The Mayan and Chibchan linguistic groups are represented by a number of distinct and mutually unintelligible languages. The Chibchans of Colombia and Central America have largely been assimilated linguistically. The Maya in Guatemala and in Chiapas and Yucatán in Mexico have been singularly successful in maintaining their linguistic identities, although Chiapas is experiencing heavy acculturation associated with both Mexican policy and the Pan-American Highway.[9] In Yucatán the Maya represent virtually the only surviving lowland Indian community lineally descended from a high civilization, and their linguistic tenacity remains high.

The many Indian communities in the western highlands of the Americas extending from Nayarit in West Mexico to Northern Chile and Argentina are numerically important, live in extraordinarily interesting terrain, and have devised fascinating means of wresting their livings from what are often the most inauspicious environments. All have long been subjected to powerful acculturative pressures and most to severe economic oppression, and they have reacted in diverse ways. We are still far from being able to make all the relevant generalizations we would like about these societies. Clearly acculturation and assimilation into the national societies are proceeding, during the last generation, at a rate faster than at most times in the past. Some communities will soon cease to be Indian, and all will acquire more non-Indian ways. The record of 450 years of endurance, however, suggests that some Indian communities will long remain a distinctive part of the highland scene. We cannot predict which will survive in terms of present theoretical understanding.

Survival as Fugitives

Scattered widely over the remainder of Latin America are a great number of Indian communities, most of them properly called tribes, who remain outside the polities, and to a large degree outside the economies, of the national states in which they live. The districts in which they live are sparsely populated, normally because they consist of unattractive and economically unproductive environments. At best such districts are undeveloped and isolated. The upper and middle courses of the tributaries of the Orinoco and Amazon Rivers and the Guiana highlands hold the largest fraction of such tribes, but some may be found in locally arid districts such as the Guajira Peninsula, the Chaco, or the Seri country in Western Sonora, the high valleys of the isolated Sierra Nevada de Santa Marta, or superhumid tropical forest areas such as the Sierra de Perija, the Choco and the Upper Sinu Valley, the Northeast Coast of Panamá, and the Miskito Coast. Some of these isolated tribes live where they have since they were first contacted. Others have been driven into their present abodes as refugees from more favored areas. Although a substantial fraction of these groups cultivate crops it is normally on a shifting, slash-and-burn basis. Permanent settlements are rare and often there is effective nomadism, though the San Blas Indians of Panamá have permanent villages on tiny offshore islets, going daily by boat to their fields on the mainland.

These fugitive tribes at present possess only primitive technology, little material wealth, and simple social organization; religion is that of the shaman rather than the priest and his organized cult. There is little evidence that the regions where they live were ever occupied by complex societies or civilizations, probably because of the poverty of their natural environments in most cases. In some instances the Indians may be displaced descendants of bearers of a high culture who could not maintain its complexity in their environmentally poorer new homes. Thus the Kogi, Ika, and Sanka of the Sierra Nevada de Santa Marta may well descend from the Tairona who once lived at its western base.[10]

The survival of these fugitive groups is invariably associated with the fact that the

Spanish or Portuguese invaders, like the Indian empires of advanced technology and political organization before them, did not find their homelands suitable for permanent occupation. They were raided for slaves, notably in interior Brazil, but in rugged or heavily forested terrain at least some could remain free if at the cost of constant movement and the consequent poverty. Patterns of hostility, effectively supported by tactics such as using poisoned blow-gun darts, and timidity offered further protection. Good examples are the Motilones of the east side of the Sierra de Perija and the Auca of Eastern Ecuador.

Whenever a technological change, discovery of mineral wealth, or the development of a market for local product occurred and induced important entry and occupation by Europeans or national citizens, the marginal or fugitive Indian tribes were doomed. The gold rush of Tierra del Fuego and subsequent establishment of sheep ranges did in the Ona and Yaghan; agricultural settlement of the Argentine Pampa overwhelmed the Puelche despite their brilliant military capabilities, and the extension of commercial grazing into Patagonia did the same to the Tehuelche; the collapse of the rubber boom in the Amazon Basin occurred before all its fugitive tribes were enslaved or exterminated, but losses were heavy. The process continues and in the past few years we have learned belatedly of exterminative attacks on Indian groups in the Orinoco basin of Eastern Colombia and in the Brazilian Matto Grosso by representatives of the expanding cattle-grazing industry. Mineral exploitation in the Guiana highlands and petroleum discoveries at the eastern base of the Andes are likely to have similar effects. The long-term prognosis for these fugitive tribes is poor.

Two groups have been especially attracted to these fugitive tribes: anthropologists and missionaries. Anthropologists seek to discover an unspoiled primitive tribe to study and record in detail its culture.[11] Only some of these students have acquired the sophistication to recognize that they are dealing with peoples who have been much affected by contact with Europeans, though in a basically negative way. Their ethnographic studies in many instances, however, included notably detailed ecologic analyses. Traditionally, anthropologists come to appreciate "their" tribe for itself and to hope that it can remain isolated and safe. Missionaries on the other hand, seek to enhance physical as well as spiritual welfare. Their explicit goal is acculturation in both material and nonmaterial culture toward their own model. As they are successful, which is less than common, a group loses its Indianness.

Semi-Acculturated Societies

Four groups, commonly referred to as Indians, require special mention because of their relative success in maintaining some vestiges of Indianness along with massive acceptance of both the technology and values of the modern nations in which they live. They are the Guajira of Colombia and Venezuela, the Yaqui of Northwest

Mexico, the Guarani of Paraguay, and the Araucanians of Chile. All groups practiced some farming but had a limited technology and relatively simple social organization. It was also one that proved resilient to powerful acculturative impacts, and biologically at least, all have weathered the storm and are becoming more numerous. Culturally, however, they are likely to become more like lower-class citizens of their respective countries whether or not they retain their Indian languages. I shall suggest what appear to me to have been critical factors in permitting these survivals.

Both the Yaqui and the Guarani experienced massive and somewhat forced acculturation at the hands of Jesuit missionaries from the beginning of the 17th century to 1767. Concerned more with religion than language and being an international rather than a Spanish mission, the Jesuit goal was to develop a Christian community rather than integrate the Indian population into the colonial culture of the Spanish Empire. They taught European arts well enough that even after the Jesuit Expulsion their former charges could compete in the greater colonial and later in the national societies. Both groups have been severely burdened since independence from Spain. The Yaquis fought so hard for rights to worship and hold land in their own fashion that post-Revolutionary Mexico has come to respect them, recognize their land tenure, and grant them some cultural autonomy.[12] Finally crushed in the Wars of Lopez and severely reduced territorially, Paraguay found its old Guarani heritage almost its only basis for survival as a state and encouraged use of the Guarani language. Except for a distinctive and wonderfully attractive musical tradition, but one that holds many European elements, there seems to be little other Indianness in Paraguayan culture.

The Araucanians south of the Río Bío-Bío fought the Spaniards and later the Chileans from the time of Valdivia to the 1880s. Neither environmental conditions nor tribal military prowess can explain their success. The vicissitudes of colonial, revolutionary, and republican history that repeatedly took the pressure off the Indian way of life just as it appeared that Araucania was pacified are relevant but not satisfying explanations. More significant was the rapid acceptance of European crops and domestic animals along with partial participation in an exchange economy by the still free Indians. The Europeans agricultural complex was more successful in Araucania than the native, as suggested by the disappearance of the two native winter-sown grains, unique to the Americas, in competition with wheat and barley. When they were not actually reacting to military pressure the Indians became more prosperous than they had ever been, engaging in a substantial trade with the lands to the north. Except for social relations and some language retention, the Araucanians are hard to distinguish from neighboring Chilean campesinos.[13]

The Guajira Peninsula, projecting into the Caribbean, was discovered and explored in the first wave of the Spanish Conquest, and Río Hacha was founded early in the 16th century. The dry thorn forest to the east, however, was singularly unattractive, and the Spaniards found little reason to go beyond the Río Ranchería. During the colonial period the Indians acquired European livestock and evolved a pastoral economy far more productive and able to support more people than had lived on the peninsula in aboriginal times. The region is now overgrazed and overpopulated, exporting people, especially to Maracaibo, rather than receiving immigrants in number. Perhaps because of matrilineal inheritance, the Guajira have been notably able to assimilate immigrants from virtually all parts of the world.[14] It is my impression that the Guajira culture is the most dynamic and vital of the four considered in this section; it shows no evidence of disappearing into the national entities. It should be noted, however, that its entire subsistence base is derived from European technology.

The Assimilated Indian or Mestizo

Biological mestization began when Europeans arrived in the New World, and mulattos and zambos began to appear as soon as Negroes were imported from Africa. Race mixing has long been so far advanced among all Indian populations, except for a few small fugitive tribes, that it now proceeds almost without reference to linguistic or cultural developments. The mixture will have a greater or lesser Indian component depending on the relative size of the Indian populations that survived the initial shock of contact in particular areas and the number of European and Negro immigrants. It has often been noted that in Guatemala, where the Indian component is large, the *ladinos* as a group are physically indistinguishable from people culturally identified as Indians. In Central and Southern Chile the Indian component is smaller, and again a large fraction of the campesinos (both *inquilinos* and *minifundistas*) are physically equivalent to neighboring Araucanians.

In Western Latin America, from Mexico to Chile and Northwest Argentina, except for a few locales such as highland Costa Rica and Antioquia, the Indian component of the several national populations is so large that they must all be regarded as mestizos. Even in the above-mentioned exceptions an Indian element is definitely present. The processes whereby these mestizo populations became identified with the developing national cultures that bear a strongly Latin and European rather than Indian character are a central theme in Latin American historiography. The transition had already occurred in Chile by the time of Independence. The same may be true for El Salvador, Honduras, and Nicaragua. In highland Colombia and Venezuela it seems to have occurred a few decades later, and in Central and Southern Mexico the shift is significantly post-Revolutionary, from 1920 to 1940. It is still under way in Guatemala, Ecuador, and Peru; rural highland Bolivia is still Indian.

There are many highly geographic components to the forces that are bringing into existence the several Latin American national cultures. Land tenure patterns, the relations between subsistence- and export-oriented agricultural production, the development of commercial markets at mines, and the creation of transportation facilities are examples. The transition from Indian to Chileno or Mexican did not occur simultaneously in all areas of Indian culture. At least official conversion to Christianity was enforced immediately after conquest. The adoption of European crops, domesticated animals, and certain kinds of technology came early and its completeness was an inverse function of how fully the Indian ways occupied the ecological niches of a particular environment. The practice of marking the transition at the time when Indian languages were or will be abandoned in favor of Spanish is perhaps as valid as using any other single index. To varying degrees in varying regions, however, Indian ways have survived the linguistic shift, often to a greater extent than in places like the Guajira and Eastern Paraguay, where Indian languages are still dominant. Dietary preferences, agricultural practices, land-holding and inheritance patterns, and familial- and village-level social organization are the sorts of culture elements that may survive from the Indian past and substantially affect the cultural geography of modern nations.

Programmatic Concerns

We are about ready for a major study that will attempt to answer, both in detail and with effort to draw significant generalizations, the question posed at the beginning of this essay: Where, to what extent, and why have Indian cultures and communities survived in Latin America? The partial and tentative answers presented here may indicate that a geographic perspective promises a sound framework for such a study. It cannot be a dissertation topic nor would it be a comfortable enterprise for one who needs regular publication to rise on the academic ladder. Its completion, however, will make a major contribution to the culture history of mankind and be of interest to historians, anthropologists, and historically oriented social scientists in general.

A great many more individuals have been and will be engaged in studying the Indians and their cultures in Latin America on the community or local region level. Few geographers have had, or are likely to have, the inclination, background, or physical durability to remain with a fugitive Indian group long enough to make a classical ethnographic study, but their ecological perspective could enhance its value. Gordon's work in the Sinú, Meigs's study of the Kiliwa, Bennett's work in Panamá, and Pennington's study of the Tarahumar are the only exceptions that come to mind.[15]

Geographers along with anthropologists have consistently made contributions in the form of community and local regional studies among the sedentary Indians descended from the high civilizations of Western Latin America. West, McBryde, Brand, Termer, and many others, pursuing a tradition that goes back to Sapper and Schmieder, have made contributions.[16] It is noteworthy that such studies may be conceptually almost indistinguishable from those of ecologically oriented anthropologists such as Tax, Foster, and Nash.[17] Both geographers and anthropologists have also worked on communities and small regions that have crossed the transition to become mestizo, as Aranda in Northern Chile, Wagner in Nicoya, and Stanislawski in Michoacán.[18]

Perhaps more significant is work by geographers focused on themes that are heavily dependent on Indian cultures. The work on domesticated plants and animals of the New World is a good example. Carl Sauer is perhaps the dean of such studies, but his students and others (J. Sauer, Gade, Johannessen, Keller, Baraona, and Bennett, for example) are maintaining a vigorous tradition.[19] Edwards's work on aboriginal sailing craft on South America's West Coast represents a parallel sort of investigation.[20] Land tenure studies involving Indian systems go back at least to McBride,[21] but they have acquired new and practical urgency with rising interest in agrarian reform in countries such as Chile.[22] At least locally the picture of impoverished Indian villages surrounded by progressive and prosperous national citizens is in error. Goins has recently observed a situation in highland Ecuador where Indian farmers, retaining their land through village endogamous marriages, enjoy far more security than the non-Indian town dwellers.[23] Through the parish priest they are actually tithing a third of their crops in a relief program. More themes and examples could be cited, but it should be clear that an active and productive area of geographic research is involved.

A final concern takes us into a realm pointed up by H. C. Brookfield in his "Questions on the Human Frontiers of Geography."[24] To what extent is it a task and capability of geographers to concern themselves with social process, social theory, and the value systems of the societies they study? The unstable juxtaposition of Indian and mestizo communities, sometimes along with Negro and essentially European ones as well, in many Latin American countries throws the question into sharp relief. As Brookfield points out, geographers have been chary about entering this field, marked by rarified theories, tautologies, and intellectual pitfalls. Some qualification may be made for certain community studies and historical treatises. Perception of environment studies seem to be our most active entry and they have just begun to extend beyond our national borders. A rich knowledge of the community to be investigated will be needed to develop valid questionnaires and protocols and to evaluate responses to them. It may still be legitimate to assert that a satisfying cultural geography of Latin America or any of its countries or major regions will not be produced until we understand not only the activities of individuals in the diverse communities but also the motivations and goals that induce them.

The Immortality of Latin American States

One can only hope that stability of now established frontiers will be the rule, since boundary adjustments that once affected little more than national prestige are likely to yield a harvest of enduring bitterness.

DESPITE THE DESIRE of the Hapsburg court to direct the Spanish Empire centrally from the Council of the Indies, the extraordinary extent of the empire required local executives with considerable freedom of action. The *Audiencias*, begun early and persisting long, were only partially effective, having little more than judiciary authority. The two viceroyalties of New Spain and Peru, with full governing powers, were firmly established before the middle of the 16th century. Despite their enormous extents they governed Spanish America until the Bourbon reforms of the 18th century cut two new viceroyalties from Peru, New Granada in 1717 and La Plata in 1776. At about that time several of the outlying Audiencias such as Chile, Guatemala, and Venezuela were converted into captaincies-general, and interior ones, notably Quito and Charcas, became presidencies. Both kinds of governments held executive as well as judicial authority.

With the independence movements all the viceroyalties and almost all the captaincies-general and presidencies formed independent foci of revolution. They became the bases of modern independent states. Despite Bolivar's and San Martín's efforts, the viceroyalties of New Granada and La Plata could not be held together, and further fragmentation ensued. The splitting of the Central American Republic, successor to the captaincy-general of Guatemala, into five independent states was deferred until 1839, and Panamá was split off from Colombia in 1903. While territorial adjustments following boundary conflicts and more general wars have been fairly substantial each of these states has survived to the present, developing its own national mythology and a self-centered economic policy, often to the detriment of its own people as well as to its neighbors.

In the past two decades we have witnessed the same fragmentation into national states of the administrative units of the British and French colonial empires in Africa, and it seems safe to predict that the economic consequences, at least for some of the smaller states, will be equally deleterious. The fragmentation of the British colonial empire in the West Indies has carried the process to an even more ridiculous extreme. It would seem that some pervasively centrifugal geopolitical force affects centrally administered colonial empires in the process of decolonization, but even more striking is the durability of the new states once they are formed. This is in notable contrast to Europe, where even in this century at least five states have disappeared into larger entities, and far more were amalgamated in the post-Napoleonic 19th century. An examination of certain events in the late 18th and the 19th centuries may throw some light on processes of state formation and maintenance in formerly colonial areas.

Why Carlos Tercero chose to detach Upper Peru, or Charcas, from the viceroyalty of Peru and attach it to that of La Plata in 1776 needs some study. The commonality of interest of two provinces that formed the heart of the Inca Empire and had been under a single colonial administration for nearly 250 years is patent. Both were heavily Indian in population, provided substantial taxable revenues from mines, and experienced a chronic deficit in agricultural products. Perhaps no more than providing the new viceroyalty of La Plata with some mineral revenue was involved.

In their efforts to maintain the independence of the former viceroyalty of La Plata after 1810 both Manuel Belgrano and José de San Martín made vigorous efforts to maintain the political unity of the area while politicians in Buenos Aires argued about whether a centralized state dominated by that city or a looser federation was appropriate. Actually the presidency of Charcas, culturally alien to Buenos Aires in all ways, remained loyal to Spain until it was isolated by the defeat of Peruvian Spanish forces at Ayacucho in 1824. It could accept independence but not dominance from Buenos Aires. Paraguay, with its long independent tradition under Jesuit tutelage, similarly could not be brought under control. The captaincy-general of Chile was an isolated frontier province of Peru and although victory over the Spaniards was gained with San Martín's support from Mendoza its deep, frontier-oriented provincial roots made full independence the only possibility. In the Banda Oriental, right at Buenos Aires's doorstep, the situation exposed all the forces of centrifugality. A substantial group favored federal but not centralized union with Buenos Aires as protection from Brazil. A loyalist group in Montivideo was supported by local merchants. The latter could easily shift to favoring an independent Uruguay when they recognized that their trading prospects would suffer if Buenos Aires controlled their commerce. Brazil's interest in annexation had historical support in terms of priority of effective settlement. British interest in fragmenting political power along the La Plata, after attempts at colonial annexation had failed in 1807, probably was decisive in securing Uruguay's independence.

It is interesting to speculate on the potentialities for economic, cultural, and political development the great state of the United Provinces of South America, proclaimed in Tucumán in 1816, might have had, but sufficient federal guarantees, for which the United States might have provided an example, were not given. The United Provinces, minus Charcas, would have had a largely European and mestizo population, been united by language and religion, enjoyed tremendous agricultural potential ranging from temperate to tropical conditions, and substantial mineral wealth.

The splitting of Gran Colombia was simpler. The viceroyalty broke up into the smaller units of the three Audiencias, each with an established metropolitan capital.

The first major postindependence effort to unify two states was carried out by Andrés Santa Cruz from 1835 to 1839 in Bolivia and Peru. The cultural and geographical basis for the union was clear and had been since pre-Columbian times, and, rather unusually for Spanish America, the boundary between the two viceroyalties, selected only in 1776, passed through fully settled districts. Argentina opposed the union, holding a residual claim to Bolivia, but it was Chile's effective military intervention that defeated Santa Cruz and blocked the union. With great internal ethnic and economic disparities unbridged, both countries have had troubled subsequent histories, and Bolivia usually manages to check in as the poorest country in South America. Whether the combined states would have fared better must be conjectural, but one suspects it might have. Chilean historians, justifying their national action, speak of the threat the combined countries would have offered to Chile's independence, and under the viceroyalty Chile had a long history of economic subservience to Lima.

At the same time the United Provinces of Central America, successor to most of the captaincy-general of Guatemala since 1823, were coming apart into the five states that persist to the present. Physical isolation of the several centers was combined with the fact that all had such limited resources that a leader in one could not long maintain control over a neighboring state, let alone over the whole area. Britain and the United States, in recent times especially the United Fruit Company, found commercial advantage in dealing with several small states rather than one larger one, and encouraged independence. The same foreign interest certainly accounts for the separation of Panamá from Colombia, the last mainland state to be formed. The reasonable success of the Central American Free Trade Association raises the possibility that reunification of the United Provinces on a federal basis might yet evolve. But differentials in living standards as well as quarrels such as the recent one between El Salvador and Honduras militate against it.

The Wars of Lopez, 1862–68, involved an intensity that was new and fortunately so far unique in Latin America. They resulted in the complete occupation of Paraguay by Argentina and Brazil and the actual death of most of that country's population. Erasing Paraguay from the map and splitting it between the victors would have been the easiest and cheapest, and a case could be made for it being the most humanitarian solution. Certainly the unwillingness of either Argentina or Brazil to let the other gain advantage played a part in Paraguay's preservation as it had earlier in the creation of an independent Uruguay. But I think by this time the ideology of the right of each Latin American "Pueblo" to permanent independence played a part. It is difficult to imagine how the collective resentment of other Latin American states could have been applied effectively on two of the largest and strongest ones, but the latter were concerned. Eventually the occupation was lifted, and while Paraguay had to make substantial territorial concessions they were limited to border areas to which either Brazil or Argentina had at least a faint historic claim. Paraguay as a state survived in isolated and economically undeveloped independence, its lands amply extensive and productive for the small surviving population.

The relative indeterminacy of administrative unit boundaries in the Spanish colonial empire left the successor states with a heritage of boundary disputes, most of them in unsettled areas. Shifts in borders on political maps have continued well into this century, the most recent major one involving the cession of most of the Ecuadorian claimed Oriente to Peru in the early 1940s. In two instances, the War of the Pacific in 1879–83 and the Chaco War in 1932–38, serious fighting preceded a substantial boundary shift in favor of the victors. In other cases compromises were reached with less violence. In no case, however, was the core of any state threatened with annexation. In general, boundary modification favored the stronger, larger, and economically more vigorous contestant. This would certainly apply to Brazil's expansion up the Amazon Basin, Chile's conquest of the nitrate-producing provinces of Bolivia and Peru, and Peru's taking over the Amazonian lands economically tributary to the Peruvian city of Iquitos.

For some four decades Chile's northern conquests seemed to yield a profit, but with the collapse of nitrate prices after 1920 they have been a considerable net drain on the national economy. An example is the attempt to create prosperity and purchase political loyalty by assembling the country's automobiles in Arica, perhaps the least suitable place in the country. One gets the feeling that maintaining a boundary dispute on a chronic basis is regarded by the governments of both states as politically advantageous. As an example, the Beagle Channel, south of Tierra del Fuego, and overlapping claims in Antarctica are always available to Chilean and Argentinean politicians for a threatening exchange of notes to make headlines when national economic news is bad. On a somewhat more serious level, an annual festival of nationalistic breast-beating fills Bolivian newspapers as they condemn the loss to Chile of the province of Antofagasta. The Chilean press ignores the event.

In much of Latin America today, in Central America and in the La Plata drainage for example, population growth and settlement have filled in the formerly nearly vacant lands that once separated the core areas of the several states. Shifting boundaries would mean transfer of people to different sovereignties. The region has become more like Europe, and one can only hope that stability of now established frontiers will be the rule, since boundary adjustments that once affected little more than national prestige are likely to yield a harvest of enduring bitterness.

To an outsider the political fractionation of Hispanic America seems to be economically debilitating, though a single government of so extended a territory would be unlikely to be effective despite unities of language and religion. Something approximating the three viceroyalties of South America plus the captaincy-general of Guatemala might make economic sense. The specific economic disability, one which has made the Hispanic American economic development look far less favorable than Brazil's in recent decades, is the insufficiency of the national market to support efficiently large assembly lines for complex manufactures such as automobile assembly and parts production. One does not get a feeling that free trade associations will permit the concentration of such prestigious and high-wage activities as automobile manufacturing in one of a group of countries regardless of the economic efficiency of doing so.

With some overlap the forces that have so long maintained these many separate entities as states can be characterized as political, economic, and ideological or philosophical. Each state provides a set of opportunities for both politicians and bureaucrats at higher levels of prestige and power, if not wealth, than most of them could enjoy in a larger though richer combined entity, and in less-developed economies both statuses are attractive. Further, the only way that peaceful unions could come about is with firmly guaranteed federalism, and the historic trend in all Hispanic American states, even more so than in the United States, has been toward ever more centralism. Not only political power but investment and industrial development have gravitated toward the capital city, which in all countries except Ecuador has taken over an increasingly dominant position of primacy. Neither political nor commercial interests would like to risk their capital's dominant position.

Although it is a relatively new development, the fact that each state holds a vote in the United Nations General Assembly is generally regarded as an asset. Exchange of a vote for some economic favor from a concerned larger state might be significant for the smaller states.

More strictly economic considerations make it in the interest of influential groups or classes to preserve the status quo of tariffs and import restrictions. Industrialists and their employees have protected access to the national market for products that may not be competitive internationally on the basis of price or quality. Conversely, when a small country does not have an industry to protect, it may admit those goods without or with low tariff charges. Uniting with countries that have such industries would almost certainly bring protective tariffs, higher prices, and lower quality, and the sorts of goods that are involved such as automobiles, kitchen appliances, and plumbing supplies are of particular interest to wealthier and thus more influential individuals. Wealth and standards of living vary considerably between the Hispanic American states. With exceptions in Uruguay and possibly Costa Rica there is a generally positive correlation between the size of a state and its per-capita wealth. The richer of two states is unlikely to seek a union that, in the short term at least, would have a negative impact on its living standard. Thus although an outside observer can envisage considerable middle- and long-term benefits from establishing larger economic entities, the short-term impact on at least an influential part of the combined population is perceived as negative.

As part of the process of establishing nationhood, immediately or shortly following independence each of the new states set up a Ministry of Education and established a school curriculum. National history was one of its important components. On the scholarly level there are many distinguished Latin American historians who have been willing to portray their nations' past with "warts and all," and more who took domestically partisan stances on many issues. As in the 19th century in the United States, however, history books for elementary and secondary school students take an unabashed nationalistic line right to the present time. This is accented in the treatment of boundary disputes and accounts of the several postindependence wars, especially in those countries that lost territory. While all of Spanish literature may be studied there is special emphasis given to that produced in one's own country. This is particularly true in regard to poetry, a form of expression that flourishes vigorously in Latin America.

At one time or another all the Latin American states have suffered from extreme internal political instability, and many do so today. Violent overturnings of governments have been frequent, with harsh reprisals often inflicted on those defeated. The sanctity of asylum in a foreign embassy, usually another Latin American one, has rarely if ever been violated, with safe exile permitted when tempers have cooled after some months. This curious form of security in a world that is dangerous for political leaders is a nearly universally accepted cultural value in Latin America. It both supports and is supported by the idea that each country has an enduring, perhaps eternal, place in the world scheme.

FLORA

AND

FAUNA

Introduction

Charles F. Hutchinson

Homer Aschmann's research interests were shaped by places familiar to him. The semiarid coastal basins of Southern California and their adjacent deserts, where he was raised and later lived and taught, provided the themes of his work and the backdrops against which they were painted. Despite his many travels throughout his career, he never strayed from these roots for too long.

Southern California during this century has been an exceptionally fertile ground for geographic research on environmental change. Its natural and cultural landscapes tend toward the spectacular and noteworthy: expansive beaches, snow-capped mountains, and endless sunshine, overlain with a phenomenal growth in absolute population and cultural diversity that have collectively influenced such things as civil unrest, disruption of native vegetation, susceptibility and costliness of earthquakes, mudslides, and fires. It could be argued that Southern California has experienced greater, faster, and more complete landscape modification than any other region in the country—perhaps more than any other region on Earth.

Aschmann's interests went beyond the obvious, superficial, and extreme aspects of the Southern California landscape. His insights tended to focus on the interesting details, those aspects that give character to the landscape we perceive but are not blatantly apparent, whether they are contained in the behavior of children toward fruit trees or the effect of management policy on plants. His consideration of the landscape was not in isolation, but stressed the way the landscape had been shaped and reshaped by a multiplicity of human actions. He worried that this emphasis is too frequently dismissed by geographers trained in a later era.

California in general and Southern California in particular offer an appropriately stylized and condensed variation on "European" settlement of North America. It begins with a small, benign American Indian population based mainly on hunting and gathering, living for millennia in relative harmony with the environment. This population is then partially subjugated by another small but slightly less environmentally benign group of Spanish—and later Mexican—pioneers intent on developing a stable, largely self-sufficient agrarian society. Despite changes in the demographic and cultural makeup of the area, many considered life in this area an idyllic existence, and it was sustained for more than two centuries. The balance was upset by the unparalleled and expansionist population growth of the eastern United States, a frontier mentality that led to the purchase and conquest of lands in the West and, ultimately, the discovery of gold and other minerals in the region. All these factors converged to attract a stampede of settlers that were to reshape the entire region.

Detecting and describing the successive traces left on the landscape by each of these groups became a major focus of Aschmann's work. To capture evidence of prior occupants and land uses, he used just about all the tools in the geographer's kit, examining everything from the deepening of arroyos to the persistence of place-names. He had a special interest, however, in the changing vegetative blanket that covers the earth, provides shelter and sustenance to the creatures that live within it, and, more than any other ingredient, gives landscapes their character.

The study of vegetation change has become somewhat of a cottage industry in the western United States. Interest in this topic may be due in equal parts to the attraction of the dramatic landscapes of the region, to the relatively short history of "European" occupation, and, compared to the eastern United States and the obvious areas of local population concentrations, the discontinuous impact of humans.

Vegetation occupies a peculiar place in the study of change. More than any other feature observed on the Earth's surface, vegetation is presumed to be determined largely by climate. Before the existence of a network of climatological stations, vegetation was used as the primary indicator of regional and global climate. Thus, many climate types were named after vegetation types, and that convention has continued because the linkage is so basic and immediately understood. Particularly at regional and global scales, this connection makes sense because the pattern of discontinuous patches of vegetation can be separated from confounding local effects of soil and land use. However, at local scales, it becomes increasingly difficult to "see the pattern for the patches," especially as the imprint of human activities continues to spread inexorably.

Because the traditional presumption is that climate and vegetation are inextricably linked, it is easy to conclude that vegetative changes reflect a change or variation in climate. But to tie vegetation change to climate change is a risky proposition. Despite the seductive force and simplicity of this argument, there are several intertwined hypotheses that might be considered before climatic change is embraced as the most probable agent of change. These hypotheses form the basis for much of Aschmann's work.

First, the dynamic history of vegetation and its relationship with climate may not be well understood or even known. As Aschmann pointed out, fire was initially judged to be a bad thing in Southern California. It destroyed vegetative cover and thus threatened the sustained yields of precious water from steep watersheds that flank the region's semiarid basins. More significant, in terms of the reaction it caused, is that the fires also destroyed valuable hillside housing. As a consequence, the first forest reserves were created in Southern California to protect mountainsides from excessive logging and also from frequent fire.

Rainfall in Southern California occurs almost entirely in the winter, and total dryness in summer is the norm. Thus, not only are temperatures high during the dry season but, consequently, so too is the incidence of fire. Not understood until recently was the degree to which many Californian vegetation types depend on fire. Most of the major vegetative types have evolved under a regime of periodic fire, on the order of every 20 to 40 years. It was eventually learned that vegetation depended on fires for renewal because fire opened the vegetation canopy, removed senescent plants, converted the nutrients they contained to a form more easily taken up by plants, and prepared seeds for germination. For the vegetation of Southern California, fire was essential.

A second—and related—reason to question climatic hypotheses for change relates to the consequences of land management policy, particularly when it is based on the aforementioned limited or faulty knowledge of vegetation dynamics. Policy is something as impalpable and elusive as light, but it carries consequences that often are too tangible.

As we have seen, fire in Southern California threatened water supplies and real estate. Thus, there was a perceived need to reduce the frequency of fires, to conserve vegetation and hence water, and also to reduce the risk of damage to real property. An unwavering policy of fire suppression was adopted in Southern California and went unquestioned for the better part of a century. As Aschmann observed, the results of this policy were not as intended. The frequency of fire was indeed reduced because there were fewer fires. However, the size and damage caused by any fire allowed to catch hold was magnified by increased fuel loads: the plant material that might have been consumed in a larger number of relatively low-intensity fires accumulated and fueled a smaller number of large conflagrations. Also, without the periodic renewal of fire, the composition of the vegetation began to change, with a decrease in plant species diversity and a reduction in faunal habitat quality.

Equally important, the perceived "control" of wildfire hazard in the desirable foothills of Southern California encouraged residential development in inherently fire-prone areas. Although fire frequency was reduced, the risk of loss was increased by the greater likelihood of catastrophic fire owing to fuel accumulations, as well as to the increased number of developments in exposed areas. This resulted in a pair of paradoxes: (1) Southern California's vegetation was altered through efforts to protect it, and (2) the risk of fire damage was greatly increased by attempts to minimize it.

The third and perhaps most critical problem that challenges theories of climatic causation of vegetation change is the general lack of understanding of preceding land uses and their selective impacts on vegetation resources. Once "natural" causes of vegetation change have been discarded, perhaps as an unintended consequence of the environmental movement, there was and often still is a tendency to establish blame rather than to understand human factors driving change.

In virtually every paper reproduced in this section of the present volume, Aschmann sought to determine the economic, social, and institutional conditions that brought about vegetation change. Rather than arbitrary, misguided, or irrational, many—if not most—human actions that might affect vegetation have an economic basis and should be understood in that context. This seems to be a difficult transition for many to make, particularly when attempting to explain the causes of recent vegetation change.

The impacts of human activity in the region are significant, despite being concentrated in a relatively short period of time. Through the diversion of an unprecedented volume of water, the deserts of the Central, Coachella, and Imperial valleys now bloom—but at the expense of every vestige of natural vegetation and creature they once supported. Such diversions permitted the growth of virtually all metropolitan areas in California.

The impacts with which we are most familiar—urbanization, irrigated agriculture, and mining—also are the most obvious and persistent. However, in considering longer-term historic impacts on vegetation, it should be recognized that there were more routine and extensive uses of land that left subtle and transient effects on vegetation, of which some traces still persist. In his later works, Aschmann was able to capture the significance of these influences by contrasting the vegetation patterns of Chile and California.

Parts of Chile and California experience the same type of climate and have comparable physical environments. The vegetation types found in each place are structurally and functionally similar but floristically distinct. Moreover, because they are each located in the New World, these two places share a broadly similar settlement history, albeit one that differs in particulars, especially in terms of settlement. In particular, their economic histories diverge, most dramatically in population size, degree of urbanization, and utilization of indigenous natural resources. Because of their physical and climatic similarities, Chile and California offer a unique opportunity to study biologically convergent evolution. However, when examined in light of the differences in their economic history, the divergence of their vegetative patterns also permits study of the impacts of human activity on vegetation.

In Southern California, vegetation currently is viewed as a resource, but one that may not have any direct economic value. Obviously, like any other natural vegetation, it has its own scenic as well as habitat value. As already mentioned, its role in maintaining watersheds and stabilizing slopes was recognized early and has held a prominent place in local consciousness. Conversely, it also has been viewed as a fire hazard. With very few local exceptions, natural vegetation plays no direct role in the local economy. Where it has not been removed by urbanization or agricultural development, those factors that have shaped vegetation have come from broad policies, notably in fire suppression.

In Chile, Aschmann found a very different situation. There, much was asked of local vegetation resources: some provided foods and medicines for humans, feed for livestock, or fuel for domestic heating and cooking, and others wood for construction, shoring in mines, and fence posts. Not surprisingly, Aschmann learned that campesinos favored different species for different uses, so there was selective shaping of vegetation patterns. The vegetation there differed from that in California because it was treated as an economic resource and, moreover, one with comparatively few restrictions on access or utilization, much as had been true in preindustrial California. It might be argued, then, that in terms of resource perception and utilization, a small rancher or farmer in the San Fernando Valley a century ago would have more in common with a Chilean campesino than with any valley resident today.

There are at least two important lessons to be learned from Aschmann's work on vegetation that have a direct bearing on current interests in global change. First in the larger part of the world, vegetation is a resource used routinely to support a host of economic endeavors. As a consequence of population increases, much of the change observed in the developing world—as seen in Chile—is the product of degradation of vegetation resources. However, the heavy use of vegetation resources has been in practice for the vast part of human history; it is only in the last century that reliance on these resources has eased in most of the developed world. As demonstrated in California, much of what we see in our own landscape is the product of earlier exploitative uses that have been rendered unprofitable or unacceptable through changes in the larger economy. Because these activities no longer exist, they are difficult to imagine. Thus, it is almost impossible for tourists in the San Bernardino Mountains to believe that the even-aged forest they are enjoying was clear-cut a century ago and that the road upon which they are driving was built along the right-of-way of a logging railroad.

The second lesson to be learned is the difficulty and dangers of establishing simple "change." As already mentioned, vegetation responds to climatic cycles. It also responds to cycles of human use. When these cycles are overlaid—as they inevitably must be—it is often impossible to determine where we might be on any trajectory of change.

Typically, initial conditions against which change is measured are established from historical documents, such as written descriptions or photographs. Subsequently, the same sites are visited, and comparable observations or measurements are made. The difference between the established initial conditions and the later observed conditions constitutes change. It usually is possible to assemble a picture of climate over this period because of relatively long-term meteorological observations and, perhaps, to correlate changes in vegetation with climate variations. However, as suggested in Aschmann's work, land use and land-use policy may be far more significant in understanding vegetation change in the western United States, particularly over the short term. For example, observations made in virtually any plant community in the Los Angeles basin will show a significant change between 1890 and 1990. These changes might include increases in total cover and, more than likely, an increase in woody shrub cover. The conclusion would be that change had occurred. Although true in the strictest sense, this conclusion would ignore the fact that there had been a tremendous population boom in the decade of the 1880s and that by 1890 these resources were under heavy use pressure and subject to regular fires. This might lead one to question whether this represented a "natural" initial condition. By 1990, re-source use had ceased three generations earlier, and a fire suppression policy had been in place for almost a century. This might lead one to question what type of "change" actually had occurred.

Aschmann showed that without an understanding of current and historic land uses and land-use policy, it is usually impossible to establish a causal relationship be-tween vegetation change and any other factor. For reasons that are unclear, the hy-pothesis of human agency in vegetation change was never particularly popular and has not yet emerged as a major theme in the current era of global change research. However, as new emphasis is granted to the "human" dimensions of global change and purely climatic explanations are examined more critically, this line of reasoning must become more common.

Man's Impact on the Southern California Flora

Fire, fire protection policy, and its implementation is perhaps the set of forces that in this century have had the greatest but least understood impact on Southern California's remaining wild vegetation.

A GROWING STREAM of archaeological evidence makes it ever clearer that man has been present in Southern California since Wisconsin glacial times if not considerably longer. In those times the local climates were undoubtedly considerably wetter and somewhat cooler. The vegetation, though probably not too different in floristic composition from the surviving native flora in a regional sense, must have enjoyed differing distributional patterns, with arboreal and mesophytic associations more widespread and xerophytic ones more restricted. Thus, when climatic patterns swiftly acquired their present characteristics between 12,000 and 10,000 years ago, man and his activities were part of the environment in which the vegetative associations would come into a new adjustment. Although there have been decades of wetter and cooler or drier and warmer weather during the past 10,000 years, there is essentially no evidence that climatic fluctuations in this long period were of significantly greater amplitude than those recorded in the hundred-year instrumental record and the additional hundred years of historical anecdote that we have.

The Prehistoric Record

The archaeological record of the first 5,000 years is distinctly thin, suggesting a low human population density. This population did control fire and it established semipermanent campsites. In the 5,000 years preceding 1769, there was a considerable population growth, and it is quite possible that by the beginning of the Christian era or a few centuries earlier, the density level encountered at contact had been achieved. By that time also almost all the cultural and technological inventory was present, and gathering and processing vegetable stuffs for resources additionally permitted a notable concentration of settlement along the shore, again matching that found by the first missionary settlers. Thus, for millennia, the vegetation was affected by and adjusted to the same, basically conservative human impacts. Only along the Colorado River was farming practiced, and even that, since it was carried on only in places left bare by receding flood waters, had minimal effects on the vegetation.

The principal element in man's impact on the vegetation was undoubtedly his initiation of fires. That the California Indians set fires at the time of contact is adequately attested in the documents of exploration. The reasons for, frequency of, and topo-

graphic and seasonal patterns of burning are far less clear. Viewpoints range from assuming no more than escaped campfires to a pattern of deliberate landscape management by burning to modify the vegetation to yield more useful plant products and
support more game, with fire drives for game and recreational burning as intermediate points. Even at the minimal level of accidental fires there would be a substantial
increment to lightning-set fires, probably at least a doubling. Further, accidental fires
could occur in any season and weather condition, being perhaps most probable at
times of strong wind and low humidity, Santa Ana conditions. Lightning-set fires, on
the other hand, while occurring in summer and fall, would often be accompanied by
rain and always with at least moderately high atmospheric humidity. Assuming deliberate burning would make likely an order of magnitude of greater fire frequency
than would occur without the agency of man.

The most absolute difference between Indian-caused fires and the modern situation is that there is no evidence that the Indians had either the will to or the capability of controlling or suppressing fires.

Under these circumstances, a probable result is that any vegetated surface was
burned at intervals one-half or one-third as long as those that now separate burn-
ings of our wildlands. Modern burn frequency in chaparral is about once in 40 years.
Lowland, naturally grassy areas now are so culturally modified that no estimate can
be given for how often they might burn, but under Indian occupance a burn frequency greater than that for chaparral is likely. A major result of the relatively high
frequency of burning suggested here is that fuel accumulation was kept down and
fires in the chaparral burned with less intensity.

It is less certain how more frequent but less intense fires affected the distribution
of vegetational associations. Perhaps the clearest variance from present conditions
would have been on the border of the chaparral and the coniferous forest associations. With frequent fires, a light burn might run through the grass and litter on the
floor of the coniferous forest without seriously damaging grown trees. Where heavy
brush interpenetrated the conifers, an intense brush fire would tend to crown in the
conifers, killing them and perhaps permanently reducing their areal spread.

In the accounts of the first overland expeditions from Baja California to San Diego
and Monterey in 1769, there are frequent descriptive references to landscapes along
the route. References to recently burned grass are not uncommon, and good pasture is described in a number of places where we now find chaparral. The suggestion
that frequent burns sustained a park landscape with grass and scattered oak trees has
been made, and its corollary is that chaparral has spread more widely subsequently.
In the northern parts of California there is an experience table based on experiments
with prescribing burning to keep down brush, to open the woodlands, and to improve grass pastures. No such experiments have been carried on systematically in
Southern California, so the impact of frequent, low intensity fires on chaparral at low
elevations and on gentle slopes remains unassessed.

Indian settlements like all human dwelling left refuse middens with high concentrations of nitrogenous matter. A number of recurrently occupied village sites at favored localities, for example near water sources, are still recognizable by their dark heavy soils. Along the coast the human population was denser and more permanently resident as well as richer in the sense of regularly having parts of marine organisms to throw away. It may be safe to say that the soil characteristics of the lower marine terraces and the edges of the estuarine embayments have been enriched in an almost continuous strip. The crumbling shell and dark soil that residents of Corona del Mar and Laguna find in their yards are products of human activity. By now, however, virtually all such terraces have been successively cleared and cultivated and then covered with residences and gardens. If they once bore a distinctive midden vegetational association, high in such weedy herbs as amaranths and chenopods as might be expected, that association is hard to recognize. There are early references to Indians planting tobacco, and village midden sites would be happy locales for that plant.

The large Indian population of Southern California sustained itself largely from gathered vegetable material. Acorns in the valleys and uplands, mesquite and palo verde pods in the deserts, and piñon and other pine seeds in the mountains were collected and consumed. The seeds of grasses and herbs, and roots found in meadows also were exploited. One gets a strong feeling, however, that this vegetal exploitation by man was not destructive of the species utilized. There were always enough seeds missed in the collection process to reproduce fully the parent generation, and digging in the meadows for bulbs and tubers probably acted as tillage to enhance the environment for just those plants.

A picture comes into focus of a long-term stable ecologic balance with many humans supporting themselves from a rich and productive wild vegetation without notably impoverishing that vegetation in terms of the products they sought, collected, and consumed.

European Entry into Southern California

The changes that began promptly following the arrival of the missionaries in 1769 and especially after the heavy imports of livestock from Sonora five years later were substantial. It was consistent mission practice to reduce the number of Indian settlements, concentrating them at missions or at a limited number of *pueblos de visita*. Further, while most of the Indian *rancherías* tended to relocate seasonally to take advantage of seasonally available resources it was mission policy to establish permanent residence. Thus, large areas that had been combed for their useful wild biota in rough proportion to their productivity were largely abandoned while right around a mission, even though the population concentration was supported primarily by agriculture, the wild biota was exploited with excessive intensity.

Until the missionization process was well advanced, that is until essentially all the

Indians in an area were under firm mission control, herds of livestock had to be protected carefully from human as well as animal depredation and were grazed only in the immediate vicinity of the missions. It was well into the 19th century before extensive land grants extended grazing activities throughout the coastal lowlands and into the foothills of Southern California. Grizzly bears kept significant grazing out of the mountains until the beginning of the American period in the 1850s. As of 1800 man's impact on the vegetation had been concentrated in roughly 10-mile radii around each of the several missions and the pueblo of Los Angeles, reduced in other parts of the coastal lowlands and foothills, and remaining in essentially its aboriginal state in the mountains and desert interior. The extent of the areas actually taken under cultivation was small, probably less than 20 square miles in all of Southern California. Close to the missions, however, overgrazing was severe as herds built up rapidly and wood cutting for fuel and construction was similarly concentrated. Until 1850, the total human population declined as immigration did not equal losses from excessive Indian deaths at the missions. But grazing activities were extended rapidly, and herds of cattle and horses came to saturate the grazing capacity of the region.

It is probable that during this period the enormous displacement of native grasses, and to a lesser extent herbs, by introduced weedy grasses and herbs took place. The inception of the event is not documented, but it certainly had been accomplished by the 1880s. The major invaders were from the Mediterranean region of the Old World and included wild oats (*Avena fatua*), foxtail barley (*Hordeum murinum*), several wild mustards (*Brassica* spp.), and in special localities *Mesembryanthemum* and *Chenopodium*. The capability of resisting heavy pasturing, perhaps an evolutionary development in the 6,000 years or more that domesticated animals have been grazed around the eastern Mediterranean, seems to have given the invading species their overwhelming advantage over their native competitors.

In contrast to exotic annual plants, which quickly naturalized themselves in much of Southern California and even became dominants, introduced perennials, while growing successfully when planted, have generally not spread into the wildlands. The eucalyptus is a particularly good example. In addition to having been planted as roadside ornamentals and around residences there were massed plantings, beginning in the 1880s, which were planned for timber production. The plan never worked out but remnants of many groves survive, often with stump-sprouting. They have just not spread.

Nicotiana glauca has established itself widely but sparsely in road-cuts and elsewhere where raw mineral soil is exposed. Castor beans (*Ricinus communis*) escape from ornamental cultivation and become part of the chaparral on frost-free slopes, but they form only a minor part of the vegetation even on favorable sites. A few other examples might be adduced, but they do not prevent the assertion that Southern California's wild perennials are natives.

Some changes in burning patterns undoubtedly accompanied the disappearance of the free-ranging Indian population and its gradual replacement by smaller numbers of ranchers. Even if man-caused fires during Indian times were only accidental a reduction in their frequency is expectable. An even greater reduction is likely since deliberate burning by the Indians is probable, and comparable action by ranchers is very infrequently reported in Southern California. The thinning of stands of grass by grazing made less inflammable and capable of carrying a fire that part of the wild vegetation which was most easily ignited. The decline in fire frequency, involving both fewer fires and less extensive burns, favored the extension of chaparral at the expense of grasslands, especially in areas of moderate slopes.

Development of Modern Land-Use Patterns

The effects on the wild landscape of the annexation of California to the United States were relatively slow to develop in Southern California. Until about 1860, the population grew slowly and there was only a modest increase in grazing pressure resulting from the market for beef in the Gold Rush Country. Threats of both Indians and grizzly bears in the mountains were eliminated and wet meadows there were grazed heavily by both cattle and sheep on a seasonal basis. Between 1860 and 1890, however, agricultural and pastoral utilization of land expanded tremendously rapidly. Except for the relatively limited areas that have subsequently been developed by irrigation the modern limits of dry-farmed and grazed land had been reached and probably exceeded by 1890.

The severe droughts of the 1860s may have played a direct part in the displacement of native grasses by cosmopolitan weeds in the coastal lowlands, and foothills, but this is hard to document. It is clear that drought, taxes, and lawsuits accelerated the breakup of the large land grant estates that held the most cultivable parts of Southern California. Government land available for homesteading had been limited to very rugged or nearly desertic localities. The purchasers of middle-sized farms cut from the large estates tended to clear and cultivate, largely in small grains, all their land that was not too steep, too rocky, or too wet, and locally such cultivation extended beyond what are present economic limits. Extensive hillslopes have subsequently reverted to pastured grasses or brushlands.

In the 1870s and 1880s there was an enormous expansion of irrigated agriculture in the coastal basins of Southern California. It accompanied the great influx of people and capital from other parts of the United States that was initiated at that time. Irrigation projects usually involved associations of relatively well-off smallholders who shared in the purchase of an undeveloped or only dry-farmed tract of land. Accessibility to water and good air drainage that offers protection to frost-sensitive crops such as citrus made the alluvial fans at the southern and western bases of the Southern California mountains the choice places for irrigation development. Deciduous trees and irrigated field crops were grown on lower ground with greater frost incidence.

The completion of major irrigation works that imported water from exotic sources into Southern California, although often initiated primarily to serve urban needs, has led to specific expansions of irrigated acreage. The earliest and still the most important of these involved the use of Colorado River water in Imperial Valley and wherever the river itself had a floodplain, as near Blythe and Yuma, that began shortly after the beginning of this century. After 1944, the All American Canal brought Colorado River water to the Coachella Valley as well. The Owens Valley Project made possible expanded irrigation in the San Fernando Valley, though generally on lands already dry farmed. This irrigated agriculture now has largely been displaced by urban developments. The Metropolitan Water District's importation of Colorado River water over the mountains into the coastal lowlands similarly permitted some expansion of frost-sensitive irrigated cultivation into hilly areas of western Riverside and northern San Diego counties. The arrival of Feather River water [through the California Water Project] is producing a modest agricultural expansion in the high desert at the northern edge of the San Gabriel Mountains, but the inability to grow frost-sensitive specialty crops makes it difficult to meet high water costs.

Almost inexorably the extraordinarily sprawling urban developments of Southern California have pressed into irrigated agricultural land. Only in the irrigated lowlands of Ventura County has there been effective resistance to this tendency. Only in a few limited areas has urbanization encroached directly on wildlands. Some instances merit mention because of their impact on fire suppression policies. In the eastern Santa Monica and along the southern edge of the San Gabriel Mountains high value, low-density residential developments have left chaparral on unbuildable slopes or elsewhere for its intrinsic attractiveness. Coniferous forests in the mountains are unusual enough in Southern California to constitute in themselves an attraction for resort development. Regulations of the National Forests, which encompass most of the areas of coniferous forest, have brought about localized, high-density resort developments. But, since it is a major part of a resort's attractiveness, as much as possible of the wild vegetation is allowed to remain, even where housing density is at fully urban levels.

In the past few decades urban developments have proliferated, generally in the form of resort or retirement communities at various points where both the low and high deserts approach their bordering mountains. Only along the northern edge of the Santa Rosa Mountains have these developments been extensive enough to threaten whole wild plant habitats.

The enormous commercial, industrial, urban, and recreational developments along the coastline have almost eliminated the wild plant habitats associated with it in Southern California. Salt marshes and lagoons have become harbors or marinas, and the coastal terraces are almost solidly covered with houses. The idea of protecting some of the coastal zone in its more or less natural form arose too late to have much effect in Southern California and the northern coast of Baja California, though there is still some hope farther north.

Modern Fire Protection Policies

Fire, fire protection policy, and its implementation is perhaps the set of forces that in this century have had the greatest but least understood impact on Southern California's remaining wild vegetation. Without much deliberate effort and as a by-product of a changing economy and land use pattern fire frequency had been notably reduced during the 19th century. The Indians who had camped over all but the most desertic parts of the area and probably deliberately and certainly accidentally set fires largely disappeared. Much of the grassland was either put into cultivation or grazed so heavily that it would scarcely carry a fire.

But some fires occurred, either set by summer thunderstorms that produced more lightning than rain or by accidents associated with human activity. Their consequences came to be recognized as destructive and threatening. Fires associated with logging activities in the San Gabriel and San Bernardino Mountains destroyed some valuable timber, but the effects of runoff-produced detritus from denuded slopes on the developing irrigation systems seemed more serious. When the San Gabriel Forest Reserve, the first in the United States and the lineal ancestor of the National Forest System, was established in 1892 its major charge was *watershed protection*. Within the next 15 years much of the chaparral and coniferous forest covered wildland in Southern California had been made part of a National Forest. In the National Forests, and by extension in their bordering brushlands, fire prevention and suppression became established policy.

Until about 1933 there was an approximate standoff between growing ability to suppress fires, enhanced by access roads and fire breaks, lookouts, and the ability to employ fire-fighting teams, and the increasing incidence of accidentally set fires resulting from growing use of wildlands. Some disastrous fires did occur, however. In the later 1930s, fire suppression was especially effective because of the availability of the Civilian Conservation Corps, but this advantage was lost during World War II and fire incidence increased.

Since World War II, fire suppression techniques such as helicopter transport of firefighters and aerial bombing of fires with retardant chemicals have made great progress. Most fires, accidentally or lightning set, are suppressed within a few acres. Total acreage burned per decade has declined, and a reasonable estimate of the average period between fires in chaparral is 40 years. Many stands are much older. Without frequent ground fires in the grass below coniferous forests, brush is establishing itself and forming a ladder up which a fire can crown with lethal effect on the trees. In recent times the season when fires occur has tended to become concentrated in August and September. Such fires are likely to be particularly intense because not only are annual plants generally dead and dry, but the nondeciduous perennials hold minimal moisture in their leaves and stems.

All these factors, of which the accumulation of dead stems in old chaparral stands may be the most critical, make the rare fire that occurs under dry Santa Ana wind conditions in late summer or fall an uncontrollable conflagration capable of jumping fire breaks and even superhighways. A number of individual burns in the past two decades have affected tens of thousands of acres, often burning out whole resorts or suburban communities. A vexing politicoeconomic confrontation arises as citizens whose properties are threatened by wildfires demand ever more effective fire protection. As the protection is more effective the amount of dead fuel continues to build up in overage stands of chaparral. When a fire does occur it is hotter, more violent, and faster moving.

Many of the wild plant associations of Southern California, notably the chaparral, have developed under conditions of frequent burning. The recovery following a fire is often vigorous as root crowns sprout and germination of some seeds is stimulated by moderate heating. What is new is the great intensification of fires resulting from dead fuel accumulation for many decades of effective fire protection. Investigation of different recovery successions following mild fires in grass or thin brush and very intense ones in overmature chaparral are only beginning. One would expect considerably different specific composition in a stand of chaparral that succeeded a mild burn where many stumps sprouted and seeds survived to germinate and one in which everything was killed, but the knowledge that would permit prediction has not been developed.

Two vegetation associations which are widely regarded as particularly attractive seem to be vulnerable to especially intense fires, the oak-grass savanna and the coniferous forest. Fully grown pines and oaks are not seriously damaged by grass and litter fires that burn at their bases. They will be killed by a hot fire if over-mature chaparral has time to establish itself among or alongside them. The upward retreat of the conifer timberline, followed by an advance of chaparral is well documented in the mountain ranges of Southern California. Several extensive losses of pine forest have resulted from severe fires of the past two decades, notably the Laguna fire in San Diego County and the San Dimas fire in the San Gabriel Mountains.

Conclusion

Man's relation to the wild vegetation associations of Southern California has been intense for a long time, but in the past two centuries it has changed in character. It may well be that his recent successful efforts to reduce the frequency of fires are provoking unanticipated changes more drastic than the deliberate burning practiced in previous centuries.

Human Impact on the Biota of Mediterranean-Climate Regions of Chile and California

[T]he combination of woodcutting and overgrazing by goats has eliminated both perennial and annual species, leaving the uplands of some communes virtual deserts.

[B]oth in California and Chile, fairly level lands that could be pastured or ploughed had their grasses and herbs replaced by introduced annuals almost as soon as Europeans entered an area.

IN THE WIDEST sense, the areas of mediterranean climate in Chile and California have similar histories in that the Europeans who came to dominate the native peoples of both regions were from Spain, another area with a mediterranean climate, and they were familiar with plants already adapted to that climate. The native peoples of the two areas, however, were different. Those of Chile cultivated crops, most of which were derived from the Andean highlands or the Peruvian coastal desert. Being more numerous and engaged in farming they had long disturbed the native vegetative cover. Further, it is likely that the comparatively recent Inca conquest had introduced large herbivores, e.g., the llama and alpaca, and probably plants such as *Nicotiana glauca* and *Schinus molle*.[1]

Significant entry of Europeans into Chile was accomplished more than two centuries earlier than in California and it was by territorial conquest. European entry into California was by the Franciscan missions and involved the establishment of distant points of control and the gradual domination of the neighboring Indian groups. Through both the Spanish and the Mexican periods, until 1846, extensive parts of California's region of mediterranean climate were not controlled or even visited by Europeans. Thus the introduction of plants into a large part of California came much later than in Chile and can sometimes be documented.

The Colonial Period

In Chile the conquistadores found a farming system that could be exploited. To this they added wheat and grapes, because of taste preferences as well as adaptability, and all European livestock. In California the missionaries and their escorts had to initiate farming and the raising of stock until the Indian peoples could be trained in these new skills. Both Mexican and European crops were introduced, but wheat did much better than maize until irrigation could be developed. As in Chile, the European livestock did very well in California.

Taking over the Inca conquests as far south as the Río Maule was relatively easy, despite a few rebellions that were crushed. Between that river and the Río Bío-Bío the native peoples showed more resistance, and in the non-mediterranean forested areas further south, the Indians were able to maintain their independence until 1880. North of the Río Maule just about all of the Indians were put in *encomiendas* almost immediately, and the Spaniards and their *yanacona* allies brought from Peru could requisition food from them. The handling of rapidly increasing livestock herds was a favored and responsible (though somewhat dangerous) job and seems to have devolved on the mestizo population as soon as one developed. The real interest on the part of the conquerors, however, was in precious metals, and the working of not very rich gold placers began immediately, using Indians from encomiendas and others drafted from villages.[2]

Disease, to some extent maltreatment in mines, and the fact that the Indians were not particularly numerous meant that by the end of the 16th century labor was scarce with respect to land. No more food was needed than in aboriginal times, but dry farming of wheat on a shifting basis probably made for more clearing of flatlands. Irrigated farming in the valleys from Santiago northward was maintained but with reduced extent and intensity. The introduction of European domestic animals, especially cattle, and their rapid increase in numbers, exposed lands covered with annuals to a new level of grazing pressure and provided a reason to burn to convert shrub-covered areas to herbaceous growth. Again, this was most effective on the relatively level land of the coastal terraces and in the Central Valley.[3]

For the next 200 years there was a slow but steady increase in the human population combined with an increase in the concentration of land in large holdings or *estancias,* the nonirrigated ones being used largely for grazing. Smallholders, often with communal land tenure, cultivated for subsistence, by dry farming in the coastal ranges, often by *curbén* or shifting cultivation, and by grazing goats and sheep in the steeper areas. In the 18th century a modest market for hides, tallow, and dried beef had been developed in Peru, and even some wheat was shipped there. The large landholders met this market, which permitted some imports and maintained living standards, which though modest were far superior to those of the smallholders or the landless. The latter were increasingly tied to the large properties as *inquilinos* in the labor force. Except for some of the northern communes, growing enough food was not a problem in the colonial period.[4]

It is not known when European grasses and herbs invaded and then replaced the native annuals on the gently sloping surfaces, but with livestock raising spreading rapidly throughout mediterranean Chile, it could have happened by the end of the 16th century. By the 18th century, with a growing population, extension of at least temporary cultivation, and deliberate burning to increase pasture, upper alluvial fans

and even some steep hill slopes in areas north of Santiago lost their broad-leaved scle-
rophyll shrubs to herbaceous plants, many of which were accidental introductions
from Europe.

Entry of the Franciscan missionaries and their military escorts into California
began in 1769. Trekking overland from the area of the former Jesuit missions in Baja
California, they brought livestock, seeds, and weeds. Missions were established at
points, first at San Diego and Carmel, with presidios at the former and at Monterey
near the latter. Filling in the system and extending it to north of San Francisco Bay
continued until 1823 with 21 missions ultimately established. All the missions were
close to the coast, although some were behind a range of coastal hills. Since the Indians knew no agriculture, their conversion to Christianity and training as farmers
proceeded together. Irrigated cultivation of a wide variety of field crops and fruit trees
in the south and dry farming further north were successful, but the Indians were
dying of disease so fast that only limited areas were needed or could be cultivated.
After 1790, raids into the interior to capture fugitive neophytes were combined with
recruitment or capture of gentile Indians to maintain the labor force.

Livestock generally flourished, especially horses and cattle, but the mission neophytes could not be trusted to herd far from the missions. A neophyte herdsman had
to be back at the mission, or at least to a permanent *asistencia* settlement, and account for his livestock daily, thereby restricting grazing to a radius of about eight
kilometers. Soldiers in the presidios and at the missions sometimes grazed their
mounts more extensively, but it was the civilian inhabitants of pueblos, beginning at
Los Angeles in 1783 and supplemented by retired soldiers, who expanded cattle raising throughout the whole coastal belt then under missionary influence. Extensive
land grants or grazing rights had been given under Spanish rule, but the rate of grant-
ing land for grazing accelerated rapidly after Mexican independence in 1821 and
through the Mexican period to 1846.[5]

Explorers' descriptions from the late 18th century indicate that the distribution
of grassland and chaparral differed little from the present, being governed by slope
and rock type and maintained by Indian burning practices that were continued by
the ranchers. But European grasses and herbs steadily displaced native annuals as
areas in the coastal valleys and the less steep hills were subjected to grazing pressure.[6]

Nineteenth-Century Developments

During the first half of the 19th-century Chile became independent, and the country was opened to foreign investment. British investors focused primarily on foreign
commerce and on silver and copper mining in the Norte Chico, an area at the dry
end of the spectrum of mediterranean climates. Demands for fuel around mining
and smelting centers devastated the local woody plants, nearly eradicating the gallery
forests along stream courses and destroying the woody *matorral* on the interfluves.

In the drier north and on the overgrazed communes it has not come back, leaving a degraded vegetation on the steeper slopes.[7] From the vicinity of Santiago southward there was an increase in wheat farming with some wheat being exported to Peru.

The economic awakening following Chile's independence, although mining offered the initial impetus, provoked substantial investment in agriculture by some, but not all, of the large landholders from the Aconcagua Valley to the Central Valley well south of Santiago. Canals for irrigation were constructed and extensive lands were converted to vineyards and other fruit trees. The extensive marshy lake of San Vicente Tagua-Tagua was converted by drainage into excellent lacustrine soil, which has been farmed intensively ever since. Rangeland was ploughed and planted to wheat. A substantial number of large *fundos* were ready to expand rapidly the area cultivated and production increased when the Californian and, shortly afterward, the Australian gold rushes expanded explosively the demand for Chilean wheat.[8]

In California, annexation by the United States was followed immediately by the discovery of gold in 1848. Two hundred thousand immigrants were attracted in a decade, mostly to the central part of the state. They came from the eastern United States and most of the rest of the world. In the Mother Lode and other mining districts there was heavy logging and increased burning, but except for actual tailings piles and valley bottoms affected by placers, and later by hydraulicking, the forest and woodland vegetation of the Sierra Nevada has recovered to something close to its original state. The demand for meat meant that grazing pressure was applied to grasslands of the Central Valley and other valleys that had been beyond mission and Mexican influence. The replacement of native bunch grasses and annuals by European annuals was extremely rapid. The market for wheat extended cultivation through the Sacramento Valley and the valleys and gentler hills on either side of San Francisco Bay.[9]

Local production could not meet the Californian demand and Chile was the country best positioned to fill the market. The Victorian gold rush in Australia in 1856 offered Chile an additional market. Both large landholders and smallholders ploughed and planted wheat wherever slopes permitted and in parts of the coastal ranges where they did not. Grazing was relegated to steep slopes even though the demand for meat was rising. The matorral of much of the Norte Chico was burned to improve pasture, and, in part because much of the land there was communally held by poor people, a common sequence was clearing, planting of wheat, pasturing the abandoned land followed by permanent degradation of woody vegetation. Except for the Araucanian-held areas south of the Río Bío-Bío (forested areas with a non-mediterranean climate), cultural land use as either cultivation or grazing, had reached its physiographic limits by 1870. Any further increase in agricultural productivity to provide for a growing population would have to come from intensification of production, especially by irrigation.[10]

Whilst gold mining in California fell off rapidly after 1860 few of the immigrants who came to mine went home. Some moved to other mining districts but many turned to agriculture. Arable land became scarce throughout the state, and hillsides were ploughed or put under orchards and vineyards in many localities.

Two mutually reinforcing developments affected land use in California during the last third of the 19th century. The completion of a transcontinental railroad in 1869, and its extension to Southern California in the next decade, opened the United States market to California agricultural products, initially wheat but increasingly to perishable specialties such as raisins, prunes, and, later, citrus fruits. The potential profitability of such activity justified heavy investment in irrigation facilities. Until the 20th century, irrigation works were limited to drainage basins, but California's topography readily permitted intensive cultivation of the eastern two-thirds of the San Joaquin Valley and the coastal lowlands of Southern California. Many of these areas were too dry for dry farming and were only poor pasturelands. The area around Riverside is a good example, it changing from a xerophytic scrub to intensive citrus culture at this time. In the coastal valleys in the central part of the state, vineyards and deciduous fruits and nuts occupied moderate slopes and needed little irrigation. Truck crops irrigated by groundwater, in part for local and in part for national markets, occupied the valley bottoms.[11]

Thus, only a few decades after Chile's agriculture and grazing had expanded to utilize all suitable land in areas of mediterranean climate, the same situation was arrived at in California. The rangelands (in both countries) were not readily irrigated, and interstices in the cultivated areas were taken over by weedy annuals, almost all of them of European origin. In Chile two wild perennial plants, one native and one introduced, have extended their ranges. The *espino* (*Acacia caven*) seems to have expanded southward in relatively flat but unirrigable lands in the lee of the coast ranges both in dry-farmed and grazed areas. As a shrub of open woodland it appears to be tolerated as a source of fuel and emergency browse. Even in the best parts of Chile's Central Valley the introduced blackberry (*Rubus* spp.) has preempted an amazing amount of good farmland. With or without human assistance blackberries are established along fencelines and road and rail rights-of-way. Probably encouraged as an effective fence in a socially insecure countryside, the impassable linear thickets have spread to widths of up to 10 meters. Whilst there are continuing attacks on individual thickets large landholders have not been willing to go to the considerable and unpleasant effort and expense needed to eradicate the weed. Native and introduced blackberries exist in California as well, but they show their land-consuming propensities only locally and they have generally been controlled adequately.

In the last part of the 19th century canals continued to be constructed on the sides of river valleys from the Aconcagua northward in Chile. The effort was to start farther upstream and with a higher canal irrigate progressively higher terraces. The more northerly of these valleys are too dry to be considered as having a mediterranean climate and they required irrigation for any cultivated crop to be able to grow. The initial impetus was to grow feed to maintain and fatten cattle driven over the Andes from Argentina, cattle which ultimately would supply nitrate mines in the desolate north. The limited flow of these northern rivers was overcommitted, and since the earlier water users had better rights, the canals often are empty and the newly cleared higher terraces are rarely planted or, if planted, do not receive sufficient water to sustain a crop.[12]

World wheat prices fell after 1870, and wheat growing remained profitable only in the more level lands of the Central Valley. These lands were generally in large estates whose owners had options of converting to improved pastures, vineyards, other fruit crops, or irrigated truck crops near the larger cities. The smallholders in the coast ranges from the Aconcagua south to the Río Bío-Bío and beyond and the communal holders of the southern Norte Chico could only try to maintain their incomes by continuing to grow wheat, something impossible in all but the wettest years. Areas on moderate slopes that were cleared of matorral are subject to strong erosion and are overgrazed when they cannot be cultivated. Annuals, including introduced species, xeric shrubs, and cacti comprise the degraded vegetation.[13] Holders of hill lands and coastal terraces, even north of the Aconcagua, who were not immediately dependent on them for income, had another option. Plantations of Monterey pine (*Pinus radiata*) from California and *Eucalyptus* spp. from Australia appeared on both moderate and steep slopes with linear boundaries that mark property lines. Free of their native predators, these trees do better in Chile than in their homelands. Whilst by world standards neither tree has very good wood, in Chile where softwoods for construction are scarce there is great demand for timber of *P. radiata*; moreover, for owners who can wait 20 to 40 years for a harvest they are profitable. Thus these plantations of introduced trees may have permanently displaced native matorral or woodland with a profitable crop which also protects the soil resource.[14]

Modern Developments

In California the early 20th century marked the beginning of the construction of great dams and aqueducts to transfer water between drainage basins, often several hundreds of kilometers. Curiously, although they made possible a great expansion of agricultural production, these great engineering works had very little effect on the wild vegetation in areas of mediterranean climate. Most of the water was destined for desert areas, such as the Imperial and Coachella Valleys and later the west side and south end of the San Joaquin Valley, or for major urban areas such as San Francisco and Los Angeles. Areas of mediterranean climate were already dry-farmed and many of them changed to more intensively cultivated crops with irrigation. Dams came later in Chile and initially were placed in the dry Norte Chico where they could stabilize irrigated cultivation in their own valleys, but they could not collect enough water in dry years to permit significant expansion of irrigated land. Recent dam construction in Chile has focused on developing hydroelectric power rather than irrigation. In both countries, of course, vegetation was eliminated from reservoir basins.[15]

In 1800 the population of Chile was perhaps 10 times that of California and it continued to grow steadily by natural increase. In California waves of immigration, beginning with the gold rush and continuing to the present, greatly accelerated population growth. Early in the 1920s the population of California came to equal that of Chile at a little less than four million people. Since then California's population has increased sevenfold and Chile's about threefold. In both Chile and California the urban population has exceeded the rural for many decades, although urbanization has been much more intense in California.[16]

"Urban sprawl" has been especially characteristic of California since before World War II and accounts for almost all of the state's population growth. It is most developed around Los Angeles and San Francisco Bay and, more recently, around San Diego, but it also occurs around middle-sized centers such as Fresno and Sacramento. The land taken over by houses, shopping centers, streets, and parking lots was pri-

Figure 16. Land clearing for subdivisions in the Santa Monica Mountains. Such replacements remove vegetative cover, destabilize slope, increase the potential for land slippage, and increase fire damage.

marily agricultural, but it included both dry-farmed areas and irrigated, intensively cultivated orchards almost indiscriminately. Citrus groves are on their way to being eliminated from the entire Los Angeles lowland. Santiago is also spreading out as it grows, but the best irrigated farmland offers more resistance to urbanization.

California's urban sprawl has long included a feature not represented significantly in Chile: exclusive and expensive suburbs have sought building sites on steep hillsides that in Southern California were typically covered by chaparral. Around San Francisco Bay both woodland- and chaparral-covered slopes are utilized. Typically, building lots are large and many owners have chosen to leave the original vegetation on much of their land. This practice creates a severe fire hazard and catastrophic conflagrations have occurred. Political pressure to improve fire protection has reduced fire frequency, but fuel accumulation makes the delayed next fire even more disastrous.[17] The original vegetation returns and people rebuild houses. Sections of the Santa Monica Mountains in and west of Los Angeles have gone through as many as three building-burning-rebuilding cycles in this century (Figure 16).

In northern San Diego County a recent entry into extremely steep chaparral-covered slopes has been orchards of frost-sensitive avocados and lemons. Enormous investments in clearing, planting, developing pipe irrigation to each tree using expensive water, and even cable systems to get to each tree for harvesting have been required. These plantings cannot be economic and may be attributed to a mixture of income tax shelters and conspicuous consumption in an affluent society. But they have made avocados abundant and cheap.

In both California and Chile slopes covered by woodland, chaparral, or matorral have resisted invasion by introduced perennials. The native shrubs could be displaced by Monterey pine plantations or houses, but even where annuals have come in after repeated burning, the native shrubs tend to return, although often in degraded form. Introduced perennials, such as *Nicotiana glauca, Ricinus communis* (castor bean) and *Cytisus scoparius* (Scotch broom), have established themselves in places with severely disturbed soil such as on road cuttings, actively gullied sites, and steep areas cleared for orchard planting. Because they occur along transportation routes, these plants seem to be more prevalent than they really are. Two trees that were introduced to both California and Chile, either for economic reasons or as ornamentals, have become naturalized enough to reproduce—viz., *Schinus* spp. (pepper tree) and *Eucalyptus* spp. Their reproduction, however, requires specialized circumstances and the trees never seem to travel far from their point of introduction.[18]

In summary, both in California and Chile, fairly level lands that could be pastured or ploughed had their grasses and herbs replaced by introduced annuals almost as soon as Europeans entered an area. The native bushy or woody vegetation in both countries has been far more resistant to invasions. Only if the native vegetation were replaced by another land use did it disappear. There is a difference, however, in the treatment of the vegetation on steep slopes in the two countries. In California, the chaparral is left alone or lightly pastured by cattle. Burning is its only disturbance, one from which it is adapted to recover, although perhaps with subtle alterations in species composition. In Chile the matorral or woodland is used, both generally for firewood and charcoal and selectively for medicinals, food, saponin (from *Quillaja saponaria*), tannin, or other uses. The quillay and the Chilean palm (*Jubaea chilensis*), which are destroyed in exploitation, have become extinct locally.[19] South of Santiago the matorral or woodland maintains its integrity, although thinned and with some alterations to species composition. To the north, especially in the Norte Chico, the combination of woodcutting and overgrazing by goats has eliminated both perennial and annual species, leaving the uplands of some communes virtual deserts.

The Introduction of Date Palms into Baja California

Modern cultivation practices would indicate that the growth of the groves was a matter of volunteer seedlings establishing themselves in moist arroyo bottoms, which saline soils rendered relatively unsuitable for most other crops.

AT SEVERAL OASES in the south central part of the peninsula of Baja California, date palms (*Phoenix dactylifera*) constitute the principal agricultural resource. Seen from the surrounding barren hills, or mesas, the settlements at San Ignacio, Mulegé, Loreto, and Comondú seem to be forests of the dark green palms. Closer observation exposes an understory of irrigated cultivation of the fruit trees, shrubs, and annual crops (Figure 17). The initial appearance of the oases, though not the indifferent cultivation practice, recalls the oases of the Old World. The date palm also figures prominently in such farming as is done in a dozen other small centers of cultivation in this dry and sparsely settled region.

The problem of when and how this Old World tree crop was introduced into this appropriate New World environment has provoked comment from several recent students of the history of Baja California. Unfortunately, the frequently repeated statement that the Jesuit missionaries brought in and planted three Arabian varieties of dates in San Ignacio in 1730 seems to be founded on literary invention instead of fact. As a recent example, Father Peter J. Masten Dunne, the noted Jesuit historian, makes this statement in his comprehensive history of Jesuit activities in Baja California without citing a source.[1] The implication is that the early introduction of date palms by the Jesuits is general knowledge. After a rather exhaustive examination of both published and manuscript literature on the history of Baja California, I am prepared to say that this statement on the introduction of date varieties first appears on page 34 of "Mother of California" by Arthur W. North, one of the first and most popular of an ever-broadening stream of travelogue histories of Baja California.[2] Though North states that he examined old manuscript records concerning the peninsula, some of which may well have been destroyed in the San Francisco fire of 1906, it is unlikely that there were any materials which antedated 1769, the year the Franciscans entered California. North was inventing a plausible and romantic story to account for an obviously long-established horticulture which characterized a group of settlements he knew to have been founded by Jesuit missionaries.

To the best of my knowledge, dates are referred to only once in all the contemporary Jesuit literature. Father Johann Jakob Baegert, a Jesuit expelled from the penin-

Figure 17. San Ignacio church from Mesa to the east with surrounding palms

sula in 1768, mentioned dates in a list of minor crops introduced by the missionaries.[3] Since he discussed most of the introduced crops at the missions at some length, it must be assumed that dates were a very recent and minor introduction. Probably none of the plants had borne fruit by the time the Jesuits were expelled.

In an exhaustive inventory of the missions, written in 1774 by the Dominican missionaries who had succeeded the Franciscans the preceding year, the presence of three date palms at Comondú is mentioned.[4] This accords with Baegert's bare mention of the plant. If they had been young seedlings in 1768, they would have begun producing, but no extensive plantings could yet have been made. The first reference to real production of dates comes from San Ignacio in 1785. In that year that mission produced 20 *arrobas* of dried dates.[5] The same source describes a rapid increase of date production at San Ignacio to 200 arrobas by 1800. Though no record has been found, it is likely that comparable quantities of dried dates were produced at Comondú, Loreto, and Mulegé during the same period. By 1857 the municipality of Mulegé, which included that city and San Ignacio, was producing 3,000 arrobas.[6] Modern date production at San Ignacio is estimated at 200 tons, or 16,000 arrobas,[7] and an only slightly smaller quantity is produced at Mulegé.

Increase of production from the late 18th through the 19th century was steady and roughly logarithmic. Modern cultivation practices would indicate that the growth of the groves was a matter of volunteer seedlings establishing themselves in moist arroyo bottoms, which saline soils rendered relatively unsuitable for most other crops. Probably by 1900 dates palms had occupied about all the areas suitable to them, and expansion of the groves ceased.

Though the time of introduction of dates to Baja California has thus been established with fair precision, the specific source of the plants brought in and whether they were seedlings or varietal offshoots is not known. Present-day cultivation is strictly of seedlings and is casual indeed. Practically no effort is made even to reduce the proportion of nonproductive male trees. Consequently there is no possibility of identifying clone varieties of known origin. There is no mention of varieties of dates in any of the mission literature, though the same literature does refer to clone varieties of grapes and figs. While date offshoots undoubtedly could have endured the trip from North Africa to the East Coast of Mexico by the 18th century means of transportation, the overland and sea journey across Mexico and on to Baja California took additional months. The possibility of keeping offshoots viable for the entire trip was exceedingly small. Plantings of seeds are much more probable.

Seedling date palms are so heterogeneous in the character of their fruits that the actual original source of the seeds of the Baja California date groves has only historical interest. The relatively low quality of the modern fruit is a natural product of any series of generations of volunteer seedlings, regardless of where the seeds came from. Three possible sources of the seeds suggest themselves. Dates have been raised at Elche in southeastern Spain since Moorish times. These palms are mostly seedlings, and male trees are almost as numerous as female. For climatic reasons the date yield could not be very satisfactory, and sale of leaves for religious decoration on Palm Sunday seems to be about as important as fruit production. The casual date culture of Spain is certainly reminiscent of that of Baja California, though in the latter place climatic conditions would justify more careful attention to the groves. A second source for date seeds, of course, is the fairly extensive Spanish trade with nearby Northwest Africa, where the fruit has long been a major article of commerce. Finally, dates may have been a secondary introduction from the coastal desert of Peru. This region possesses a climate which is entirely appropriate to date culture, and there are now some palm groves though they do not constitute a major crop. To the best of my knowledge the time of introduction of palms in this region has not been investigated, but Peru was intensively exploited by the Spaniards more than a century and a half before the first settlements were established in Baja California. A fairly steady trade between Mexico and Peru operated in the 16th and 17th centuries.[8] It may also be noted that modern date cultivation practices in Peru resemble closely those of Baja California,[9] and a modern introduction from there would certainly be in the form of seeds rather than of offshoots.[10]

The low level of care received by the date gardens of Baja California is symptomatic of the minor interest in agriculture in that region. The owners of agricultural land are few and generally persons of relatively great wealth. A garden means prestige, and dates will grow wild in arroyo bottoms where the water table is high, require little investment in labor, and yet yield a modest return. Braceros returning to Baja California villages from the Coachella Valley tell of the better-quality fruit and higher productivity of the date gardens there, but they have never persuaded the local owners to undertake the investment of adopting the higher standards of cultivation.

The commercial date plantings of the California and Arizona desert, though only a few hundred miles away and established after the Baja California groves had attained roughly their present extent, seem to have had no historic connection with their southern neighbors. This fact can be considered as fortunate. Although the first successfully propagated dates introduced into the United States in 1890 may not have been the supposed varieties, the intent was to bring in desirable offshoots. Most of the offshoots selected and shipped by Walter T. Swingle from Algeria in 1899 proved to be the sought Deglet Noor and Rhars varieties. Subsequent commercial and U.S. Department of Agriculture imports attempted to introduce all the desirable established varieties grown from Morocco to the Persian Gulf.[11] The standards of cultivation in the United States are based on, and developed from, the most advanced systems worked out over millennia in the Old World drylands. It is doubtful that Baja California can make any contribution to the United States date industry, but it might greatly increase the production and improve the quality of its own date gardens by adopting some of the cultivation techniques practiced in California and Arizona.

Recovery of Desert Vegetation

[D]estruction of perennial plants, as is occurring through concentrated use of off-road vehicles in the southwestern United States ...
will scar the ecosystem [for] generations or centuries.

Aɴʏᴏɴᴇ ᴡʜᴏ ᴏʙsᴇʀᴠᴇs a particular patch of desert vegetation over a period
of years will be struck by the slowness of change in individuals of perennial species.
In some years there is no perceptible growth at all in the stems though flowers may
bloom and dry up, but an apparently dying plant or branch may take a decade to die.
The reverse situation characterizes the herbaceous plants which can carpet an area
with wildflowers and then disappear in a few weeks, with different species dominat-
ing the display in sequent years.

We are probably correct in concluding that destruction of perennial plants, as is
occurring through concentrated use of off-road vehicles in the southwestern United
States, for example, will scar the ecosystem so that recovery, even with full subsequent
protection, will require decades, generations, or centuries. The factual basis for this
conclusion, however, is singularly weak. The arid and semiarid regions south and east
of the Mediterranean Sea have shown an appallingly degraded wild vegetation for a
long time, but the abuses, notably overgrazing by goats, have persisted for millennia
and continue.

Experimental investigations, where the initial conditions, the extent of disturbance, and degree of subsequent protection are known, are indeed necessary. Fairly
extensive plots will be needed to gain insight into invasion or colonization rates for
various species, but the most critical limitation is time. Many decades, perhaps more
than an individual's career, must elapse, with a consistent level of protection being
continuously maintained, before fully valid observations can be collected.

When, through accidents of history, a locality's vegetation experienced major
damage that terminated at a known time, and then it was protected from almost all
man-induced disturbance, that locality can acquire the properties of a formal experiment, particularly in desert areas where the vegetation is too sparse to sustain extensive fires. The stated conditions seem to be unusually well met at the Mina de
San Fernando, a now-unworked copper mine near the center of the Baja California
peninsula just south of the 30th parallel.

The Mina de San Fernando produced copper ore from shafts and drifts along a
vein during the 19th and early 20th centuries. Located in a roadless area it could ship
ore containing 25 percent or more copper only some 55 km by muleback to the coast.
The mine was discovered about 1870, and beginning in 1882 there were alternate periods of operation and abandonment until 1907.[1] The maximum labor force, includ-

ing muleteers, is estimated at 45 and the resident population 150 to 200 in periods of full operation.

After 1907 the mine itself was never worked, but the stored protore, ranging from 7 to 25 percent copper, attracted interest with the rise in the price of copper during World War I. From 1917 to 1919 a smelter was operated utilizing local vegetation for fuel. A slag deposit of about 1,000 cubic meters indicates the extent of the operation, but with the collapse in the price of copper in 1919 the whole operation was suddenly abandoned. Piles of ore ready for the smelter and vegetable material, to be used as fuel, still lie around the principal mineshaft.

The mine lies at about 475 m in a hilly district of complex lithology about 5 km south of the former Mission San Fernando Velicatá. The terrain has always been difficult for wheeled vehicles, and the district is nearly unvisited. Brief and somewhat dubious rainfall records from four stations in the region indicate an average annual rainfall of 100 mm.[2] Following good winter rains the ground is covered by a herbaceous vegetation that briefly affords good pasture for cattle, but the perennial plants are meager forage indeed. The cattle population, which does range freely, is sparse, limited by the normally poor perennial pasture. In the vicinity of the Mina de San Fernando significant grazing occurs only when there is herbaceous vegetation, and the perennial shrubs are not affected. Thus the perennial plants in the area of the mine have had more than 50 years to recover from the depletion induced by mining activities.

My own observations of the area around the mine began in 1949. Then and in 1971 a definite aspect of impoverishment characterized the district, fading proportionally with distance for about 1.5 km away from the centers of mining activity. The operational structures for the mine were on a slope west of the southerly and most active mineshaft. They occupied an area about 200 square meters on which remnants of adobe walls still exist. Some 2 km to the west, on the point where two arroyos join, a number of large ruined adobe buildings and some stone fences and corrals indicate the place where the pack mules and their gear were kept.

Apart from places where mine waste was actually dumped the most disturbed and impoverished area is that where the mining administration buildings stood. *Franseria chenopodifolia* is the most abundant shrub, with plants up to 40 cm high constituting more than half of the perennial individuals in the area. Most of the other perennials are thorny succulents. Two species of *cylindropuntias* (chollas) had certainly invaded since abandonment, as resident miners would have removed their threatening masses of thorns. Clumps of two species of Agave had also come in. A giant *cardón*, *Pachycereus pringlei*, which can be identified in a 1911 photograph, still stands. Its massive stems, weighing many tons, were too difficult for the miners to remove. The young *cardones* that have developed since mine abandonment are single stems up to 2.5 m tall. This would represent their growth in perhaps 50 years.

In the radius of a kilometer from this center of maximum disturbance there is a progressive increase in the number of species of perennials, gradually bringing the flora up to par with that of the surrounding terrain in diversity. The chollas in particular seem to have recolonized actively in areas from which they are likely to have been eradicated.

In the area of the mule corrals there is a notable thinness to the vegetative cover with some recolonization by chollas, cardones (up to 2.5 m tall), and some viznagas (*Echinocactus*). Inside a stone corral about 30 m square, the ground is covered with remains of the annual *Mesembryanthemum cristallinum*. This naturalized South African plant is abundant on the foggy coastal terraces of Baja California's west coast, and also covers formerly cultivated fields in the same area. Evidently the disturbance plus the addition of animal manure in the corral was sufficient to establish this annual plant for more than 50 years, though it does not occur in the region on undisturbed sites.

With its thick succulent rind a freshly cut cirio would weigh more than 100 kg, so exploitation was a direct function of accessibility and an inverse one of distance from the smelter. The radius of complete extirpation is about 400 m, a bit less behind a steep hill just west of the smelter. The availability of mules near the corrals caused a secondary center of extirpation. Probably when mules were taken to the mine for other purposes they dragged cirio trunks with them. In the 52 years of subsequent recovery the following pattern has developed. Within 150 m of the smelter there are no cirios at all. In some places out to the 400 m limit of cutting there are groups of small individuals 1.5 to 3 m tall. In one place within this zone there are half a dozen individuals about 50 cm tall.

Two conclusions can be derived from this distribution pattern. I conclude that there have been only two or at most three seasons in the past 52 years in which climatic conditions were propitious enough for seedlings of *Idria columnaris* to establish themselves, even though the plants put out seed nearly every year. In my judgment the 1.5 to 3 m tall plants all were established in a single season, growing at varying rates because of differing immediate edaphic conditions. No records of observation are available to permit identification of the years of establishment and thus rate of growth under wild conditions. Secondly, *Idria columnaris* is an extremely slow colonizer. Another half-century is likely to elapse before there are even small cirios throughout the affected area and perhaps another century without disturbance before they recover.

A more thorough formal sampling of the perennial vegetation around the Mina de San Fernando, hopefully when the vegetation is in a less-desiccated condition so that species identification may be more secure, is called for. There may be other, less conspicuous species whose distribution has been seriously altered. A rough temporal conclusion can be drawn, however. Some 200 years are required for the recovery of one conspicuous exploited species in this desert, even though the area of intensive exploitation is only about a square kilometer and the surrounding region is available for natural restocking. Were a larger area disturbed, the time needed for recovery would be much longer.

Man's Impact on the Several Regions with Mediterranean Climates

[The California Indians] may have developed the most humanized landscape any nonagricultural society ever created.

Eucalypts will long be a prominent feature in the man-modified landscapes of all the mediterranean lands.

THERE IS TREMENDOUS variation among the several regions of the world with mediterranean climates in the nature and intensity of the modification of their ecosystems as the result of human activity. The duration of occupation by man is not a significant variable. It exceeds 10,000 years in all cases and may well be longest in South Africa, one of the less-modified regions. Through his use of fire in an area that is seasonally inflammable even primitive man's influence on the ecosystem was not negligible, but it was probably comparable in all parts of the world, favoring some organisms and disfavoring or even exterminating others. Areas fully free of fire would undoubtedly carry a heavier forest than they do now. Natural fires would still obtain in some regions, but perhaps with much less effect on the vegetation. The island of Madeira is unique in having the only area of mediterranean climate untouched by man until historic time. The tremendous fires that burned for years shortly after the first settlement early in the 15th century altered its vegetational pattern from forest to brush and grass, evidently permanently.

The next great technological innovation that generally affected local ecosystems, the domestication and propagation and protection of some species of plants and animals, however, occurred perhaps 10,000 years ago at the eastern end of the Mediterranean Basin and reached Western Australia less than two centuries ago.

There are good reasons to believe that horticulture, the manipulation of individual plants, commonly vegetatively reproduced clones, antedates broadcast sowing of seed grains.[1] Our earliest archaeological records, however, come from the uplands at the eastern end of the Mediterranean Sea and seem to represent a society that planted grass seeds in late fall in fields that had been cleared of brush by burning at the end of the summer dry season. They would sprout in the winter rains, survive the mild winter cold, and be harvested in late spring. Hahn's inferences that domestication of the cow, sheep, and goat, respectively the fiercest and wildest of animals in their wild forms, could only be accomplished by a rich and sedentary, i.e., farming, population look ever more probable in the light of new archaeological discoveries.[2] That the motivation for domestication was religious rather than economic is also likely.

Before the dawn of written history, some 5,000 years ago, a complex set of variant farming patterns, adapted to distinctive climatic and topographic situations, had been developed in the Middle East, and had begun to spread at varying rates in all directions including westward on both shores of the Mediterranean Sea. Irrigation probably began with small streams easily diverted over alluvial fans. It fluoresced in the great riverine oases of the deserts to the south and east, where evolving societies could direct masses of human labor in water control. Plowing with oxen perhaps also originated in these great oases, but it could spread into dry-farmed areas that had extensive level surfaces. True pastoralism, often accompanied by slash-and-burn farming with the hoe rather than the plow, led the way westward through the Mediterranean Basin as well as into Central and Northern Europe. Herdsmen could exploit effectively lands too dry, too rough, and too cold to plant, but in the early stages a single community usually carried on both activities, individuals and families only being specialized. This pattern persists to the present in the rougher and less-developed parts of Anatolia, Greece, Italy, Iberia, and North Africa. Specialized and somewhat transhumant communities composed entirely of pastoralists are associated with the drylands and the high mountains rather than areas of mediterranean climate.

These agricultural activities all had profound effects on the native biota. An important element is that farming and grazing permit a great increase in the human population, often by an order of magnitude or more. Horticulture and agriculture involve the more or less complete replacement of the natural vegetation by the cultivated one over limited or extensive areas. The cultivated land may be maintained as such more or less indefinitely or may be abandoned for a number of years after one or a few years of cultivation. In either event the disturbances of cultivation also prove to favor a set of plants called weeds, which may actually evolve by selection along with the deliberately bred crops. Pastoralism similarly puts heavy grazing and trampling pressure on the natural vegetation, favoring some species but reducing or eliminating others. In the regions of mediterranean climate with their characteristically broken topography only the smoother surfaces are cultivated, at least until there is severe population pressure on the land, while almost any place is accessible to and affected by goats.[3]

The modification of an ecosystem by cultivation or grazing is a relatively slow and cumulative process, progressing over time measured in millennia. In general human populations and their capacity to affect the landscape tend to increase with time. Disequilibria such as erosion of cleared slopes that tend to decrease a resource will force a concentration of activity and perhaps cause additional disequilibria in other locales. Further, the way in which the general ecosystem is affected is modulated by cultural practices and values that vary in detail among societies. Such features as the preferred grain for breadstuffs, the species of domestic animal herded, or whether agriculture is a commercial enterprise or a subsistence way of life will strongly affect the choice of surfaces for cultivation or abandonment and the intensity of implantation of a cultural landscape in place of the natural one. Because of these particularities it is appropriate to characterize briefly the histories of human occupation of each of the several regions of mediterranean climate over the past 10,000 years, attending especially to varying patterns and intensities of land use.

The Mediterranean Basin

From the beginnings of the neolithic with its domesticated plants and animals, which was present at the eastern end of the Mediterranean Sea about 10,000 years ago, nearly to the beginning of the Christian era, that region or its immediate neighbors recurrently produced both technological and social innovations that spread at varying rates in many directions. A fairly steady spread was to the western end of the basin. The neolithic may have reached the Atlantic by 2000 B.C. or a bit earlier, but by that time highly organized and populous states, specialized cities, and long distance commerce, even in bulky goods, were present around the eastern end of the Mediterranean Sea. The cultivation of fruit trees, especially grapes, figs, and olives, was an eastern introduction, and new fruit crops such as citrus kept coming from the east into Roman times.

The brilliance of the Middle Bronze Age around the Aegean Sea in the middle of the second millennium B.C., and its fall into decadence shortly thereafter, as well as the political turmoil that began even earlier in the irrigated deserts to the south and east, suggest that locally populations and land exploitation had peaked and were pressing on resources. The stable and unified control of the Roman Empire had the effect of bringing the western basin up to the technological level of the eastern, with major and modern urban centers developing in Iberia. For a time population growth was rapid in the west, and land use and abuse of the eastern sort developed. Though there were local florescenses in Iberia and Morocco from the 8th to the 12th centuries A.D., evidently associated with specialized crops and irrigation techniques, the Mediterranean Basin had by later Roman times established a land-use system that was not significantly modified until the present century.

The political stability or law and order maintained by the Roman Empire seems also to have favored the extension of farming into drier areas than are now cultivated in Syria, Cyrenaica, Libya and, Southern Tunisia, areas of ancient civilization even then. Both dry farming and the maintenance of drought-resistant perennials, olives, grapes, and figs were involved, and areas of ancient cultivation can be traced archaeologically beyond the modern border between the steppe and the sown. Full famine relief for the populace could be provided by the empire in years of drought. In the politically disordered centuries following the fall of the Roman Empire and extending into the 19th, intensive pastoralism, often associated with overgrazing and the obliteration of all shrubs but the toxic oleander, pressed into areas of mediterranean climate, especially in North Africa.[4]

Through this long period level land was planted with small grains, sometimes with long fallow periods; in small alluvial basins intensively cultivated and irrigated gardens of fruits and vegetables were developed; the olive and the vine found places on steep lower slopes, and locally where population pressures had developed some artificial terraces on similarly steep slopes were planted to small grains; domestic animals, especially the goat and sheep, were pastured in the roughest and poorest country, putting severe pressure on the native vegetation. Native forests, however, never abundant or fast growing, had come to be recognized as a resource that needed care and protection. Pines, oaks, and chestnuts were preserved to yield food and fodder, and selectively their wood. The intensive care and exploitation of native trees, at about the same level as olive cultivation, is a peculiarly mediterranean complex that has spread very little to other parts of the world. The live oaks of the Sierra Morena of Southern Spain are a part of an ancient wild ecosystem that is preserved as a crop. They are pruned, stripped of their cork, and yield a regular acorn crop for fattening hogs.

This system of land exploitation could support stably a large but not infinite population. Pressures of overexploitation were most acute in the upland grazing areas, but erosion of overgrazed slopes could clog drainage systems and create marshes on some of the best alluvial lands. When the social fabric was rent by wars and political instability, burning and cutting of forests and degradation of terraced fields and groves for the expansion of poor pasture tended to destroy resources.

Although the areas of mediterranean climate remain the poorest parts of Europe, a product of rural overpopulation, an atomistic family-oriented social organization, and the retention of an ancient land-tenure system in Iberia and Southern Italy, as well as of man-caused damage to the land-resource base, the past century has witnessed significant changes in agricultural practices. With the improvement of transportation facilities winter vegetables and subtropical fruits can reach markets in Northern Europe as well as local urban centers. Although dependent on risky international luxury markets the producers gain enormously higher returns than from subsistence farming. This specialty production is concentrated in irrigable lowlands, and even paddy rice is cultivated on marshy tracts. Israel and the North African states of the Maghreb are participating in the development. Where they are practiced and extended these intensive cultivations effectively eliminate the native biota, or create a new ecological system. There is a parallel, perhaps indirectly related development in the abused and eroded uplands. Returns are so low that villagers who have raised grain in small patches of terraced or sloping land and grazed sheep or goats where the terrain is more rugged are abandoning their ancestral homes and moving to cities or the irrigated lowlands. Reforestation or rationally restricted grazing has become politically feasible, though substantial effects on the long-denuded upland slopes may take some generations to appear. In the poorest areas with populations that continue to grow explosively, as in North Africa, the prognosis is for even more destructive exploitation of the uplands.

Chile

When peoples from the western Mediterranean Basin set out to explore and exploit the world in the 15th century, the land-use system of their homeland was what they knew and tried to spread. In Central Chile in the mid-16th century the Spaniards encountered a similar physical environment. Hunting, fishing, and gathering Indians had lived in Chile for at least 10,000 years,[5] but the spread of horticulture to that region from the north probably was much more recent, perhaps starting about the beginning of the Christian era. The local domestication of three grain-bearing plants adapted to winter rains and summer drought, unique to the New World, suggests a considerable antiquity for agriculture in Central Chile. *Teca* disappeared in the 18th century and cannot be identified; *mango* was a grass, *Bromus mango madi* was a composite produced for the oil in its seeds. Llamas and alpacas were herded throughout Northern and Central Chile at the time of contact, but numbers seem to have been too small to put serious pressure on pastures.[6] Population densities had evidently not been built up to near saturation except in the irrigated valleys from Aconcagua northward. From Santiago south to Chiloé a neolithic shifting cultivation was practiced on gentler slopes. Summer burning for clearing is not readily controllable in a mediterranean climate, and it is likely that the vegetation of the gentler alluvial surfaces of the Central Valley had been greatly modified by Indian practice as far south as the Río Bío-Bío, even though only a small fraction of those surfaces was under cultivation at any one time.

The Spaniards introduced winter-sown small grains and the Old World grazing animals. The Old World grains were enough superior to eventually eliminate the native ones. Chile was extraordinarily isolated, and its only realizable wealth for a long time was from gold placers, and the manner of their exploitation had a destructive effect on the native population. When the placers or the labor force to work them were exhausted a colonial society developed that could feed itself with a combination of Indian and European crops, especially wheat, beans, and potatoes, and European grazing animals. With labor scarce and markets small, extensive haciendas exploited the best land, the gentle slopes of the Central Valley, primarily for grazing cattle, each growing enough crops to feed itself. Indian communities found shelter in little pockets of the coast ranges, joined by Spanish and mestizo smallholders. Both the latter groups were more likely to graze goats and sheep than cattle.

The formative period for the modern Chilean landscape is really the mid-19th century. The sudden development of a cash market for wheat to supply the California gold rush, and shortly thereafter one in Australia, led to an expansion of its cultivation into any place that wheat would grow, including steep and vulnerable slopes in the Coast Ranges.[7] The *fundos*, or large estates, on better land in the Central Valley produced more and profited more. Wealth from the silver mines of the Norte Chico was invested in irrigation systems as well as in the best land, and an intensification of cultivation, focusing on the vine, proceeded. With European weeds and ornamental trees as well as crops there is probably no more completely exotic vegetation in the world than that which covers irrigable level land in mediterranean Chile.

The slopes of the Andes and those of the Coast Ranges seem to have had consistently different histories of exploitation. Andean slopes were parts of large, lowland-based estates or were government land and were grazed with cattle, on an extensive basis. Large parts of the Coast Ranges were planted in wheat; grazing was by smallholders, commonly of goats and sheep. The combination of severe soil erosion and a progressively less-favorable ratio between costs and returns continues to cause the abandonment of wheat land. In some cases planted forests have been introduced; in others a wild regrowth that continues to be grazed is the replacement.

The recently instituted (1964) and now-accelerated land reform program focuses primarily on breaking up large estates of irrigated and intensively cultivated land. Many such estates also contain even more extensive acreages of rough land or notably poor soil and have been used for grazing or forest plantations. It is too early to tell whether such holdings can be managed as large units on a sustained yield basis or will be subjected to overgrazing as smallholders seek to increase their incomes. Intensely political rather than economic pressures are likely to determine government policies. As in Southern Europe, however, rising aspirations of the rural population and migration to the cities have a tendency to reduce grazing pressure and some recovery of the mediterranean scrub vegetation, now thinned or destroyed in extensive areas, may be foreseen.

California

Although discovered by the Spaniards at the same time as Chile, California was not settled by them until the late 18th century. The Indian population consisted of hunters, fishers, and gatherers of vegetable stuffs, but despite its lack of horticulture it was numerous and adept at exploiting the wild resources. A conservative culture had achieved a delicate balance with its environment.[8] The 150,000 humans were a major element of the biota and strongly affected it with fire and collecting pressures on particular species, but they seem to have been in dynamic equilibrium for more than a millennium. They may have developed the most humanized landscape any nonagricultural society ever created.

The establishment of Missions, though humanitarian in aim, caused a catastrophic collapse of the Indian population, especially through introduced epidemic diseases.[9] By the time of the accession of California to the United States, and the almost coincident gold rush, mediterranean California had an extensive grazing and limited farming economy and perhaps one-third the population it had supported with the more primitive system. Beginning before 1850 around San Francisco Bay, and three decades later in Southern California, the level areas shifted rapidly into the production of small grains. On the pastured grasslands Old World weeds rapidly replaced the native grasses and herbs, possibly as early as Mission times. Strongly after 1880, areas subject to irrigation were devoted to specialty fruit and vegetable crops, favored by newly available railroad transportation and free access to the whole United States market. At about the same time a retreat of the dry-farming frontier began, associated with rising living standards that made cultivating slopes too steep for mechanized farming uneconomic. Most such abandoned land has reverted to chaparral or scrub or has been maintained as pasture dominated by the weedy Old World grasses and herbs. Grazing has remained extensive and highly commercial with only big ranchers who are able to rationalize their operations and avoid overgrazing participating. Thus much steeply sloping land carries a wild vegetation that is very little abused by grazing. Fire, as in Indian times, has important effects on the extremely vulnerable vegetation. In Southern California, where fire protection is assiduously sought, fuel accumulation may cause the very intense fires, when they occur, to have an impact on the chaparral, oak grasslands, and coniferous forests different from that of the lighter more frequent burns of earlier times, but the relationships are complex and as yet little understood.[10]

The tremendous sprawling urbanization that has afflicted California in the past few decades has been concentrated in areas that had been devoted to irrigated specialized fruit production except in hilly sectors immediately adjacent to the major metropolitan centers. New irrigated plantings have tended to occur in desert rather than mediterranean climates. Thus despite its very large population California retains extensive rough areas with their natural vegetation only moderately modified. These chaparral- or brush-covered slopes constitute a growing set of management problems to land managers and planners. Can their recreational potential be increased without increasing fire hazard? Could a grass cover be artificially maintained to reduce fire hazard, and would it afford adequate erosion control and slope stabilization? At the present time the direct agriculture or pastoral productivity of these wildlands seems to be their least relevant capability.

South Africa

Only a few centuries before the Dutch brought European land-use patterns to the mediterranean part of South Africa in the mid-17th century the pastoralist Hottentots had entered the area, displacing the hunting and gathering Bushmen. The pastoralist tradition had diffused slowly southward through Eastern Africa from its ancient hearth near the eastern end of the Mediterranean Sea. Hottentot grazing activities do not seem to have developed much intensity in the area they had newly occupied. After the Dutch settlers had slowly gained a foothold they spread inland into drier areas with extensive grazing activities. In the Cape District itself small grains and garden produce and fruits found a market in passing ships. These crops were produced on the alluvial lowlands, and with the growth of a modern economy in the Union of South Africa the specialized irrigated crops, including vines, have tended to displace the small grains. On the steeper slopes neither commercial grazing nor dry farming became important, and the native subsistence grazing has been eliminated within the region of mediterranean climate, though it persists in the drier Karoo to the east. There was ample flatland too dry for anything but grazing in the interior for the European commercial grazers to exploit. Consequently the chaparral-like forests on steep slopes in the mediterranean part of South Africa are perhaps the least modified of any in the world.

The Australian Lands with Mediterranean Climates

The aboriginal populations of both Western Australia and South Australia were hunters and gatherers with cultures as conservative as those of the California Indians, but their population density was considerably lower, possibly because of a less food-rich native flora. European settlement came late, 1828 in Western Australia and 1836 in South Australia. The early settlers were few, but they looked on the extensive, relatively level lands of mediterranean climate strictly in terms of their capacity to produce commercial products for export. Local subsistence needs were and remain small. The characteristic chaparral-like forest of other mediterranean areas was largely absent. In the wetter areas there were dense forests of giant eucalyptus trees. In drier locales large shrubs, primarily other species of eucalypts, were scattered over a grassland. With very extensive lands and little labor available, sheep grazing became the dominant land use. What labor there was was devoted to eliminating the scrub or forest and upgrading the pasture with European grasses, especially in the wetter areas. On level lands toward the drier interior huge wheat fields were planted, and still further into the drier interior sheep grazing again became dominant. Limited areas of specialized fruit and vine production in the hills behind Adelaide and Perth satisfy local markets.[11] The extensive mallee scrub of Kangaroo Island and the Ninety Mile Desert southeast of Adelaide remained a meager pasture, despite a sufficiency of rainfall, until well into this century, when it was discovered that adding traces of copper, cobalt, and other trace elements to the soil could both upgrade the pasture and protect the sheep from deficiency diseases.[12]

Despite the fact that the extensive areas of mediterranean climate in Australia support the lowest human population densities of any region of the world with that climate, and even then more than half of the people live in the two large cities, the natural ecosystem may well be the world's most disturbed. The relative lack of relief coupled with uninhibited concentration on extensive commercial grazing and cropping are the determinants.

Interregional Dispersals

It may be appropriate to conclude with a few comments on how certain plants native to one of the world's mediterranean climate regions have successfully established themselves in others after being transported, deliberately or accidentally, by Europeans during the past four centuries. Platyopuntias (*Opuntia* spp.), or prickly pear cactuses, possibly from California but more likely from Central Mexico, have been established in all the mediterranean regions. In Southern Italy and Sicily and to a degree in Chile they are cultivated for their fruits, but they also tend to establish themselves on overgrazed, eroded, and rocky slopes, where they prevent pasturing and are hard to eradicate. Their clumps offer refuge to small animals and can form exotic bits of enduring wilderness in otherwise heavily grazed terrain.

The Monterey pine, *Pinus radiata*, is endemic to restricted areas along the cool coast of Central California. It has been established in extensive plantations in the moister parts of the Chilean mediterranean area, both in the Central Valley and the Coast Ranges. Its rapid growth there allows it to be the most important source of lumber in the country.

The almost complete replacement of native grasses and herbs in California and Chile by weeds of Old World origin such as wild oats, foxtail millet, and mustard has already been noted. The importation was almost certainly accidental, but the immigrant annuals are generally believed to have been successful because they could tolerate, better than the native plants, pasturing and trampling by the introduced domesticated grazing animals. One is tempted to regard the Old World weeds as having evolved and adapted in their eight- to ten-thousand-year association with man and his domesticated grazing animals.[13]

Eucalypts from Australia, many of them species from the parts of that continent with mediterranean climates, have in the past century and a half been given a remarkable worldwide distribution and propagation. They are now abundant in all areas of mediterranean climate, as well as being the most visible tree in the drier parts of the highlands of Western South America and East Africa. Even in humid Southeastern Brazil they are strikingly abundant. Almost invariably they were planted deliberately, and often hopes for their commercial utilization as lumber were unrealized. But they survive without care and withstand repeated cutting for firewood or lumber by stump sprouting, growing slowly on dry slopes and fast where their roots have access to groundwater. I have seen few places where eucalyptus groves are expanding by the dispersion of seedlings. This may take a long time. But I suspect that eucalypts will long be a prominent feature in the man-modified landscapes of all the mediterranean lands.

LINGUISTICS

Introduction

William G. Loy

LANDSCAPE AND LINGUISTICS were intertwined in the mind of Homer Asch-mann. The publications that follow document decades of concern about both. He sought linguistic evidence, and he used evidence generated by anthropologists from such techniques as lexicostatistics and glottochronologies to help explain the movements of people across the landscape. These people brought their languages but also the remainder of their cultural baggage, including their habits of material culture. The study of language and geographic names allows us to trace migrations, as well as understand various influences in the development of cultural landscapes.

People feel the need to verbalize; the need to encode the landscape and to give names to places. We name physical landscape elements such as hills and valleys but also places we create—roadways and towns, for instance. Ever curious, Aschmann must have questioned every geographic name he encountered, searching for the clues each held about the use and development of the land.

He felt, as I do, that geographic names are an unappreciated category of geographic evidence about the land. A few modern academic geographers besides Asch-mann, notably Wilbur Zelinsky and John Leighley, have considered geographic features and noted their value in giving clues to the history of landscape development.

In the city where I live (and probably where you live) there is a Mill Street and a Ferry Street. Not surprisingly, these streets once terminated at a flour mill and a river ferry. Historical geographic names provide far-flung clues about now-defunct economic activities such as the mining or ranching sites throughout the arid areas preferred by Homer Aschmann. Maps show cultural incursions by the Spanish as missions and by the Anglos as forts. The very language of the geographic name distribution indicates that most geographic names were recorded and fixed by the cartographers of the dominant group at that time. The lack of extant names by other groups known to be present testifies to their lack of influence. While these principals hold in general, it must be emphasized that the process of naming is complex and that the cautious scholar does not jump to conclusions. The geographic name *Cambridge* does *not*, for example, refer to a bridge over the river Cam. Naming is seldom a simple process.

Aschmann's dissertation fieldwork in Baja California in the late 1940s and early 1950s apparently figured in his publication 40 years later on the origins of the geographic names *Coromuel* and *Pichilingue* near the Bay of La Paz. To support arguments for the most likely origins of these names, he used a wide range of evidence about pirates. Although the origins of these names will probably never be known with certainty, Aschmann made well-reasoned arguments. One wonders if he pondered these exotic names and kept an eye out for related evidence for four decades until writing the essay included here.

Realtors hope to influence the use of a place by coining an appropriate name. The oldest known example of the phenomenon is the giving of the name *Greenland* to a forbidding piece of real estate. Apparently a more enticing name was needed than the previously named country Iceland. There is probably some truth to the matter. Almost anyone would choose to settle in Happy Valley over Stinkingwater Flats, especially when facts are few. One of the most famous of the "real estate" names is *Miracle Mile*, a portion of Wilshire Boulevard in Los Angeles. In 1957 Aschmann responded to an inquiry published in *American Speech* concerning the origin of the term *Miracle Mile*. He went to the newspaper archives to find a 1938 key article; then he rounded his answer out with a cogent explanation of what happened to the term since its inception.

People of different cultures name the landscape in different ways. American Indians generally give practical and descriptive names, whereas whites tend to apply names that merely identify, often commemorating a person. Different peoples also encode the landscape features in varying ways. Europeans seem to conceive of a river from its mouth to its headwaters as a unit, whereas most American Indians assign a name to a place on a river, and then another name to another place upstream or downstream. The knowledge of how peoples conceive of the landscape differently has been gained largely by anthropologists working in the American West. This was the area of Homer Aschmann's longest and most intimate experience. Western North America, especially the zone from west Texas heading west through New Mexico, Arizona, northwestern Mexico, and on to Southern California seemed to fascinate Aschmann the most, and this region served as the stage for many of his investigations. His interest in, and knowledge of, Southwestern anthropology and history was exceptional. These were among the landscapes he knew the best.

By understanding naming processes, the cultural geographer can gain insights that assist in reading the landscape. Studying the distribution of geographic names provides clues to the perceptions of the people, their values, and their influence at the time of naming. Even today, American Indians use geographic names based on the availability of natural resources important to them. White society is generally oblivious to these names. With an astute power of observation and knowledge of the ways of western American Indians, Aschmann used geographic name evidence in an implicit as well as an explicit way.

Few people have been as perceptive and as active as Aschmann. Of some 100,000 prisoners of war held in Germany during World War II it was he who recognized the uniqueness of speech terms such as *kriegie*, a contraction of the German term *Kriegsgefangener* (prisoner of war), and *Goon* for a German. Attentive and curious, even under POW conditions, he collected and was able to analyze about two dozen terms that he described as "Kriegie Talk" in a professional article; this article served as a benchmark of a decades-long fascination with names on the land.[1]

Calendar-date street names are rare to the point of being unknown in the United States. Within a short time of arriving in Asunción, Paraguay, Aschmann noticed several calendar-date street names. He suspected language was an important element in this bilingual country.[2] Soon he was demonstrating some of the best qualities of the professional geographer—curiosity, fieldwork, and careful map reading. He discovered that most of these names related to historical events of national significance and that most of these events occurred during the cool season, not during the hot and muggy days of summer. Almost 40 years after publishing "Kriegie Talk," he published his calendar-date street names research in the journal *Names*.

No landscape in the country has been more dominated by the development and use of the automobile than Southern California, where Aschmann lived most of his life. Perhaps Aschmann's exposure to this phenomenon had something to do with his investigation into the use of the term *model* in regard to Ford cars. His succinct article tells the story of Ford car model designation from the 1903 Model A through to the famous Model T Ford, to the reincarnation of the Model A in 1928, and then the designations that followed in the 1930s. These cars produced a mobile element in the landscape of Southern California and brought about the development of permanent structures such as freeways, parking lots, and sales centers that give Southern California so much of its automobile-centered character.

On other occasions, Aschmann traced linguistic distributions as a device for understanding origins of current patterns of settlement and landscape development. In the longest essay reprinted here, that on Athapaskan expansion in the Southwest, Aschmann examines how language—even when it is not recorded in written form on maps—can suggest the movement of peoples, their customs, and their impacts on the land.

Homer Aschmann had an abiding interest in explaining landscapes, answering questions, and setting the record straight. One of his preferred kinds of evidence was anthropological and, more specifically, linguistic. He was uniquely able to combine these lines of evidence into cogent explanations of how the land got to look the way it did and how its usage changed over time.

Athapaskan Expansion in the Southwest

At present the Navajo Reservation, though displaced westward, is fully
occupied by Athapaskan speakers and actually involves more area than
the Navajos ever occupied prior to American conquest, as well as 10
times as many people

COMMUNALITY OF LANGUAGE is perhaps the most important factor in the integration of human social groups into potentially cooperative entities that can share
and exchange ideas, both those of a technological and resource-exploiting nature and
those of a less material sort such as value systems, aesthetics, and understanding of
the supernatural. Within small groups such as the family or primitive band, the regular communication among all members sustains the linguistic unity even though
linguistic drift causes the language itself, something that is always learned by the next
generation but never learned perfectly, to change through time measured in generations. Larger groups, tribes, and ultimately nations can be sustained as operational
entities only if there is sufficient intercommunication among the component interacting groups to maintain mutual intelligibility. Time and distance, the latter emphasized by sedentary living habits, have served through the long sweep of human
history to fission linguistic units into an intricately branching set of languages of
varying degrees of relatedness.

Only with the development of strong centralized political systems, written languages, and revered literary documents such as the Bible, the Koran, or the *Divine
Comedy* did it become possible for really extensive and expanding speech communities to maintain themselves as units for long periods of time. In his insightful essay
Borderlands of Language in Europe,[1] Vaughn Cornish showed how the linguistic
boundaries of Central and Eastern Europe were established shortly after the fall of
the Roman Empire by the accession of barbarian tribes to Christianity and the accompanying literation of their languages. Until the present generation these boundaries have been substantially fixed. The speech communities then crystallized are in
considerable measure the ancestors of the modern European nations. With modern
media of mass communication, beginning with the printing press, single languages
have been able to sustain their unity among hundreds of millions of people in separate continents.

For most of human time and all or almost all of it in most of the world, however, neither writing nor highly structured political systems were part of man's cultural equipment. The distribution and articulation of linguistic entities, unified and dialectized languages, language families, and linguistic stocks that were present in particular areas at the beginnings of their recorded histories are products of cultural processes that affect language working over tremendously long periods of time, forming and erasing patterns and sometimes leaving enduring residuals. Among the potential products of the study of linguistic distribution of nonliterate peoples, two stand out. This is one of the most effective means of reconstructing the histories of tribal groups and their relation to cultural evolution and the diffusion of other culture elements. Also, it is possible to gain some insight into certain social processes concerned with intergroup relations. How does one group displace another or overwhelm it culturally in effective assimilation? What are the cultural concomitants of success in intergroup competition? From abundant examples in both simple and complex societies it is clear that technical superiority or economic advantage often results in failure rather than success in such competitions.

The Indian populations of North America north of the high-culture areas of Mesoamerica afford an enormously interesting set of examples of linguistic displacements. Especially in the western part of the continent a vast number of distinct languages are grouped in families and stocks in intricately intermixed geographical distributions. Related languages survive as isolated residuals, in some cases probably the result of migration of parts of a speech community, in others a continuous speech community split by the interposition of an alien group. In a fragmentary way the language distributions known at the time of European contact reflect tribal movements, expansions, and contractions over at least five and probably ten thousand years. Unraveling the earlier part of the history and relating the linguistic to the archaeologic evidence is of necessity speculative and uncertain. What we can learn of the mechanisms of linguistic shifts from more recent and better-controlled examples will enhance the validity of earlier reconstructions, both in America and elsewhere.

The area comprising Southwestern United States and northern Mexico is a particularly favored locale for such studies. Although European contact and documentary records began as early as 1540, some of the visited areas did not come under effective control until after the middle of the 19th century. Notable linguistic expansions and displacements occurred during that span of more than three centuries, and we have at least fragmentary documentary records of them. It cannot be denied that the establishment of Spanish missions and settlements within the area and the introduction of domesticated livestock had a special impact even on those Indian societies that remained independent of Spanish control. But in their competition with each other the several Indian groups probably followed patterns basically established in pre-contact times. For the modern student the study of North American Indian language displacements has an advantage over similar studies on other continents. As a fact of history, all the Indian tribes have been overwhelmed by the American society of European origin. The tribes are not ancestral to modern nations, and there is infinitely less danger of the student coloring his understanding by himself identifying with one group in its contest with another.

Within the past two decades lexicostatistical techniques have been developed which permit an estimate of the amount of time that has elapsed since two related languages were part of the same speech community.[2] While the precision of the instrument is subject to question, the results of glottochronological studies of the histories of American Indian linguistic groups are consistent with, though more detailed than, earlier historical interpretations made on other grounds, largely those of geographical distribution, by such students as Powell, Sapir, and Kroeber.[3] Glottochronology seems especially secure when dealing with a linguistic family that has in part experienced extensive migration coupled with territorial expansion in the relatively recent past. The splitting of a speech community because of increasing distance between its elements is an objective event. The technique provides a time for the occurrence of the event, probably accurate within 20 percent or two centuries within the last thousand years.

The Athapaskan linguistic family, at least in its Southwestern representatives, is a singularly appropriate group through which to examine the modalities of linguistic displacement among peoples organized at the band—certainly no higher than the tribal—level. From west-central Alaska to northeastern Mexico a single linguistic family has been recognized since the time of Powell. More recently this family has been combined with the Tlingit, Eyak, and Haida of the Northwest Coast into the more extensive and more distantly related Na-Dene stock. The Athapaskan family proper falls into three subgroups of languages, each widely separated geographically from the others (Figure 18). The most extensive section is the Northern one, which occupies all of interior Alaska and northwestern Canada as far south as the Sarsi on the prairies of Alberta. A second section occupied areas from northwestern California [NW Coast Athapaskan on the map in Figure 18] to southwestern Washington in interrupted fashion. Jacobs pointed up the singular fact that the several Athapaskan-speaking tribes in this region seldom occupied the coasts or the banks of the major rivers, that is, the sites which in that part of the world and to the native technology were economically the richest and supported the highest population densities.[4] His explanation that the Athapaskan speakers were relatively recent immigrants who had moved into the essentially unoccupied stream headwaters and were gradually expanding into more favorable terrain downstream is eminently reasonable. It is supported by the glottochronological calculations by Hoijer that the several Pacific Coast languages are separated from those of the Northern section by between 700 and 1,300 years.[5]

The Southwestern section of the Athapaskan family is appropriately identified as the Apachean group of languages. It is a more integrated section linguistically, showing glottochronologically a separation of only 400 years between its two most geographically distant elements, the San Carlos in central Arizona and the Lipan in

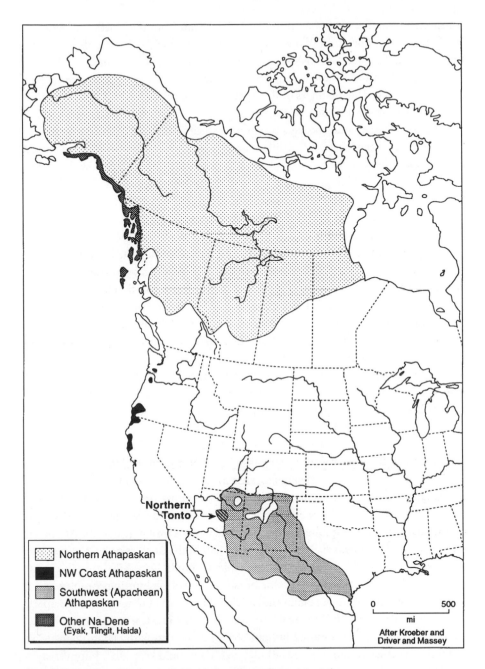

Figure 18. Na-Dene languages in North America at late contact times

south Texas. Further, the separation of the Apachean from the Northern section is smaller, between 600 and 1,000 years, than its separation from the California section or the California sections's separation from the Northern. Finally, within the Northern section itself there is roughly the same separation among its several languages as between the members of the Apachean group.[6]

A relatively unforced interpretation suggests itself. About 1,000 years ago, perhaps less, a group of Apachean speakers began migrating southward from central British Columbia along the Western Cordillera and the Cascades. They must have passed rather rapidly through the sparsely inhabited lands of the interior Salish, without permanently displacing that group. Establishing themselves in the Cascades and Coast Ranges in California and Oregon, they slowly moved downstream toward the coast and larger river valleys. The Galice, an interior group in southern Oregon, probably left central British Columbia no more than 700 years ago.

The groups ancestral to the Apachean section probably left the Northern section from the area of northern Alberta no more than 600 years ago; 500 years might be a better estimate. Navajo shows only a little over a 600-year separation from Chipewyan in Hoijer's glottochronological calculations, and it is improbable that the ancestors of the Apachean speakers were specifically part of the Chipewyan speech community. It is more likely that they came from the nearer Sarsi area, but that group shows a wider separation—more than 900 years. One can assume greater modification of the Sarsi language due to intimate contacts with their fellow Plainsmen, the Algonkian-speaking Blackfoot, and this problem of language modification through contact with adjacent alien speech communities suggests caution in interpreting the apparently precise dates available from glottochronology. In any event, by the mid-19th century a distance of 900 miles separated the most northerly Apachean group from the Sarsi, the most southerly language in the Northern section.[7]

The fact that the Apachean languages are most closely related to those of Northern Athapaskans who lived well east of the Rocky Mountains as well as the inherent difficulties of terrain farther west that would have required a sequent series of economic adaptations for exploitation and survival make it probable that the Apachean migration southward was either along the east face of the Rockies or farther east, out in the High Plains. The latter is more probable. It should be recalled that during at least the early part of the migration the horse and the bison hunting complex based on it were not available to the Plains Indians. With permanent water sources far apart, vegetable foods sparse and seasonal, and the bison, the greatest resource, mobile at a rate far faster than unmounted human groups could travel, the High Plains were certainly unattractive and sparsely populated. If they could keep alive, impoverished bands would not find their passage southward disputed unless they pressed east against the river-bottom farmers such as the Mandan, Hidatsa, or Pawnee, or toward the Rocky Mountain face, where better water and fuel supplies and more diversified resources afforded a home base for the Arapaho. Champe identified the

Dismal River Aspect in western Kansas, an archaeological complex, with early 18th-century Lipan Apaches.[8] Lipanan might be a better designation, since the historically known Lipan Apaches lived much further south in Texas. They had some pottery and practiced a little river-bottom farming—culture elements that could have been borrowed from groups like the Pawnee to the east or the Pueblos to the southwest.

At about Latitude 35° N and in 1540 this migration begins to enter history. In the large corpus of documents concerning the Coronado expeditions, 1539–42, when both large and small groups crossed and recrossed the area from the San Pedro River in southern Arizona to the valleys of the Rio Grande and Pecos, there is virtually no reference to Indians other than those residing in pueblos.[9] Castañeda does mention some barbarous seminomadic *ranchería* dwellers near Chichilticale, and they may have been Southwestern Yavapais or even Papagos. The term *despoblado* appears frequently, especially in the lands that later were the home of the Western Apache. East of Pecos on the plains of the New Mexico–Texas border, however, rancherías of nomadic bison hunters called Querechos and Teyas were encountered. The Querechos, with whom the Spaniards were to have sporadic contact for more than a century, have consistently been recognized as ancestral Apaches. The age separations calculated by Hoijer for the several Apachean dialects in 1956 are from 150 to 400 years.[10] It is completely reasonable to conclude that at the time an essentially undifferentiated Athapaskan speech, ancestral Apachean, was spoken among these nomadic rancherías.

The excellent quality of the topographic and ethnographic information in many of the Coronado documents must be emphasized. These explorers were looking for Indians, and those on the Plains were not hard to find. In the bulk of the area later to be dominated by Apaches, there was no resident population. Further, individuals and small groups from Zuni and other pueblos traveled far and regularly from their permanent settlements—hunting, perhaps trading, and especially visiting. Accurate reports on the next town and several beyond it always seemed available. There was more intertribal hostility on the plains and news was faultier. Of all groups the Querechos seem to have been the most timid and peace seeking, as might be expected from newly arrived impoverished migrants. The destruction of a pueblo near Pecos in about 1525 is not attributed to them by Castañeda but to the Teyas.

Then or shortly thereafter the Querechos came under pressure from the north from the Comanches who were drifting southeastward from central Wyoming. When the Spaniards returned to New Mexico at the end of the 16th century, New Mexico as well as west Texas held nonpueblo populations which were variously called Querechos, Cocoyes, and Apaches de Navaju. The first reference is from Espejo's expedition in 1583 and mentioned Querecho mountaineers west of the Rio Grande in the vicinity of Ácoma;[11] identification as Apaches is not certain, but they could have been ancestral to the Gila Apaches of a century later. In 1598 the Cocoyes, a ranchería people practicing a bit of agriculture, were reported in the headwater area of the Rio Grande,[12] across the Sangre de Cristo Mountains from the Plains. The Navajo report from 1626 is clearest and locates these people in the headwaters of the San Juan, that is, in *Dinetah*, the legendary Navajo homeland.[13] At this time Spanish missions and settlements were just being established in New Mexico, and one is struck by the attitude of the chroniclers that these nomads were timid and tractable, less likely to offer resistance to Colonization than the organized societies with formal religious systems of the Pueblos.

Seminomads in areas farther west who were reported in the same period, however, were not Apachean. Espejo went from the Hopi villages toward the southwest and on the Flagstaff plateau encountered a "warlike mountain people" who were in communication with the Cruzados of the Verde Valley.[14] The cross symbol, probably related to the cardinal directions and sacred four, continued to be a characteristic of the Yavapais (a Yuman-speaking group) when Corbusier reported on them in 1875.[15] Farfán in 1598[16] and Oñate in 1605[17] encountered the same cross-wearing groups in that area; curiously, the former found occupied rancherías on the high, snow-covered but fuel-rich plateau in midwinter. The peoples west of the Hopi villages who ranged over the high but largely unforested plateau are called Cosinas or Coninos and are pretty clearly the Yuman-speaking Havasupais.[18] Athapaskan speakers had not yet entered the modern territory of Arizona.

In the 17th century Athapaskan speakers seem to have established themselves widely throughout New Mexico and perhaps in east-central Arizona. They were no longer so tractable, raiding livestock and killing an occasional missionary. On the plains of eastern and southern New Mexico they were mounted and participated, at least peripherally, in the flourishing Plains culture. Mixed among the sedentary Pueblos, they seem to have found an ecological niche and were multiplying rapidly. West of the Rio Grande there is a clear preference for rough and generally wooded country. It also seems clear that contact with the Spaniards greatly reduced the extent of the areas exploited from the Pueblo settlements. Extensive hunting was discouraged, since it made mission control harder to enforce. Domesticated animals provided animal products in more concentrated terrain, and new crops, especially fruits, probably made gathering wild plant foods less essential. Also, populations in the pueblos were definitely reduced for a variety of reasons.

The Pueblo Revolt of 1680 and the Reconquest 12 years later afforded far-reaching advantages to the newly arrived and only marginally participant Athapaskan speakers. The abandonment of the Piro Pueblos opened a passage westward for the Apaches, and Apache threats are certainly one of the reasons the pueblos were never reoccupied. In the dry plains with isolated ranges of southern New Mexico, the west Texas panhandle, and northern Chihuahua there were a number of seminomadic groups, largely nonfarmers and with no more than ranchería political organization. Conchos, Jumanos, Mansas, Sumas, Janos, and Jocomes seem to be distinct groups, distributed roughly from southeast to northwest. The Southern groups had been under some pressure from the Spaniards and missionaries, but within a few decades after 1680 essentially all disappeared or were absorbed into several Apache groups, from east to west the Mescalero, Mimbreño, and Chiricahua. Despite the establishment of Spanish presidios from El Paso to Tubac, these Apachean groups raided successfully into Chihuahua and Sonora for more than a century, effectively cutting off the reestablished New Mexican colony from all but heavily convoyed communication.

Fleeing the persecutions that accompanied the Reconquest, many people from the Rio Grande Pueblos fled westward to live among the Navajo of northwestern New Mexico, contributing numbers to what was doubtless a small population and, perhaps more significantly, enriching their culture with agriculture, herding, and crafts such as weaving and pottery making.

In both the above cases of contact between Athapaskan speakers and other Indian groups, although the forms of the encounters and the respective levels of technology and social organization were quite different, one result was consistent. Athapaskan speech prevailed, spreading its domain over hundreds of thousands of square miles.

There was also Apachean expansion westward into Arizona into the rugged area, noted as a *despoblado* in Coronado's time, between the Gila River and the Mogollon Rim as far west as the mouth of the San Pedro River. Whereas in the 16th and most of the 17th centuries there had been contact between the Sobaipuri Pimas of the San Pedro Valley and the Hopi, Zuni, and even the Spanish settlements along the Rio Grande, Velarde made the specific statement in 1716 that the trade was cut off by Indians who had moved into the intervening area.[19]

The 18th and the first half of the 19th centuries were characterized by Apache raids against enclaved and bordering sedentary peoples—Indian and Spanish and Mexican. Livestock taken in these raids came to be a major factor in the economies of many Apache bands. It is noteworthy that only the Navajo paid much attention to herding the captured stock; the cultural influence of the incorporated Pueblo refugees is apparent. Some Apaches rode captured horses, though they paid little or no attention to breeding them. Other stock were driven to home camps and were consumed. More could be captured later. Territorial expansion was not sought; perhaps it was consciously avoided, as it would take sources of loot farther away. But the sedentary folk suffered and contracted their grazing ranges. In one instance, abandonment is thoroughly documented. In 1762 the Sobaipuri Pimas abandoned the San Pedro Valley, moving westward into the Santa Cruz Valley and into the deserts of the Papago country.[20] Apaches moved into the Aravaipa Valley but were in no hurry to occupy the entire vacated area.

The history of Apache and Navajo raids and Spanish, Mexican, and U.S. reaction thereto from 1700 to 1870 is inordinately complex and disgustingly thorough though not necessarily accurately or coherently documented.[21] Until the Americans were able to bring overwhelming wealth and numbers into the contest, the Apachean side had the better of it. Though the Apaches sometimes suffered losses from retaliatory raids, no other groups made permanent settlements in their now-vast domains, with the qualified exception of the Santa Rita mines, and the sedentary folk contracted their ranges and abandoned some exposed settlements. Gálvez's policy instituted in 1786 had some success in protecting northern Sonora and Chihuahua. It was to pay tribute or doles to Apache bands to induce them to keep the peace, and to demoralize their social structure with liquor and trade goods. After 1811 the revolutionary disorganization of New Spain cut off funds, and Apache raids were renewed. The policy of buying Apache scalps later introduced in Mexico provoked the expectable atrocities on both sides. It freed no territory from Apache occupation or raids but probably reduced population growth among Athapaskan speakers, notably the Chiricahuas and Mescaleros. Against other nomadic Indian groups, however, the Athapaskan speakers were less successful. Comanche and Kiowa Plains warriors were driving them westward and southward into less hospitable and productive parts of the plains, and the impoverished Utes were pressing the Navajos southward in the San Juan Basin. By this time the herding-oriented Navajos should be considered economically more advanced than the Utes.

Two quite different expansions in territories occupied by Athapaskan speakers occurred very late, even after American power had become the dominant force in Southwestern history. As the Navajos became more involved in herding activities, their demand for grazing land increased at an accelerating rate, and they expanded westward as far as the Hopi villages. While the scorched earth policy of Kit Carson's 1863 campaign that forced most of the Navajos to surrender for internment at Bosque Redondo to avoid starvation nearly vacated the Navajo homeland in northwest New Mexico and adjacent Arizona, many bands remained free by moving with their herds as far west as the Colorado River and even beyond the Little Colorado, completely enveloping the Hopi lands.[22] After the Navajos were allowed to return from Bosque Redondo, the combination of demographic vitality and land hunger induced by a grazing economy gradually persuaded the U.S. government to expand their reservation westward into Arizona. At present the Navajo Reservation, though displaced westward, is fully occupied by Athapaskan speakers and actually involves more area than the Navajos ever occupied prior to American conquest, as well as ten times as many people.[23]

The story of the Northern Tonto of the Middle Verde Valley is more mysterious. At the beginning of the 17th century the area was occupied by a single people, called Cruzados and pretty definitely Yavapais. In the 1860s and until 1875 when the Indians were moved to the San Carlos Reservation, the district was occupied by both Yavapais and Apaches. The two linguistic entities were in full alliance; intermarriage was frequent; even individual bands included speakers of both languages. Each individual, however, felt that he belonged to one group or the other, although bilinguality was normal. Even today bilingual individuals who claim either Apache or Yavapai as their native language can be found on certain reservations.[24] That this complete mixing of two linguistic groups could be a stable system of long standing is improbable. The Apaches were almost certainly recently—and peacefully as far as the Yavapais were concerned—intrusive. It cannot be fully documented, but a reasonable explanation is that Southern Tontos moved into the Verde Basin after 1860 to share in the raiding of the newly opened mines in the vicinity of Prescott and the pack trains that supplied them. Final subjugation of both groups occurred before the Yavapais could assimilate the Apachean immigrants or the Apachean linguistic area could be further expanded by the reverse development.

In sum, then, the period of a little more than three centuries following 1540 witnessed some displacement of Apachean speakers from the Great Plains but at least a threefold expansion of their total area in a southerly and southwesterly direction. Some of the original resident groups were just absorbed by the Apaches, others were driven from their homes by repeated raids; in other instances the Apaches established themselves in essentially vacant districts.[25] It is now appropriate to inquire as to what cultural characteristics the Athapaskan speakers possessed that gave them competitive advantage over long-resident and frequently richer and more highly organized societies in the Southwest. The following suggestions are necessarily tentative.

*   *   *

1. Although in earlier times pueblo-type settlements were widely distributed in the uplands of the Southwest, by the 16th century most of the sedentary farmers had abandoned the uplands. The Apaches readily found essentially unoccupied ecological niches, for example, in the Gila Mountains of western New Mexico and later in the rugged country at the foot of the Mogollon Rim toward the west. One suspects that a residue of skills for coping with cold weather in a forested land developed in the Athapaskan homeland, the taiga of northwestern Canada, made such an environment quite acceptable to them as it had become progressively less attractive for the later Puebloan peoples. At the time of their arrival in small bands they were in no position to contest with large groups of the established peoples for mutually attractive sites. They sought and largely found a refuge zone, from which their incessant raiding made life ever more unpleasant for the sedentary peoples.

2. Their very poverty in material culture made the Apaches aggressive and expansionistic. They were always hungry and looking for a place to make a living. If conditions got bad, for ecological or sociomilitary reasons, there was no place to retreat in which a living was assured. The lands they came from were likely to be occupied by similarly hungry congeners. Although it was accompanied by raiding and fighting, Apache expansion was really an *Unterwanderung*, the poorest people seeking ecologic niches that established residents had neglected because of their unattractiveness.

3. Either as a direct product of the simplicity and unadaptedness of the material culture they brought with them or because of a fundamental style or focus within

their culture, Apachean groups were notably willing to borrow, adapt, and improvise with ideas and techniques for making a living that came from their neighbors. Thus the Western Apaches seem to have gotten from the Yavapai an agricultural system of irrigating small plots from upland springs and dry farming more extensive ones if the summer rains were good. They did not take up the highly organized system of canal irrigation of valley fields from the Pimas. The Navajos got their dry-farming pattern from their Pueblo neighbors and from the fugitives who sought refuge with them during the Spanish Reconquest of the 1690s. Their knowledge of and interest in stock raising came from the same source. There is a singular lack of evidence for rejection of any locally developed economic technique because of its incompatibility with their culture. Even their uneconomic fish and water animal food taboo was borrowed from an ancient and widespread Southwestern pattern.

4. The simple social organization in which a band of 100 or so persons was the largest unit of political control was an advantage to people who were usually on bad terms with their neighbors. They could never be defeated in war except individually. A successful Pima or Puebloan punitive raid—of which there were many—would at best knock out a single band, and for centuries Spanish, Mexican, and American military forces similarly found the Apaches' sociopolitical fluidity completely frustrating. On the other hand, the pattern of strict clan exogamy required the maintenance of loose but friendly interband contacts which could on an ad hoc basis unite much more numerous forces for concerted action if there was a clear and apparent immediate advantage in so doing. The most typical pattern was that of recruiting a raiding party of young men by visiting a series of bands en route to the area to be raided.

5. Despite their widespread and deserved reputation for military prowess, the Apaches did not have a war pattern in their culture. In contrast to most Indian tribes in North America a man did not seek or achieve glory through fighting. There was little ceremonial associated either with going to or returning from war. Nor was there any book of prescribed tactics, as the Colorado Yumans fought with clubs, the Plains people with lances, etc. In a fight the Apaches used what they had and found convenient. Most important, the Apaches really did not go to war except to the extent of preparing ambushes against pursuing or punitive expeditions. They went out to raid property, especially livestock after domesticated animals had been introduced by the Spaniards. Earlier they presumably raided other tribes for grain stores. The most successful raiding party was one which obtained the property without contacting its owners. Fighting occurred only if it could not be avoided. It often was vigorous, but the goal was to get away with the property without losing any men. If there were losses, it was not necessary to retaliate except on a personal or familial level. Another raid would be organized only if there were good prospects for booty.

Few societies of the past or present have ever had such a rational approach to intergroup hostilities. Their system worked so well that on several occasions various Apachean groups would for decades gain a significant part of their subsistence from

booty. Warfare (really raiding) was able to show a net profit for the entire community, a rare phenomenon indeed.

6. Recruitment of captive children into the tribe was not uncommon among
North American Indians, but the Apachean groups seem to have been particularly
active in this regard. Nakai, a recent president of the Navajo Council, has a name that
means Mexican and is the family name of a whole clan derived from assimilated Mexican captives. Earlier a Jemez clan had developed among the Navajo from the
Puebloan refugees of the 1690s. Since the Apaches were expanding under terms of
hostility with their neighbors, this steady recruitment of captives was surely important in compensating for heavy battle casualties. Apache culture seems to have enjoyed both a high degree of egalitarianism and a respect for an individual's personal
accomplishments rather than his ancestry. Despite its simplicity and physical poverty,
an Apache community was one into which a child captive could and would want to
be assimilated.

7. The Apache abhorrence and avoidance of the dead results in property destruction and loss through burning of the deceased's residence and effects and sometimes
killing his livestock that is indeed uneconomic and tends to increase poverty. This
death avoidance, coupled with a dispersed settlement pattern, however, would tend
to restrict the spread of epidemic diseases. One of the most decisive causes for the
loss of the North American continent to Europeans by the Indians was their vulnerability to newly introduced diseases. The Apaches were probably as vulnerable phys-
iologically as any Indian group, but their customs may have given them a survival
advantage over their sedentary neighbors during the centuries between the first European contacts and their final conquest and pacification.

8. When speakers of two languages are living in close and nonhostile contact,
bilingualism will develop. The situation, however, tends to be unstable and one of
the languages is likely to displace the other. Obtaining data on such movements
among preliterate peoples such as the American Indians is singularly difficult, but
one tends to suspect that the Athapaskan tongues tended to win such competitions.
Suggesting reasons why these languages, notoriously difficult to learn, should have
this capacity is far beyond my linguistic competence. We can note, however, the vigorous survival and growth of Navajo in the past century. The language is more widely
spoken by more people today than it ever was in the past. The complete absorption
into Apache of the Jano, Jocome, and Suma societies and languages during the early
18th century is even more striking. One gets the impression that the recently intrusive Apaches were winning the peaceful linguistic competition with the Yavapai in
the Middle Verde Valley in the 1860s and 1870s.

*   *   *

The Athapaskan speakers were not invariably successful in their competition with
other Indian groups. The diffuse social organization to which they clung so tena-

ciously was not so effective on the Plains once the horse was available and the Plains war pattern had developed. Notably, the terrain failed to afford refuges to small groups. The Pawnee, Kiowa, and especially the Comanche, in gradually pushing the Athapaskans into the least attractive part of the Plains, provided some of the impetus for Athapaskan expansion southward and westward. During the 18th century the Utes were applying effective pressure on the Navajo and the Jicarilla from the north. Here the case is again the simple one of Unterwanderung. Farther removed from pastoralists and sedentary farmers, the Utes raided the Navajos as soon as the latter had acquired by looting enough livestock to make them vulnerable, and until the mid-19th century the Navajo were tending to withdraw from the Upper San Juan Basin. If not to the meek, at least to the impoverished the advantage tended to fall.

In their generally successful competition with neighboring Indian groups and Spanish and Mexican frontier outposts, the Southern Athapaskans were contesting with peoples who were richer and possessed more complex cultures. But the differential was only a modest one. Furthermore, the sedentary peoples exploited significantly only a small fraction of the terrain they nominally occupied, leaving ecological niches open in which the invaders could establish themselves.

When a much richer, more numerous, and technologically more advanced society such as the European population of the United States swept into the Southwest, all of the Indian groups were engulfed by it. There are abundant examples over the world of conquest and expansion by rich and powerful societies such as the Roman and Latin overwhelming of the Celtic populations of Gaul and Northern Italy and the displacement of the aboriginal inhabitants of the Red Basin of Szechuan by Chinese. This paper endeavors to illustrate and show the historical significance of another sort of linguistic and cultural expansion, one that normally is harder to document but which has probably played an equally important part in creating the world's cultural and linguistic mosaic.

Miracle Mile

It represented one of the earliest efforts of the more exclusive specialty stores to evade downtown traffic congestion and to create an appropriate neighborhood in which to display their wares.

THE QUESTION ASKED by Mamie J. Meredith concerning the origin of the expression *Miracle Mile* prompts this answer.[1] The flamboyant commercial district in Los Angeles on Wilshire Boulevard, from La Brea Avenue westward toward Fairfax Avenue, has borne this name at least since the 1930s. A Los Angeles realtor, A. W. Ross, is given credit for creating the flourishing development. His efforts to get the appropriate commercial zoning and to control the kinds of buildings that went into the area are said to have begun in 1923.[2] Mrs. Dee Davis, the executive secretary of the Miracle Mile Association, gives 1929 as the date when the expression *Miracle Mile* was first used orally. Mrs. Davis reports that "Mr. Ross was so enthusiastic about the development that a friend one day remarked that he spoke as if the area were really a miracle mile." The earlier name for the district was the Wilshire Boulevard Center.

The first printed record I have found of the expression was the lead phrase in a full-page advertisement in 1938 by the A. W. Ross Realty Company, describing developments in the district and inviting further investment.[3] An intensive search of advertising copy in Los Angeles newspapers probably would produce a citation a year or two older.

The development has enjoyed a high level of prosperity. It represented one of the earliest efforts of the more exclusive specialty stores to evade downtown traffic congestion and to create an appropriate neighborhood in which to display their wares. Their business success has, of course, made the attempt to get away from traffic a failure. In 1941 the Miracle Mile Association, a chamber-of-commerce-like organization of the merchants of the district, was incorporated,[4] and it continues to operate vigorously. The association claims a property right to the name, but the right has never been established in court.

Construction of the new stores began near La Brea at Sycamore Avenue, moving westward across undeveloped land, but was restricted for a long time to property facing Wilshire Boulevard. The term *Miracle Mile* was in use long before Fairfax, which is precisely a mile away, was reached. It may be recalled that during the 1930s, when vacant store space was easy to find, such an outburst of commercial construction was a major event in the real estate world. In this sense the real estate promoter's choice of a name was apt, and the success of the Los Angeles development has certainly not discouraged copying in other parts of the country. By now Wilshire

Boulevard has commercial establishments, many of them of the most pretentious sort, along most of its 20 miles from downtown Los Angeles to the beach at Santa Monica. People still speak of the *Miracle Mile*, but its precise limits are less clearly recognized.

Calendar Dates as Street Names in Asunción, Paraguay

My impression is that streets named with calendar dates refer to events so deeply involved in Paraguay's history that in general such names have been and will continue to be more durable than the name of the average street.

Taking my first 30-minute walk after arriving in downtown Asunción I crossed three streets with calendar dates for names. Riding buses around town brought another half dozen similarly named streets to my attention. Perusal of available large-scale maps showing street names brought the total up to 34, including one that has been suppressed.[1] There may be more because many streets that show on large-scale maps are unnamed. Paraguayan towns follow closely the street-naming patterns of the capital city. With their smaller numbers of streets to be named my impression is that calendar dates form an even larger fraction of the street names. I

Figure 19. Looking down Azare to the center of Asunción

have noticed at least one date that is not represented in Asunción, *15 de Mayo* in [the suburb of] Luque. It is an alternate date with *14 de Mayo* for celebrating Paraguay's independence. Many an individual of local distinction rates a street name in his hometown but not in Asunción.

Naming streets is an official act in Paraguay. In the 19th century placing a name or altering one was by presidential decree. More recently the act has been carried out by a municipal ordinance or a municipal resolution in Asunción. Thus there is official documentation for the history of almost all street names, and it has been compiled by the municipality.[2] Osvaldo Kallsen, an amateur historian, has published a substantial volume in which he attempts to account for the origin of every street name in the city, providing, as appropriate, a discussion of the event or person commemorated.[3] As stated in the book's introduction, these discussions constitute a concise history of Paraguay; actually the best-balanced history I found in that country of intense historical partisanship. From these sources as well as discussions with informed long-time residents I have prepared the following annotated list of calendar date–named streets. The classes of events commemorated are N for an event of national significance; R for a religious holiday; RN for a religious holiday of special national significance; and I for an event of primary significance to other countries.

31 de Enero	Battle of Corrales, 1866. Paraguayan over Argentina.	N
3 de Febrero	Day of San Blas, patron saint of Paraguay.	RN
28 de Febrero	Undetermined, but on modern 1:25,000 topographic map.	—
1 de Marzo	Battle of Cerro Corá, 1870. Death of Francisco Solano López and effective end of the War of the Triple Alliance.	N
9 de Marzo	Battle of Tacuarí, 1811. Defeat of a force from Buenos Aires making possible Paraguay's declaration of independence two months later.	N
1 de Mayo	International socialist holiday. The street name has been suppressed.	I
2 de Mayo	Battle of Estero Bellaco, 1866. Paraguayan defeat.	N
3 de Mayo	Day of the Holy Cross.	R
14 de Mayo	Independence of Paraguay, 1811.	N
24 de Mayo	Battle of Tuyutí, 1866. Paraguayan defeat.	N
25 de Mayo	Important date in the Argentine independence movement, 1810.	I
8 de Junio	Battle of Picada Diarte in War of the Triple Alliance, 1869, last Paraguayan victory.	N
12 de Junio	Signing of protocol to end the Chaco War, 1935.	N
14 de Junio	Cease-fire ending Chaco War with Bolivia, 1935. Replaced 12 de Junio on the same street as the more significant date and 12 de Junio was applied to another street.	N
22 de Junio	1876. End of the occupation of Asunción by Brazilian forces, six years after the end of the War of the Triple Alliance.	N

25 de Junio	Undetermined, but on modern 1:25,000 topographic map.	—
4 de Julio	Beginning of second attack on Nanawa (Fortín Presidente Eligio Ayala) in Chaco War, 1933. Paraguayan defensive victory. (Not the United States holiday.)	N
14 de Julio	Bastille Day, French holiday.	I
18 de Julio	Culmination of the Battle of Boquerón y Sauce, 1866. Paraguayan defeat.	N
20 de Julio	Governing Junta sent letter to Buenos Aires declaring independence, 1811.	N
15 de Agusto	Day of the Holy Virgin of the Assumption, founding of Asunción, 1537.	RN
25 de Agosto	Proclamation of independence by Uruguay, 1825. Also noted as the date of the new Paraguayan constitution, 1967.	I, N
8 de Setiembre	Undetermined, but on modern 1:25,000 topographic map.	—
11 de Setiembre	Founding of the Republican (*Colorado*) Party, 1887.	N
22 de Setiembre	Battle of Curupayty, 1866. Paraguayan victory.	N
29 de Setiembre	Capture of Fortín Boquerón in Chaco War, 1932.	N
12 de Octubre	Columbus Day, discovery of America.	I
23 de Octubre	Commemorates the deaths of student protestors, 1931.	N
1 de Noviembre	All Saints Day.	R
3 de Noviembre	Birthday of President Stroessner, 1912.	N
27 de Noviembre	Date associated with Fortín Gabino Mendoza, established in 1934 during the Chaco War and now on the Bolivian-Paraguayan boundary.	N
8 de Diciembre	Day of the Virgin of Caacupé.	RN
11 de Diciembre	End of the 50-day battle of Zenteno-Gondra (Campo Vía) of the Chaco War, 1933.	N
25 de Diciembre	Date chosen in 1842 as the de jure date of Paraguay's independence. Earlier independence under Dr. Francia was from Buenos Aires, not necessarily from Spain. The coincidence with Christmas is not mentioned in the official records.	N
Año 1811	The year of Paraguay's independence is the only year commemorated in a street name.	N

Three dates are noted as undetermined because I found no official record of what they commemorated. *8 de Setiembre*, for example, could refer to the beginning of the Battle of Boquerón in the Chaco War, 1932, but it might celebrate another event.

There seems to be no record of officially naming streets in Paraguay's long colonial history, though undoubtedly the residents identified passageways with descriptive terms. In 1788 the governor commented that houses were randomly placed and streets or paths wound around them. The official grid plan described in the Laws of the Indies, and almost universally applied in Spanish America, was not followed in colonial Asunción. A map of the city prepared about the same time by the distinguished Spanish surveyor and naturalist Félix de Azara showed only two regularly aligned rows of buildings, parallel to the northern waterfront, that could define a street.[4] The dictatorship of Francia, which kept Paraguay isolated from the world until his death in 1840, similarly did not bother with street names. The government of President Carlos Antonio López, as part of its bringing the country into the modern world, named 45 streets in Asunción by presidential decree on April 1, 1849.[5] The first efforts to pave these and other streets did not begin until 1874.[6] Of these named streets, eight bear their original name although one of them, *El Paraguayo Independiente*, was known for most of the past century as *Buenos Aires*.

Of the initial complement of street names, two consisted of calendar dates, *14 de Mayo* and *25 de Diciembre*. The former survives unchanged, but the latter was renamed in 1927, and *25 de Diciembre* was given as a street name in another part of the city.

Using dates for street names and for other features or institutions is not uncommon in Latin America. It is supported by the fact that the Gregorian calendar was used through almost all of Latin America's postconquest history, a situation not available to the English-speaking lands. The Paraguayan distinction is that such usage was so common and goes back to the very first street naming in what was and remains the center of Asunción. A recent street map of Caracas, Venezuela,[7] a much larger city, shows only five streets with calendar-date names, two of them being *1 de Mayo*. All are in outlying barrios and three such barrios themselves have calendar dates for names, something that does not occur in Asunción. Santiago, Chile,[8] again a city four times as large as Asunción, has seven date-named streets, but in this case two of them are near the city center.

In Buenos Aires there are three date-named streets in the city center and four more were identified in outlying suburbs.[9] Both in Buenos Aires and in provincial cities, parks or plazas are likely to be named with calendar dates; *25 de Mayo* and *9 de Julio* are commonly chosen for this purpose. In the provincial cities of northwest Argentina—Salta, Tucumán, and Jujuy—date-named streets figure more prominently, but in these smaller places only two or three streets are involved.[10] In contrast calendar dates have been popular in Argentina as names for settlements. Seventeen such settlements were located on the ESSO road map and some have grown to be substantial towns.

Montevideo, Uruguay, which like Asunción refrains from having streets designated by numbers of letters, also comes closer to the Paraguayan pattern of calendar date–named streets. On a map showing about two-thirds of the city,[11] still a larger and more populous area than all of Asunción, 18 date-named streets were identified. Two pairs, however, were duplicated in different parts of town, suggesting that street naming in Uruguay is a less rigorously monitored process than in Asunción. A number of major avenues hold dates as names as well as some short streets in close-in and outlying areas. It may be speculated that both Uruguay and Paraguay, squeezed between two much larger powers, have felt more need to keep their national histories constantly in public consciousness. Like Argentina, Uruguay uses calendar dates to name settlements.

In Paraguay dates are used to name infantry regiments, but the most popular usage is as names for provincial football clubs. Perusal of the sports pages of the Asunción newspapers yielded no less than 38 clubs with such names playing in at least 15 relatively small towns. Sometimes all four teams in a small town's league have calendar dates for names. Curiously this practice has a faint reflection in professional sports in the United States with the San Francisco 49ers and the Philadelphia 76ers.

Since the naming of streets in Paraguay is an official act and not one left to the discretion of real estate developers, names are bestowed with great political consciousness. Street signs offer a means to establish national identity and honor figures deemed meritorious in national history. Even foreign relations may figure in the process, with friendly countries or chiefs of state having important streets named after them. The opportunity to thus use a street name is too valuable to squander on a logical system of numbered or lettered streets. Battles and participants in the War of the Triple Alliance (1865–70), even Paraguayan defeats, began to be commemorated as soon as Paraguay regained its sovereignty. The location of a battle, its date, and sometimes one or more Paraguayan participants would each get a street named for them. As Asunción began its rapid expansion after 1945 it was decided to honor with a street name every officer who died in the Chaco War (1932–35) under honorable conditions. A modest number of enlisted men received the same honor for especially meritorious service, but only senior officers got a street name while still alive.

These practices put a severe strain on a finite number of streets (now about 1,200), and one approach to relieve this pressure has been to apply different names to different sections of the street. This may be done where the street takes a slight bend at an intersection or at the crossing of any of several axial thoroughfares or at any intersection. Since a downtown street or a principal avenue is more honorific than a residential street in a poor neighborhood, the municipal council has frequently bumped a central street name to the outlying district to permit a more worthy person or event to have a higher honor bestowed on it.

As the record of the 45 streets named in 1849 by Carlos Antonio López suggests, a street name in Asunción has an expected half-life of about 50 years. My impression is that streets named with calendar dates refer to events so deeply involved in Paraguay's history that in general such names have been and will continue to be more durable than the name of the average street. There are almost certain to be exceptions, especially when the name refers to an event or idea associated with a political party or ideology. The suppression of *1 de Mayo* is a good recent example. *11 de Setiembre*, *3 de Noviembre*, and *23 de Octubre* could well have their names changed should the Colorado party lose its long-time political ascendancy. Dates of significance to foreign countries rather than to Paraguay may be in a similarly precarious position.

Looking over the whole calendar of date-named streets, an instance of real geographic or climatic influence, if not determinism, is apparent. The winter half-year (May through October) provides the names for 23 streets, the summer half-year only 11. The difference is statistically significant at the 0.02 level. Eliminating from the list dates of undetermined significance or of northern hemisphere association gives a ratio of 17:7, with a slightly stronger statistical significance. The bulk of chosen dates commemorates battles or political assemblages. Paraguay's hot, muggy summers are a very poor time to carry on a military campaign or to conduct a political assembly.

Coromuel and Pichilingue

These two exotic place-names open a window on the considerable part Baja California played in the history of the Pacific during the more than a century and a half between its discovery and permanent settlement.

On the west side of the rocky little peninsula that projects into the Gulf of California just north of the Bay of La Paz in Baja California, Mexico, are two exotic place-names, *Pichilingue* and *Coromuel*.

In current usage the term *Coromuel* is not primarily a place-name. It refers to the cooling westerly breeze that blows over La Paz every afternoon. Its meteorological basis is straightforward. Waters of the Pacific Ocean are notably cooler than those of the Gulf of California on the other side of Baja California, so a mild thermal low pressure develops regularly in the latter area. Frontal passage and tropical cyclones that might alter this pressure gradient are infrequent, and, most significantly, the area west and southwest of La Paz is the only place in the entire length of the peninsula where a mountain or upland spine does not impede the air flow. *Coromuel* is also a place-name applied to a *balneario* or swimming resort on the little peninsula that extends north from La Paz and is just outside its somewhat polluted bay. The Coromuel blows right against it. The resort was well established when I first visited it in 1952, but the place-name is probably only a decade or two older and clearly is named after the wind.

All the long-term residents of La Paz speak of the Coromuel, and it is a significant amenity in their environment. Even in summer, when all other coastal places on the Gulf of California swelter in stifling heat and humidity, the La Paz climate is tolerable and attractive to tourists who exploit the fishing opportunities of the Gulf. There is also a widespread agreement that Coromuel is a hispanicization of the English name *Cromwell*. The fanciful stories of how the name was implanted in Baja California are more varied.

One of these stories was made into a radio feature on *Bob Ferris News,* KNX (Los Angeles), on December 5, 1955.[1] Ferris obtained it from the Russo family, important and well-established merchants in La Paz. Cromwell, a great and clever English pirate, hid in the Bay of La Paz and used the regular wind to sally forth to the Cape to attack laden Manila Galleons as they sailed by. He claimed several prizes and buried them somewhere in the sands around the bay. A final feature is that early in this century a great *chubasco,* or tropical cyclone, altered the character of the bay, obliterating all landmarks and losing the treasure forever. Other tales have Cromwell becalmed in the bay and being threatened from land and sea by Spanish forces. The afternoon wind permitted him to get out through the narrow channel and elude his pursuers in the darkness. The investigative journalist Fernando Jordán, in his 1951 book *El Otro México*, discounts this legend but from old residents obtained another one, that the name comes from the sailing vessel *Cromwell*, which used the wind to exit the bay.[2] Jordan gives no date and no one has found a record of a ship named *Cromwell* in those waters.

Among the scores of English pirates and privateers identified by Peter Gerhard in *Pirates on the West Coast of New Spain, 1575–1742* the name *Cromwell* does not appear at all.[3] One suspects, though I have found no documentation for it, that the reference is to Oliver Cromwell. In actuality, Oliver Cromwell never left the British Isles, but during his rule in the mid-17th century, Britain was particularly active in interfering with Spanish shipping and in making incursions in Spanish territories around the Caribbean. That the British leader should become the English pirate incarnate to colonial officials and mariners attempting to defend the long Pacific Coast of the Spanish Empire is not unreasonable. This extended, poorly defended frontier, with slow and interrupted overland communications and sea travel often actually blocked by real pirates, was repeatedly swept by rumors of buccaneering raids that were completely fictitious, probably with fictitious captains.

Furthermore, in the 16th and twice in the 18th centuries English privateers (Spaniards would have identified them as pirates), lying off the southern tip of Baja California, did intercept the Manila Galleon as it stayed close to shore en route to Acapulco. Twice they were successful. It is likely that the fearsome British leader became the bogeyman to Spanish mariners in the Pacific that Sir Francis Drake was in the previous century.

Documentation for 17th- or 18th-century use of *Coromuel* has not been discovered, however. Permanent Spanish settlement of Baja California began with the Jesuit mission in 1697 at Loreto. This was too far up the Gulf, an impoverished region, to be of interest to pirates. No mission was established in La Paz until 1720. Nonetheless, the Cape region of Baja California was not an unvisited country. Although unsuccessful, Cortés's attempt at settlement in the La Paz area in 1534 and 1535 had obtained some pearls of high quality. Over the more than a century and a half that followed, licensed and unlicensed pearling expeditions worked the Gulf Coast at least as far north as latitude 28°. Although crossing from Sinaloa in small boats was risky, small entrepreneurs, who were willing to risk criminal penalties to avoid the royal *quinta* (20 percent tax) and the hassle of getting licensure, were more numerous and regular visitors than the dozen official expeditions. The latter often had the mission of founding a permanent settlement but clearly focused their energy on getting pearls.

In addition, beginning with Thomas Cavendish at Cabo San Lucas in 1587, English and Dutch privateers and pirates in undetermined numbers used any embayment in the Cape region to take on water and wood and careen their ships, badly in need of such attention after the voyage around Cape Horn. Though Cabo San Lucas was the place from which to ambush the Manila Galleon, the protected Bay of La Paz was a favored place for careening ships. These foreigners must have interacted with the pearlers; there were reported instances of their relieving them of their pearls.[4] Less hostile interactions, especially with illegal pearl seekers, would not have been reported to Spanish authorities, but they would become part of the mariners' lore of the West Coast of New Spain. It is my conclusion that the Coromuel legend arose from these interactions.

As an aside it may be noted that the English and Dutch visitors consistently reported friendly receptions from the local Indians. They wanted peace and quiet to attend to their repairs and limited reprovisioning and could make minor gifts to the Indians. The pearlers, however, were concerned to induce or impress the Indians into the laborious and dangerous activity of diving for pearls. Spanish authorities and, especially after 1697, the missionaries regularly complained bitterly that abuses of the Indians were impeding missionization and making the latter hostile to Spaniards in general. The temporarily successful revolt by the Pericú in 1734 must have stemmed in part from this hostility.

The earliest written reference to Coromuel that has been found is in the Los Angeles *Star* in 1857: "At the commencement of summer rain squalls gather about the mountain tops ... and the coromoel [*sic*] comes off the mountains, cooling the air."[5] In Southern California the wind described would now be called a Santa Ana, a descending and warming wind, but at its inception air movement could make it seem fresh and cooling. It can be noted that when the United States, in the treaty of Guadalupe Hidalgo, returned Baja California to Mexico in 1848, those residents in La Paz who sided with the United States found it expedient to be evacuated and many wound up in Los Angeles, then the most comfortably Mexican town in California. It is likely that they attached their own wind name to one prominent in their new homes. In the Southern California site the name did not last long. The reference, however, would document *Coromuel* in La Paz to before 1848.

A stranger reference comes from the French novelist Gustave Aimard. His adventure tales, mostly set on the American frontier, were popular and most were translated into English with later editions abbreviated into dime novels. In *The Freebooters*, *Coromuel* appears several times, but it has become a violent gusty storm wind, and the locale is shifted to the Texas Gulf Coast.[6] This is far from La Paz, but Aimard had spent a youth collecting adventures which he would later use in his novels. One of his adventures involved participating in the filibustering expedition of Gaston Raoux Raousset de Boulbon who, simultaneously with William Walker in Baja California (1852), attempted to set up a state in Sonora and Sinaloa. Raousset de Boulbon was executed in Guaymas, but Aimard was finally freed. The exotically named wind, probably picked up from Sinaloan pearlers, was inviting to use creatively in any dramatic context. Airmard also invents a topographically impossible hill fort on the Gulf Coast of Texas.

Santamaría, in his *Diccionario general de americanismos* associates *Coromuel* with Baja California and derives it from "the famous English pirate Cronwell [*sic*]."[7] But he makes it the prevailing northwest wind that blows along the Pacific Coast from San Francisco to the Cape. This normally careful lexicographer evidently attempted to extrapolate rationally from fragmentary information.

Immediately north of the Balneario Coromuel, just outside the Bay of La Paz, is another place-name that by a similarly shadowy route came from Dutch buccaneers to be attached to the Baja California Coast, presumably a place where ships were careened: *Pichilingue*. In an exhaustive study in 1944 Engel Sluiter derives the term (his variant is spelled *Pechelingue*) from *Vlissingen*, a former seaport in the southwestern corner of Zeeland in the Netherlands.[8] The strange set of sound shifts came from the name entering Spanish by way of French, with the shift from [v] to [p] coming from the lack of an [f] or [v] sound in Basque; it is only slightly more bizarre than *Vlissingen* becoming *Flushing* in English, as in New York. The small, moated town of Vlissingen in the southern Netherlands was one of the principal ports from which Dutch "sea beggars" and later privateers set forth to harass Spanish shipping in the 16th and 17th centuries.

In contrast to *Coromuel*, *Pichilingue* became a general term applied to Dutch ships sailing in violation of Spanish law wherever they were encountered. It is mentioned as a place-name in the Venegas manuscript written in 1739. Gerhard, in a manuscript by Sigismundo Taraval, who left the Cape region in 1734, found mention of Pichilingue Island near La Paz.[9] Taraval stated that the name came from a pirate who careened his ship there many years before 1697. Sluiter discovered a Dutch map, published in Amsterdam in 1765, on which Puerto de los Pichilingues appears close to the position of the present place-name.[10] The map title refers to recent discoveries by the Jesuits in California. These discoveries probably were made around 1720 when the Cape region was explored and the mission at La Paz founded. *Pichilingue* shows up as a place-name also in Sinaloa and Ecuador and, with variant spellings, possibly near Cartagena in Colombia and on Chiloe in southern Chile.

These two exotic place-names open a window on the considerable part Baja California played in the history of the Pacific during the more than a century and a half between its discovery and permanent settlement. It is also remarkable how these names, casually attached because of real or imagined events, have persisted in a sparsely populated, at times uninhabited, locality for two or more centuries.

DESERTS

Introduction

William E. Doolittle

MANY PEOPLE FIND arid and semiarid lands strange, even foreign, places. Some feel uncomfortable in these lands, and, hence, tend to avoid them. Indeed, the term *desert* is a noun derived from a verb meaning depopulated. We actually know very little about drylands because we have, over the course of human history, emigrated from them and sought the comfort of more humid environments. Indeed, today only 15 percent of the world's population resides in drylands, even though these lands comprise about 35 percent of the total land surface of the globe.

We in the United States have a fairly limited knowledge of drylands worldwide in part because of our Western attitudes. The social, economic, and political centers of the world are located in north-central Europe and eastern North America—all humid areas. A common Western view of deserts is as "wastelands," good only as defensive frontiers, sources of minerals to be exploited for their wealth or, possibly, places to be preserved for their exotic and majestic natural beauty.

In the same vein, native peoples of these lands are often viewed as second-class citizens, inconsequential, cheap labor to be used, or cultural artifacts to be preserved, depending on one's motives. It was no coincidence that when Indians in other parts of the country were being isolated during the last century, they were forced to occupy the arid lands of the western United States. Today, some of these peoples have successfully developed supposedly useless lands for mineral extraction (Navajo), skiing (White Mountain Apache), and resorts (Mescalero Apache). For these people, despite the continuing impoverishment of some of their number, the deserts have been evolving into economic assets.

An irony of desert landscapes is that the places in which people live in the deserts are really places rarely deficient in water. Most desert dwellers actually agglomerate in oases. This is as true in the southwestern United States as it is in Arabia. Although we have used air conditioning to make the hot deserts more habitable, most of its cities started out with enough water. El Paso, Albuquerque, and Phoenix are all on rivers, though substantially reduced in volume by modern upstream diversions.

The study of drylands is conducted by an odd group, though not odd in a sinister or threatening sense. Homer Aschmann and I fall into this category, taking great delight in working in the openness of xeric landscapes. There we rejoice in being able to discern not only the soil, landforms, flora, fauna, and sky, but the effects of human agency on each. This leads desert scientists, perhaps more than others, to share a common appreciation of the importance of the connections, linkages, and relationships among them all. This appreciation for the interconnections among disparate phenomena predisposes arid lands scholars toward the benefits of inter- and multidisciplinary research.

When he was not in his office, Aschmann often resembled a 19th-century Death Valley prospector who has left his equipment-laden burro tied up out front. The earthy appearance, however, did not mask the intellect. Through even a brief initial conversation with him, any new acquaintance could immediately sense the authority of his knowledge on virtually every aspect of drylands and life therein.

He acquired his expertise, in large part, out of doors. He felt that only in the field can someone come to grips with its environmental elements and come to "know" a landscape. He knew what it was like to be "attacked" by jumping cholla (*Opuntia fulgida*), entangled in the branches of cat's claw acacia (*Acacia greggii*), and to have a foot punctured by the thorn of an ocotillo (*Fouquieria splendens*), all within an hour. He had witnessed first-hand, and sometimes even experienced, the excruciatingly painful and literally life-threatening retches produced by the loss of critical amounts of body fluid. His appreciation of the harshness of arid landscapes and of the hardships experienced by local residents is most evident. His writings also reveal a scholar knowledgeable of virtually every aspect of the natural and social sciences such that they pertain to water-deficient environments.

While in the field, Aschmann took great comfort in others' company, in part because so few others were nearby. He would share his view that drylands are not necessarily places of social isolation, especially when food becomes available. Social isolation is much more common in forested areas, probably because of the ease of isolation compared to the relative openness of deserts. When he was by himself, however, he was comfortable conducting research alone, enjoying the solitude of empty quarters. Perhaps this solitude lends itself to the absence of pretense, because Aschmann was the least pretentious of our lot.

Although his studies of drylands explored many paths, their common core is singularly geographical, and regional descriptions underlie each of them. Apart from his classic book on the Central Desert of Baja, "The Head of the Colorado Delta" best illustrates this facility of regional description. As in all his work, this description is not a mere cataloging of facts, but an insightful, first-hand portrayal of those things of immediate, but not always apparent, importance to the region's inhabitants. He addresses the natural environment and then enlightens us about how it has been changing over time under the influence of different cultures, particularly through the application of irrigation water. The patterns he saw three decades ago in southwestern Arizona have continued, abetted by a great explosion of "snow birds" who come each year seeking warmth and spring baseball.

Aschmann's arid lands research was comparative in nature, if not explicitly. He would comment frequently on similarities or differences between the area of current study and an area he studied previously, comparing landscapes and dryland adaptations among them all as he did in his discussion of the *turno* in Chile and the use of irrigation in the southwestern United States, coastal Peru, and the Indus Valley.

As a scholar he was fair and evenhanded, reflective, rational, and balanced. After reading any individual work, the reader might come away thinking that Aschmann was a proponent of either unlimited development or conservation at all cost. Neither interpretation is right. If anything, the corpus of his writings reveals a moderate scholar. He was opposed to people who sought short-term profit. Similarly, he was not favorably disposed toward those who made impassioned but unsubstantiated pleas for restraint. Perhaps his most notable caution against extreme polemics is his reminder that although drylands seem to restrict development, some rather large cities have emerged and continue to prosper in deserts.

His studies reflect a curiosity about both the past and the future. Like so many students of drylands, he was struck by the antiquity and persistence of human occupance there and prospects of what is to come. One of the reasons for the interest in the former is that aridity tends to preserve archaeological remains that are nonexistent in humid places. In the field, one is so often jolted by evidence of past occupance and manipulations of the landscape. These discoveries inevitably lead to questions of how those people survived the harsh conditions and what happened to them. These were questions central to Aschmann's understanding of how landscapes changed with time and the extent of influence attributable to humans.

This fascination with landscape change in areas that were sparsely inhabited or perhaps completely uninhabited led Aschmann to concentrate much of his attention on the use of ethnographic data, investigating indigenous people living in either the same or similar areas today. He also vigorously studied the archaeological data and archives in pursuit of information pertaining to days gone by. He persisted in these interests and contributed many articles and reviews to publications in the field of anthropology.

His studies reveal an honesty that was refreshing not only among arid lands scholars but also among all scholars concerned with the past and with native peoples. He felt that one of the most important lessons he could pass on to others was that development should be based on lessons learned from what went before. When looking at the present, he was always aware of antecedent landscapes and how they influenced what was before him.

Aschmann is considered by many to be a cultural or human ecologist, although I cannot recall his ever using the terms or labeling himself as such. He never seemed too keen on labels, preferring instead to pursue questions through whatever means was necessary for a better understanding. Such intellectual rigor was the basis of his wide-ranging curiosity and expertise, including a great respect for native folk and their communities. He was obviously fascinated with how native inhabitants of drylands *interacted* with their environments, and understanding such interactions, he believed, was the key to understanding the evolution of the landscape itself.

In more than one place Aschmann illuminates the distinction between herding and agriculture, and between dry farming and irrigation. These obvious distinctions demonstrate the effects of different food production strategies and different water distribution technologies. What is done where and when is not simply a matter of some innate and unchanging physical geographical quality. In fact, contrary to what most people think, the biophysical environment is not static. Rather, it changes, and some human-induced changes can be for the better as well as for the worse. This is not to say, of course, that people are always successful in their attempts to control nature. Aschmann illustrates that few such human endeavors stand the test of time; technology may override environmental limits, and it may do so according to economic principles, but it does so for relatively brief periods in the overall course of history. The desert, he said, will reclaim itself or even expand.

The economics of ecology, or in a cultural sense, adaptation, involves management of the lifeblood of drylands peoples—water—but also of other renewable and nonrenewable resources. Although Aschmann touches lightly on the issues of crop domestication and the rise of civilization, these topics held a special fascination for him, and his comments always carried uncommon insight. He articulates the issues, not necessarily the problems, of overgrazing, timber and fuel wood depletion, and soil erosion. Ever present is an acceptance of human numbers, and how this fluctuating demand can best be satisfied. Similarly, the reader is never allowed to forget that mineral resources are not only finite but also that historically, extractive activities usually have very short lives. Though they may allocate decision making to government planners, people today are faced with the same problems people have always faced in drylands, and no one has yet come up with a durable and viable solution. The obstacles that must be overcome by investments in infrastructure are expensive and have limited long-term payoffs. Aschmann concludes, albeit subtly and unobtrusively, that tourism may hold some hope in a few places, and that preserving some drylands as wildlands makes more economic than ecological sense. All we need to do is come to our senses.

Today we read and hear a great deal about environmental degradation, especially at the hand of humans. Aschmann never accepted this party line, and reiterated repeatedly that we should think in terms of environmental dynamics. Soil fertility is not an inherent quality that is fixed in nature and decreased by people. It can vary over time due to factors that have nothing to do with the actions of farmers.

Environmental change is not unique to drylands. It is, however, manifest most clearly there. Rainfall, for instance, is far more variable in arid regions than in humid areas. Indeed, an inverse correlation exists between the total amount of precipitation and its variation over space and time. This irregularity and unpredictability does not lend itself to very accurate regional generalizations, and it does not accommodate seasonal differences. Unlike humid regions, arid lands tend to have marked and distinct rainy and dry seasons. Vegetation responds accordingly. During part of the year it is leafless, brown, brittle, and sparse. At other times it can be green, dense, and luxuriant. With such pronounced seasonality, the vegetation itself and the soil beneath it become fragile and not very resilient.

The ecology of drylands is easily disrupted, and once disturbed, the environment remains scarred for a very long time. It is also evident that these environments are dynamic in their own right. Floodplains such as along the Nile or the Colorado are highly productive environments, and, as a result, have long been attractive to people. Paradoxically, these oases are also very unstable. Droughts one year can result in decreased discharge and an insufficient supply of irrigation water. Floods the next year can destroy a crop and significantly damage long-term productivity by removing valuable topsoil. Similarly, a flood can remove long-cultivated and nutrient-depleted soil one year and replace it the next with a thick layer of fresh sediment and organic detritus. Throughout the world, efforts have been made to even these swings of nature, often through the construction of dams and irrigation projects. The result is the most noticeable and lasting example of landscape change. Therein lies the inherent fascination that so attracted Aschmann's attention.

Desertification—A World Problem

*In Libya, and in southern and eastern Syria . . . extensive Roman ruins
stand out in unfarmed and overgrazed landscapes*

Tʜᴇ ᴡᴏʀᴅ *desert* has a long history in the English language, being well established at a time when few if any Englishmen had ever seen any of the world's drylands. Its etymology shows clearly that the noun is derived from the verb and its
primary meaning refers to a land occupied by no or few people. In American and especially western American usage, however, desert has come to mean a region that is
sparsely populated because it is dry. In fact, geographers have defined deserts as a climatic region, bounding them by specified relationships between precipitation and
temperature, the latter term a surrogate for evaporation or precipitation effectiveness, which are hard to measure. Such definitions do some violence to the basic meaning of the word when intensively occupied areas with access to exotic irrigation water
such as the Nile Valley, Imperial Valley, or the middle Rio Grande Valley are called
deserts. Only a limited part of the latter area in New Mexico could be called a desert
on climatic grounds, and it is not extreme. Even the driest portions of the state can
support some population by grazing.

Desertification is a word of recent French coinage and refers to the process and
action by which a region becomes unsuitable for human occupation. It may have
physical causes such as climatic change, be initiated by human action, or be some
combination of the two.

There is abundant geologic and paleontologic evidence of expansion of dry areas
into formerly moderately humid regions in the Western United States at the end of
the last glaciation, perhaps 12,000 to 10,000 years ago. Similar evidence is present in
other parts of the world. This expansion of the deserts clearly resulted from basic climatic changes, alterations in the world's atmosphere's circulation patterns. Since that
time there is no evidence of comparable climatic change. The evidence of smaller
variations in climate is equivocal. [As evidenced in various records] . . . there have
been both unusual seasons and periods of one or more decades that were warmer or
colder, wetter, or drier than the long-term norm. Some students feel that they can
identify periods experiencing different climatic conditions that lasted for centuries
or millennia, but there is little agreement on timing or even the direction of the
changes. It seems clear, however, that postglacial climatic changes have been at least
an order of magnitude smaller than that represented by the end of the last glaciation,
and there does not seem to be any trend. A major long-term climatic shift beginning
now or at any time in the future is not precluded. It will begin with a year and then

a series of years with unusual but not unexperienced weather. I know of no theory capable of predicting such an event.

In the Old World, particularly in the lands south and east of the Mediterranean Sea and in Central Asia, however, there is clear evidence of desertification. In Libya, and in southern and eastern Syria, for example, extensive Roman ruins stand out in unfarmed and overgrazed landscapes. Regular rows of pits in the ground have been identified as olive orchards and vineyards where no crop now grows and poor brush or ephemeral herbs are the only vegetation. Comparable situations elsewhere in North Africa and the Middle East are numerous. Their causes must be sought in human actions and the normal characteristics of semiarid climates.

Extending outward from all major dry areas, deserts, if you like, are gradients that extend into progressively more humid regions, except on West Coasts, where the desert may front on the ocean. There is abundant evidence that the domestication of the small grains and the major grazing animals, thus the beginnings of agriculture, occurred in subhumid foothill regions that form a crescent to the north of the Arabian Desert. Neolithic farmers expanded outward, both northward into more humid lands and southward into drier lands, met frequent years of crop failure, and became more dependent on their grazing animals for subsistence. A long time ago true pastoralist societies who did not plant became established in even drier districts.

Since wealth is equated with the number of animals held, pastoralists tend to overstock their ranges rapidly, competing with other pastoralists and nearby planters for pasture, especially in dry years. Because of their mobility, and frequently a warlike tradition, the pastoralists, though normally fewer in number, tend to have the advantage in conflicts. Both kinds of occupants tend to extirpate the sparse woody vegetation of dry regions in seeking fuel. After one or a series of dry years has caused crop failure in agriculturally marginal districts the planters will have to move out. They will be replaced by pastoralists and are likely to be kept from returning to plant; their olive trees and grape vines likely to be cut for fuel. Erosion of soil bared by overgrazing will reduce forage and further intensify overgrazing.

It would seem that the might of the Roman Empire could protect the farmers' fields and settlements from pastoralist raids. In the unsettled centuries after Rome fell, the agricultural frontier retreated and extensive regions were notably reduced in their capacity to support a human population. It will be interesting to see whether the recently independent states of North Africa and the Middle East, many of them rich with oil, will be able to reverse this pattern of degradation.

In the present decade a similar pattern of land degradation has been observed in the vast Sahel region stretching across Africa, south of the Sahara Desert. The trigger was a three-year drought affecting the entire region, but an unstable situation had been developing for some decades. The effective pacification of the region under French and British colonial tutelage, coupled with increasingly effective disease control, had resulted in a buildup of the human population and a concomitant increase in attempts to cultivate land, the number of grazing animals, and pressure on woody vegetation to serve as fuel. Deep bores to provide water for livestock permitted animals to be kept on ranges throughout the year rather than moved when surface water sources dried up in the long winter dry season.

The drought-induced pasture failure began in the drier north, and, as they could, herdsmen moved their animals south, putting additional pressure on the vegetation of a region that was also drought-stricken. The sparse and slow-growing shrubs and small trees were stripped of their foliage for fodder and their woody parts burned for fuel. What protection these deep-rooted plants afforded the soil has been lost for many decades, if not permanently. Crops failed completely in the drier northern parts of the dry-farmed zone and herdsmen moved in, forcing the belts of land use farther south.

Eventually, of course, most of the animals died and widespread human starvation was alleviated, at least in accessible areas, only by an international food relief effort. Reasonably normal rains have occurred in the past two summers. Those pastoralists who had preserved some fraction of their herds and flocks will return to the north, building up their herds as fast as possible to repeat the same cycle in lands with less vegetative resource than before the drought. Those who lost their animals must remain in the south, either on relief or attempting to cultivate submarginal land in agriculturally overcrowded districts. Without climatic change, with only cyclical variability in weather, the uninhabitable Sahara Desert has expanded southward.

The chains of social causation that can lead to desertification as described above seem to be of two types. The classic one is based on poverty. With its technology, typically one associated with pastoralism, a given population finds that the resources of the area to which it has access do not sustain it. Resource depletion occurs even in normal times but is accelerated by recurring droughts. A further acceleration will result from either an increase in the human population or an effort to raise living standards by maintaining larger herds. The situation becomes acute when the national economy or the technical skills and interests of the population afford few or no alternative means of making a living.

Fortunately, it is possible to be confident that this causal change is not now operating in the United States. Something very close to it was developing on the Navajo Reservation until the mid-1930s. There, a rapidly growing population that had reason to stay on a finite territory was forced to increase the size of its herds and flocks, its main economic resource, just to maintain its living standard. Severe overgrazing was general. The stock reduction program provoked bitter resentment and resistance that persists, but the wealthy United States economy was able to substitute various relief programs to sustain the still explosively growing Navajo population. That these relief programs grow ever more expensive is a problem, but at least deterioration of the Reservation's range has been largely halted, perhaps even reversed.

A second chain of causation, however, is a threat in the western part of the United States. It involves the overexploitation of vegetative resources or the mining of groundwater in search of short-term profits. Overgrazing on public lands, where the pastoralist with an insecure lease or permit has no self-interest in maintaining a range and must be controlled by policing, is a problem but one that may lead to range degradation rather than true desertification. Before the latter is reached the lease or permit will be given up and livestock removed because animals kept on the most meager pasture have little or no value on the American market. This, however, does not justify ignoring range degradation, erosion, and gully formation that lowers water tables, problems that are real in New Mexico and elsewhere in the Southwest.

Pump irrigation in semiarid regions has increased explosively in the last two decades, with a concomitant fall in groundwater tables. Most of the groundwater of the High Plains of Eastern New Mexico and elsewhere is of ancient, probably Pleistocene, origin. At present rates of withdrawal it will not be replenished in the foreseeable future. The green circles that dot the Plains every summer will be brown, eventually showing a somewhat impoverished version of the sparse original vegetation. Here again we are not talking about true desertification. The lands were grazed before and will be again. The social malaise that will occur when expensive wells run dry or must be deepened at excessive cost and productivity of the land will support only a small fraction of the population that is building up both on farms and in revitalized small towns will be significant.

Salinization of lands irrigated either by surface water or brackish groundwater is a local problem in many parts of the American West. If it progresses far enough it can produce true deserts and complete abandonment. In general, salinization results from attempting to irrigate with a minimum of water. Groundwater or stream water with a moderate content of solubles repeatedly evaporates or is transpired from the fields, leaving the salts behind. The only reasonable way of abating salinization is flushing with an excess of water so that the salts are carried away, but if the amount of water available is finite at least some of the formerly irrigated land must be abandoned, now salty and useless.

It is generally believed that deserted lands of the Indus Valley and Mesopotamia that show the ruins of great civilizations, such as Mohenjo Daro, were abandoned because the farmlands were ruined by salinization. The Rio Grande and parts of the Gila and Lower Colorado Valleys as well as more localized areas have some vulnerability.

The salient aspect of these several sorts of land degradation or desertification due to overexploitation of a vegetative or hydrologic resource in the American Southwest is that they are unnecessary. We can feed ourselves more than adequately without them and the profits gained by overgrazing, exhausting groundwater, or irrigating too many acres with a limited water supply are small in total and affect very few people. They cannot compensate for permanent damage to any significant amount of land, and it would seem that society is justified in proscribing such activities.

Divergent Trends in Agricultural Productivity on Two Dry Islands: Lanzarote and Fuerteventura

The extensive deposits of subsurface lime concretions afforded a valuable export, but fueling the lime kilns led to the absolute destruction of all woody natural vegetation.

IN TERMS OF their physical environments and early history of human occupation, Lanzarote and Fuerteventura, the eastern islands of the Canaries, are remarkably similar. Both are fully volcanic in origin, and their construction has involved a number of periods of volcanic eruption, beginning at least in the early Tertiary, and continuing into very recent times. Lanzarote experienced major eruptions in the 18th and 19th centuries, and fields of scarcely weathered lava and undissected domes of lapilli on Fuerteventura show that its most recent eruptions were only slightly prehistoric. On neither island is there a major relief feature. The highest elevation on Lanzarote is 671 meters, and the highest on Fuerteventura is 807 meters. The most recent volcanic peaks and craters in both cases reach lower elevations, generally below 500 meters. Where the most recent volcanism is ancient, deep valleys have been incised, but in general both islands are long low domes above which recent cones and older dissected plateaus rise almost at random.

The absence of significant elevated areas prevents both islands from extracting locally extra amounts of orographic precipitation from the trailing cold fronts of the cyclonic storms that pass over the islands in winter. Nor do they enjoy the fog drip that contributes substantially to the water resources of the higher Western Canaries. On the north side of the latter islands a bank of clouds, borne by the northeast trade winds, abuts against steep slopes between 700 and 1,500 meters and sustains the monteverde, or moist broadleaf evergreen forest, as well as replenishing the groundwater supply. No such resource existed on Lanzarote or Fuerteventura.

Brief and fragmentary rainfall records from a number of stations on both islands suggest average annual rainfall values of 5 or 6 inches at the lower elevations, where most stations are located. Localities above 300 meters which are exposed to the north may get 10 inches. Rainfall varies greatly from year to year; less than 2 and more than 15 inches have been recorded at several stations. Almost all of it falls between November and April. Temperatures are usually moderate, under the influence of the incipient tradewinds; frost never occurs, and really hot, dry, Saharan weather may come for a few days several times in a decade. At other times the relative humidity is quite high, averaging over 50 percent in summer and over 70 percent in the winter, with steady tradewinds blowing most of the time.

Both islands were characterized by a low bushy vegetation in which certain endemic *Euphorbias* were prominent. There may well have been more palatable herbs present in earliest historic times, but overgrazing has reduced their occurrence. Orchilla (*Roccella*), a lichen formerly used as a dyestuff, grows on new lava and low bushes along coasts exposed to the tradewinds. It constituted the major export crop in the early history of both islands. The Canary palm (*Phoenix canariensis*) was the only large tree, though small tamarisks and willows grew along watercourses.

For a century, prior to the conquest of both islands at the beginning of the 15th century, the islands had been visited and raided with some frequency by European and North African mariners, and there had been erratic contacts back to classical times. On the smaller island of Lanzarote the native population seems to have been reduced by slave raids and perhaps disease to a few hundred and offered little resistance to feudal conquest by Juan de Bethancourt, a Norman vassal of Castile. Slightly more resistance was put up by the larger population on Fuerteventura, but it too quickly came under feudal lords who established European vassals in many seats to govern and exploit the islands.

The native population had raised goats, cultivated barley with hoes formed of goat horns, fished, and gathered such wild fruits as the poor dates of the Canary palm. They continued to do so under feudal rule. Figs and grapes were introduced early and could survive without irrigation in certain river-bottom sites. Somewhat later the dromedary was introduced from Africa and fields were plowed and sown to small grains; wheat was introduced but barley dominated, being better adapted to the low rainfall.

Only in the more extensive cultivation of small grains, aided by draft animals, was the content of the aboriginal economy significantly modified by European conquest for the first two centuries. The economy's focus, however, changed in that the feudal lords of the *señoríos* sought to maximize production for export. Some Moors were brought as slaves or serfs from North Africa to provide more abundant labor and make up for emigration losses to the richer western islands during and after their conquest at the end of the 15th century. Extensive areas were planted in grain, and in wet years, typically one in three, substantial surpluses were available for export, especially to Tenerife and Madeira. Orchilla was collected as a laborious feudal duty, and all its cash value went to the landowners. Goat herds were built up to maximum size, and the export of hides and occasionally some cheese was of importance.

In what was basically a difficult environment the expenditure of rather more than one-fifth of total production for the benefit of feudal rulers, who as time went on began to live away from the islands, plus another tenth to the Church, placed a severe strain on the islands' economies. Emigration of a substantial part of the resident population occurred whenever one of the periodic droughts developed, and, perhaps more significantly, only an exploitative economy could support the demands placed on it.

With its more extensive area, and large relatively level surface, Fuerteventura was initially the more productive island. In wet years the extensive grain fields produced great surpluses, but they were exported to the benefit of absentee landlords. In dry years serious overgrazing by goats, and the subsequent wind and water stripping of bare fields, however, began a permanent depletion of the soil resource base. The extensive deposits of subsurface lime concretions afforded a valuable export, but fueling the lime kilns led to the absolute destruction of all woody natural vegetation except on the inaccessible peninsula of Jandía at the south end of the island.

By the middle of the 18th century destruction of soil and vegetation had progressed so far on Fuerteventura that the island entered a period of depression from which it has never recovered. Its major export has been people, and in dry years, scrawny, starving goats. Population growth has been very slow, and living standards are the lowest on the Canary Islands. The island is reported to have held 7,344 people in 1744. By 1857 the number had risen to 11,412, and by 1950 to 13,506. In the same period Lanzarote's population grew from 7,210 to 29,388, although it too was a source of substantial emigration.

The bulk of the steady stream of emigrants that has flowed from Fuerteventura has consisted of young adults so that, even for Spain, the fraction of dependent children and old people is disproportionately large. A chronic shortage of able-bodied labor parallels an acute lack of capital on the island. These shortages have made Fuerteventura minimally able to take advantage of fortuitous economic opportunities that have presented themselves. Most of the island is well adapted to the *Opuntia* cactus and might have been a leading center for the cochineal industry that boomed during the mid-19th century. The lack of capital and labor prevented a development of the industry anywhere near as extensive as occurred on Lanzarote, or on Gran Canaria and Tenarife. While the cochineal trade collapsed after 1870, because of competition from the cheaper aniline dyes, capital accumulation among smallholders from its profits permitted them to shift into other kinds of specialized agricultural production. Very little such capital ever came to Fuerteventura.

One specialized form of dry farming has developed on Fuerteventura, in fact characterizing its economy. This is the *gavia* system, whereby a series of perhaps six bordered, level plots averaging an acre or less in extent are laid off along minor sloping watercourses. A narrow stream channel is preserved along the side of the gavia plots. During a rain sufficient to cause any surface flow, all the runoff is directed into one gavia plot, filling it to a depth of perhaps one foot. If there is enough rain and runoff, more of the plots are filled in sequence. When the water has soaked into a gavia that has "*bebido*," i.e., "drunk," the heavy mud will be planted, and from that single irrigation will yield a crop of grain or leguminous seeds.

While the ability to get a crop of any sort under these conditions is notable, the gavia system has great limitations. Only a limited number of places, most now exploited, have the requisite physical characteristics: small watercourse, deep heavy soil, and moderate slope, to permit constructing a series of flat terraced plots. The number of gavias that can be filled with water, and thus planted, is enormously variable from year to year, and unusually heavy rains may wash out the whole system. Although per-acre yields may be good, this pattern of farming depends on hand labor and per-capita returns must be low, in addition to being uncertain. The focusing of agricultural interest on the gavias has tended to encourage overgrazing of the slopes above them. Bare surfaces seem to be desirable so that there will be quick runoff into the gavias. The long-term results of such a land-use pattern are probably irreversibly deleterious, and many surfaces that, at least occasionally, yielded grain crops by dry-farming methods in earlier centuries are now stripped to barren caliche surfaces that offer the poorest of pasture for goats.

On Lanzarote, developments since the mid-18th century have followed a different course. The catastrophic volcanic eruptions of 1730–53, which covered much of the best agricultural land on the island with thick layers of lava or lapilli, seem to mark a turning point and to have resulted in a developing rather than a declining economy. Two social factors, as well as the obvious physical factor, may have been coincident. With the volcanic disasters the feudal lords of the señoríos may have found it necessary to relax their rent demands and in some instances grant something approaching freehold rights to prevent the working part of the population from abandoning the island completely. I have not yet been able to document fully such a development, but it is clear that on Fuerteventura, where there was no such shock, vestiges of feudal land tenure persist to the present day while they do not on Lanzarote. The Canary historian Viera y Clavijo, writing in the 1770s, characterizes the peasants of Lanzarote as energetic and those of Fuerteventura as indolent. If the description was then accurate the situation may have been due to peasant populations of different cultural origins or to differently developing land tenure patterns. The rural Lanzaroteño of today seems more industrious than his counterpart on Fuerteventura, but this can be explained as a result of the subsequently developing relative hopelessness of the agricultural economy on the latter island, and to the deficiency of able-bodied adults in the emigration-depleted population.

The posteruption physical developments in Lanzarote agriculture are unmistakable. It was discovered that by digging to the bottom of the lapilli layer and planting grape or fig slips in the underlying soil a remarkable yield could be obtained. The lapilli cover reduced water loss from the soil by both runoff and evaporation so that from as little as four inches of rainfall a good crop was assured. Furthermore the *malvasía* grapes growing in the hothouse of a black pit acquired an inordinately high sugar content, yielding a heavy wine of great distinction. Great effort was needed to dig these pits, some over a meter deep and three meters across, to get the grape roots into the underlying soil, but there was an assured reward for the effort. By now these strange vineyards have been extended over many square miles.

Where the lapilli layer was only a few inches thick, grains such as maize or barley or beans, peas, or garbanzos could be planted in shallow furrows and root in the underlying soil. Whether the black volcanic ash layer encourages moisture from the humid trade wind air to condense at night or whether it merely makes rainfall more effective by minimizing evaporation and runoff, four or five inches of rain is enough to make a crop under the ash.

The development of grape plantings deep in the lapilli occurred first, though there may have been some grain plantings through the lapilli right after the volcanic eruptions, where the layer was thin near the edges of the ash fall. Sometime during the latter half of the 19th century the practice of covering ordinary soil with three or four inches of lapilli developed. The additional practice of clearing raw lava fields for planting probably developed during the cochineal boom of the mid-19th century as well. This *enarenado artificial* was observed and well described by Karl Sapper in 1905. Once accomplished, it afforded assured though modest yields, for example, 20 bushels of maize or 12 bushels of peas per acre, in an area where perhaps only one year in three would get enough rain for any crop cultivated with ordinary dry-farming methods.

Tremendous labor is involved in covering a field with lapilli. Even today much of it is transported on camelback, but it does not have to be carried far. In addition to the area of historic volcanism, there are recent ash domes in other parts of the island, notably in the north near Haría, an area where the enarenado artificial is proceeding particularly actively. Near Yaiza, in the southwestern part of the island, raw lava fields are being leveled, a layer of soil brought in, and then covered with lapilli. The principal crops produced from such extremely expensive fields are onions and tobacco for export. Even with the extremely cheap rural labor available it cost up to $500 per acre to create a lapilli-covered field, and after about 10 years another $100 per acre must be spent in scraping up the lapilli and cleaning it of the dirt that gets mixed in by plowing.

In the strip of calcareous dune sand (*jable*) that runs across the middle of the island another specialized dry-farming technique has been developed. Little fences of rye straw protect against wind-borne sand, and quite respectable crops of sweet potatoes, melons, and tomatoes are obtained despite chronic conditions of drought.

The returns for the labor investment from these specialized forms of dry farming are certainly meager. But they are real and relatively secure. The underemployed peasant with little capital can, by enormous effort, add to his productive land holdings, and at the same time he increases the total productivity of his home island. Emigration still takes away a substantial fraction of the natural increase in population, but it does not result in rural and insular depopulation. The farmers have the option of leaving home and seeking their fortunes or encouraging their children to do so, or of expending great effort to improve their land capital, and remaining in their home communities. Both choices are reasonable. Because of the substantial local market and labor supply, Arrecife, the island's metropolis, has been able to develop as a major fishing port, exploiting the rich waters off West Africa. There is no comparable development at any port on the equally strategically located island of Fuerteventura.

By now over 12,000 acres on Lanzarote have been made many times more productive by the laborious process of covering them with lapilli. The area covered is being slowly but steadily expanded. The enormous labor cost prevents this land development program from attracting speculative capital and the development of large lapilli-covered holdings. It is an activity for the local peasant to pursue during slack farming seasons, perhaps hiring a few of his neighbors. Rural underemployment makes this enarenado artificial feasible, but the enarenado artificial reduces the economically and socially debilitating effects of rural underemployment.

There are domes of lapilli scattered over the island of Fuerteventura, and physical conditions are suitable for even more extensive application of this land improvement method. Despite considerable government stimulation, however, less than 500 acres had been covered with lapilli by 1960. There is just not enough energetic labor available among the local rural population to get the program under way, and the low yield does not attract outside capital. The most vigorous farming development on Fuerteventura at present involves producing tomatoes for export, using mined groundwater for irrigation in the vicinity of Gran Tarajal. This activity attracts local labor and outside capital, but, since the groundwater supply is small and not being replenished, it is surely transitory.

In summary, then, the agricultural economics of the two eastern Canary Islands followed completely distinct courses after the middle of the 18th century. On Fuerteventura agriculture was traditionally exploitative, and the land resource base has been steadily depleted. Lack of investment in social amenities (for example there is no secondary school on the island), has been a further incentive to continuing emigration of the able. Without inordinate investment by the not too prosperous Spanish government there seems to be no way to prevent the island from remaining a depressed area, perhaps becoming more so. The island is now certainly a drain on the national economy.

On Lanzarote, however, vigorous local experimentation to develop more productive means of farming was continued, at least since the great volcanic eruptions. Enough success was achieved, in producing malvasía wine and temporarily in cochineal production, to permit some capital accumulation and the maintenance of an active and slowly growing resident labor force. Life on the island is difficult, but it is not hopeless, and productivity is increasing at least as fast as the population without depletion of the land capital.

From this history some generalization concerning agricultural settlement in physically marginal areas throughout the world may be derivable. Unless deliberately and heavily subsidized by governments or groups based on inherently richer lands, such settlement can proceed only by depletive exploitation of some stored resource: minerals, initial soil fertility, groundwater, timber, or other elements in the natural vegetation. The wealth accumulated through such exploitation may be exported or it may be used to build homes, roads, and schools and to provide other amenities. At some point, however, the depletive exploitation must become less rewarding and a decisive juncture or threshold will have been reached.

On Fuerteventura, unemployment, emigration of the able, decay of facilities, and continued and even accelerated land depletion through overgrazing and soil erosion have been the rule. There is no assurance that the pattern can ever be changed, and a fraction of the earth is less useful to mankind than it was 500 years ago. On Lanzarote, however, at the critical juncture, for reasons not yet entirely clear, experimental probings of agricultural potentialities by the local population yielded real if modest returns. A slow but steady increase in production has been achieved, and individually many small farmers are imbued with an urge to invest their basic capital, which is their labor, in further experiments which may give further gains.

On the world scene there are myriad examples of both trends, larger in scale and importance if less clearly defined. The progressive impoverishment of Appalachia in the United States might be contrasted with the Swiss Jura, where the labor resource was reinvested in hand industries. The spread of paddy rice cultivation in Southeast Asia may be regarded as an older and far more extensive parallel to the enarenado artificial of Lanzarote, while the overgrazed and eroded hill landscapes of the borders of the Mediterranean result from decisions like those made on Fuerteventura.

Evaluations of Dryland Environments by Societies at Various Levels of Technical Competence

The barrenness of the overgrazed landscape around Jarmo makes us forget that it receives annually about 18 inches of rain.

The utter barrenness of the landscapes south and east of the Mediterranean does not accord with the recorded precipitation values.

Anyone who surveys the entire history of human cultural development, most of it known only archaeologically, must be impressed by how much of the early record comes from drylands in the lower and lower middle latitudes. It is in such areas that we find the earliest evidences, in both the Old World and the New, of the discoveries and inventions that created neolithic cultures and led on to city building, metallurgy, writing, and eventually the formation of states and empires. The accident of preservation does not provide a fully adequate explanation for the preeminence of the records of civilization in the drylands. We know historically that Egypt and Mesopotamia in the third millennium b.c. had cultures far more elaborate than those in nearby humid lands, both north of the Mediterranean and in East and Central Africa; and at the beginning of the Christian Era the arid coast of Peru was enormously more advanced than the humid eastern two-thirds of South America.

Although they are not capable of supporting as many people as certain humid lands, the world's drylands seem to have made a heuristic contribution to man's cultural development, to have stimulated discoveries that increased man's control over both his physical and his cultural environment. This paper endeavors to identify sets of physical conditions peculiar to lands of deficient precipitation that may have had this stimulating effect, or that seem to correlate causally with the characteristic features of desert civilizations as we know them, now or historically. It also shows that, once a new level of technology or social organization had sources of these favored localities, they might be available to support an augmented population during such emergencies and not be dissipated in ordinary times. To what extent this problem was recognized and faced by preagricultural societies has not been examined widely, to my knowledge, but many aspects of totem and taboo in central Australia seem, when viewed from this standpoint, to possess a rationality that would enhance prospects of group survival.

The marked seasonality of vegetative growth in drylands also was likely to present a brief period of abundance and easy availability of some foodstuff. The fall ripening of the fruit of the *pitahaya dulce* (*Lemaireocereus thurberi*) in southern and central Baja California, at which time as many as 1,000 people are reported to have assembled for ceremonial and social purposes, is a good example.[1] Thus, both of necessity and because of opportunity, the dryland environment, despite its sparse total population, did not produce social isolation, but at certain times of the year it placed large numbers of people in fairly intimate social contact. Where a major oasis was created by an exotic river, such as the preagricultural Nile Valley, we may infer intensive social contact among quite large numbers of people. This is a situation that is conducive to cultural elaboration and appears to give the drylands an advantage over forested ones where social isolation seems to be the rule among primitive groups.

The ecology of contagious disease in drylands as opposed to humid ones is a problem that merits investigation from many standpoints, and the following comments are intended to be provocative, rather than conclusive. Three distinct situations can be distinguished: open country without permanent water sources, small isolated springs or waterholes, and major extensive oases such as a slowly flowing exotic river.

In open country a disease could be maintained during the annual dry season only through its human vectors. Should they die, the disease would disappear promptly.

At waterholes, however, insect vectors and aquatic worm parasites or vectors could persist for longer periods, and diseases could be introduced to new and distant waterholes by man and, perhaps, by other agents. A particular water source might acquire a reputation as a pesthole and long be avoided on any of several sorts of rationales. On the other hand, if the disease organism required a human host in its life cycle, it would ultimately disappear unless it were mild enough in its effects not to exterminate or even scare man away from the water source.

Where a major oasis existed, man and his diseases would stay together and achieve some level of mutual adaptation, perhaps lowering man's efficiency, biologically and otherwise, but never eliminating the human species. From his diseases the surviving dweller at a major oasis received a sort of protection. He could be conquered but not replaced by outsiders. The 5,000-year stability of the Egyptian physical type can be explained better in these terms than in any other I know. Until modern sanitation and antibiotics were available, involving perhaps the last 100 years, immigrant conquerors had little biological future in the Nile Valley. Each major oasis—the Nile, Mesopotamia, and the Indus Valley, for example—would tend to develop both its own disease varieties and a resistant human population. With the rise of historic empires and interoasis travel and conquest, some exchanges of infection could occur. Such developments have explanatory potential in accounting for cyclic declines in oasis civilizations known historically and, at the same time, help to account for the long-maintained cultural identities in each of the great Middle Eastern oases and the several coastal sectors in Peru.

In humid tropical areas the environments for diseases and their vectors were generally favorable, and there was a tendency for diseases to have broad and continuous distribution. Humid areas with cold seasons were less favorable to diseases, but again disease-spread was likely to be continuous and not to serve as a mechanism for isolating a moderately extensive culture area.

Beginnings of Agriculture

The logic of environmental interpretation, the distribution of the postulated ancestors of the earliest domesticated plants and animals, and, most recently, dated archaeological finds, all indicate that small-grain, seed agriculture had its earliest beginnings in a steppe region or regions, probably in the more humid sectors near the forest margin, between the Indus Valley and Anatolia.[2] Jarmo still looks typical of the initial stage of neolithic agriculture, although it may not have any temporal priority. The system spread quickly over extensive areas. Digging-stick or hoe preparation of land for planting was involved. Soil fertility and yields had to be high on a per-acre basis to maintain the neolithic farmer, so slash-and-burn seminomadism was more appropriate to lands of adequate rainfall where soils lost their initial productivity quickly. The drier steppes, where crop loss from drought threatened regularly, were not suitable for dry farming until the plow permitted the planting of extensive fields and the accumulation of large surpluses in good years. The barrenness of the overgrazed landscape around Jarmo makes us forget that it receives annually about 18 inches of rain, concentrated in winter, which is just the right time for the small grains grown there.

Although the shift from hunting and gathering to neolithic farming undoubtedly increased the total population that an appropriate area could support, and although the areas under cultivation might be shifted regularly and fairly rapidly, we would err to imagine that the scope of social contacts was increased for the early farmers. It probably decreased. The neolithic farming community, although not fixed in residence, was less mobile than its gathering predecessor. It also needed to, and was able to, exploit less territory. Finally, it was more completely self-sufficient, and in regard to its prime resource, space for farming, it was directly competitive with its neighbors.

To leave the forest steppe or forest areas with their reasonably assured moisture supplies from seasonal precipitation and carry neolithic farming into more arid lands, new techniques of varying difficulty and generality of applicability had to be developed. These techniques may be grouped into two sets: dry farming and irrigation. Successful dry farming depends on substantial surpluses being produced in good (that is, wet) years and stored for drought years. Although excellent soil characteristics in some dryland districts might provide greater than normal yields, hoe or digging-stick tillage, especially difficult in steppe sods, sets such a severe limitation on the amount of land an individual can plant that surpluses sufficient to counteract major and frequent climatic risks could not be accumulated. Dry farming could become important only when domesticated draft animals and the plow permitted an order of magnitude increase in the area cultivated and planted. The great florescence of agriculture in the American, Argentine, and Russian steppes is entirely modern and is based on mechanically derived power.

Irrigation affords more immediate benefits, especially in truly arid localities that are watered by exotic rivers. Although the requisite preparatory work might have been greater than that for farming based on rainfall, crop yields that were both bigger and more secure could be obtained, and, barring local difficulties with soil depletion or salinity, a given plot could be tilled indefinitely.

The quality of the soil resource is the most important asset common to irrigated valleys that have become centers of desert civilizations. Because of seasonal variations in riverflow that have a ratio of 10:1 in a normal year and much more in extreme years, the deposition of alluvium at flood stage will be extensive wherever the valley is not an actual canyon. Where there is a long river, heading in a humid region and entering or passing through a dry one, the alluvial material will have been sorted and will consist of a mixture of sand, silt, and clay, rather than the coarse cobbles and boulders that cover the upper slopes' alluvial fans on short steep watercourses fed by local rainfall. The riverine alluvium in drylands possesses a full supply of the soluble minerals needed for plant nutrition, and these minerals are renewed by the regularly recurring flood deposits. If an adequate water supply could be assured, crop yields of unprecedented size and security, and available on a continuing basis from the same field, could be obtained from soils of nearly ideal texture, having high soluble-mineral plant-nutrient content, and natural restoration of fertility.

The economic advantage to be derived from irrigated farming in a limited but considerable number of favored desert valleys was revolutionary in its scope. Neolithic shifting cultivation, although it undoubtedly increased the number of human beings a given area could support, rarely yielded a substantial food surplus. Virtually the entire labor force of a community was engaged most of the time in producing the sustenance for that community. In a favored irrigated valley a true surplus of food—that is, more than the producers could eat—could easily be garnered. Julian Steward has indicated that two-fifths of the labor force was able to feed the whole population in irrigated valleys of coastal Peru, and this was without energy from other than human muscles.[3] A similar situation probably existed in the predynastic Nile Valley, before population pressure required cultivation so intensive that labor was uneconomically used.

In all probability flood farming was the earliest irrigation activity, and the historically known system of the Yuma and Mojave Indians of the lower Colorado may have existed for a long time prehistorically in several of the Old World valleys.[4] The Yuma and Mojave effectively dissipated their potential for producing a food surplus by a systematic pattern of individual indolence, coupled with a war pattern that was sufficiently violent to prevent population growth and to forestall any pressure to develop a more encompassing scheme for exploiting their agricultural opportunities.[5] Elsewhere canal systems for diverting and controlling irrigation water were developed, thereby expanding the area available for planting and making its yield more certain. The utilization of schemes like this, varying slightly in their technical details, characterizes every culture area where what we can properly call a desert civilization flourished.

The self-renewing character of dryland soils, which made possible fixed agricultural settlement, encouraged investment of a community's labor, which was unneeded for food production, in immobile property. The investment could go into dwellings and monuments to enrich the quality of life, or into capital structures to increase the productive capacity of the locality, such as canals for irrigation and transportation, land clearing, leveling, and draining, and embankments for protection against uncontrolled flooding. In an arid environment such investment does not decay or erode as rapidly as it would in a humid land. The capital investment by one generation could benefit, and be added to by, subsequent generations for an indefinite period into the future. In desert oases, for the first time in all of human history, a generation received from its parent something more than life with a cultural inheritance of knowledge and a few mobile tools and art objects; it received productive capital in the form of a developed agricultural landscape that needed only maintenance to yield ample food for less than the maximal labor investment.

Although the Nile Valley, once cleared and canalized, has proved its almost indestructible agricultural potential for 5,000 years, serious deterioration afflicts most of the other major developed oases. Where heavy local rainfall occasionally occurs, as in Mesopotamia, the 100-year storm may wreck a canal system, and lesser storms may damage it by siltation. A vigorous society could repair the damage promptly, but one in social and political decadence might lose, even permanently, the assets that had been laboriously constructed by preceding generations. Even with modern chemical understanding, the problem of increasing salinity in irrigated desert lands remains intractable. In the simplest terms, the problem will arise whenever irrigation water with a moderate content of dissolved salts—and this is almost a universal property of the waters in dry parts of the world—is applied to fields and allowed to evaporate or be transpired by plants. The dissolved salts will be left behind and will accumulate to intolerable concentrations. The problem can be avoided only if an excess of water is applied to afford some runoff, which will flush out the excess salt.

Except in the delta, the sharply demarcated valley bottom of the Nile does not provide enough accessible flatland to use all the water at flood stage. Some has always flowed to the sea and served as a flushing agent. In the Indus Valley there is more potentially irrigable land than water to irrigate it. Waters have been led from the river to distant fields and never returned to the river. Hundreds of thousands of once-irrigated acres are now alkali wastes, and the expansion of these wastelands and the forced abandonment of farmland is a vital problem in modern Pakistan.[6]

Grazing Activities and the Drylands

Although it is developed on theoretical and deductive grounds, Eduard Hahn's 65-year-old thesis that the domestication of large herbivores was carried on by the rich societies of the irrigated river valleys for ceremonial purposes, for milk and blood sacrifices, rather than for utilitarian ones, maintains its general validity.[7] The economic utilization of such animals was derivative and came later.

In truly desert Egypt, plowing with oxen contributed little to the economy, but in steppe lands like northern Mesopotamia and the Punjab the possibility of thus multiplying the acreage planted made marginally drylands capable of being farmed and yielding a surplus. This was especially true if large and highly organized societies could store surplus grain and distribute it to localities that had suffered crop failure from draught. With their more efficient digestive systems, the domesticated herbivores could also be maintained by grazing still drier steppe lands or rocky uplands that yielded no crops for man. In broken terrain with small plots of land of varying quality, a given community or village might both farm and keep animals economically, and such a pattern spread from Persia and Anatolia to the Mediterranean peninsulas of Europe. But when broad plains graded from steppe to desert far from large oases with irrigable land—as happens east of the Syrian littoral, south and west of the Tigris Valley, north of the Caucasus, and east of the Caspian Sea—specialized pastoral economies were developed. None of these were fully independent of planters, but their trade relationships were infrequent, coming after long treks from the pasturelands. An inevitable concomitant of a grazing economy is that ranges become overstocked in no more than a human generation.

Military and political power tended to gravitate to the herdsmen whenever they competed with steppe farmers. Their wealth was as mobile as they were; the size of their herds could be expanded in favorable seasons, and at times of drought they might protect at least some of their animals by migration. Direct and violent struggle for grazing lands was a normal part of their lives. Ordinarily a major center of intensive irrigated agriculture could protect itself against the herdsmen, although conquests of great civilizations by small groups of warlike pastoralists are a familiar theme in history. Most such conquests occurred when internal political disorganization had afflicted the irrigation civilization. The steppe lands that were marginally suited for dryland farming, if it was supported by the plow and draft animals, tended to be conquered by pastoralists and used for grazing. This clearly constituted a regressive form of land use, since it permitted fewer people to be supported by a given area. The regularity of this development is perhaps best emphasized by noting one major exception. At the height of Roman power, when reasonably effective order was being maintained in the steppe margins south and east of the Mediterranean, the frontier of tillage was advanced in places like Cyrenaica and central Syria far beyond its present limit.

The ability of pastoralists to take over the nonirrigated steppe lands of the Old World had direct environmental consequences that their occasional conquests of irrigated areas did not have. In the latter case the agricultural system persisted or was soon restored, and it is possible to consider the military conquests as beneficial in that they introduced new political ideas. In the steppe lands, however, overgrazing and the deliberate removal of brush and timber produced a cycle of soil degradation and removal by erosion, a phenomenon especially apparent in the limestone hills and plateaus that rim much of the Mediterranean. The utter barrenness of the landscapes south and east of the Mediterranean does not accord with the recorded precipitation values. Mosul and Kirkuk get more rain than the Sacramento or Salinas valleys of California, and the bleak, rocky Judean hills are fully humid lands, receiving annually 25 to 35 inches of rainfall.

In the irrigated valleys of the Old World, domesticated animals were of ceremonial and aesthetic significance but of relatively little economic importance. On the forested lands of northern Europe, with their relatively poor soils, animal-oriented agriculture and careful use of manure made possible a permanent, nondepletive, and reasonably productive agriculture. But throughout the lands of steppe climate, which they might have benefited, grazing animals and their herders have proved to be destructive of the earth's resource capital. In the warmer Old World drylands, the term *steppe* is relatively insignificant. They contain irrigable farmland and overgrazed pastureland that is little more productive than true desert.

Dryland Civilizations

Karl Wittfogel's theme of hydraulic civilizations and the rise of oriental despotism assumes that the management of complicated irrigation systems required a central authority to control water for the welfare of the entire community dependent on irrigated agriculture.[8] This bureaucratic, often priestly, authority then arrogated to itself absolute power over the community. The sociopolitical implications of this theme are beyond the scope of this paper. The rigidity of the socioeconomic patterns in the several great irrigated valleys of the Old World, however, and the capacity of these patterns to endure almost unchanged for millennia—facts that Wittfogel points up—are enormously significant elements in human culture history. I should like to apply a less particular theory to the economic structure of irrigation civilizations; namely, the classical theory of rent operating in a specific sort of environment.

Once the gallery forest was cleared and canals constructed for irrigation, the land that could thus be farmed was enormously productive. The 5,000 square miles of irrigable land in the Nile Valley alone probably supported more people than the whole of contemporary Europe in the third and second millennia before the beginning of the Christian Era. Furthermore, to do this, only about half of the labor force had to be engaged in food production. Roughly comparable situations existed in Mesopotamia, the Indus Valley, and the coastal valleys of Peru. In other words, the productive energy of nearly half of large populations, living so close together that effective organization was easy, could be diverted to other activities. These activities ultimately produced the domestication of large herbivorous animals; monumental constructions, both utilitarian and ceremonial; complex priestly and religious traditions, as well as governmental bureaucracy and law, institutions that often led to a system of writing, metallurgy, and magnificent refinements of arts and crafts: in sum, the objective attributes of civilization. The surplus productive capacity of irrigable land that supported these activities may be thought of as rent.

During the initial stages of land clearing and canal construction, the procurement of productive land was the reward to labor, applied in an organized and rational fashion to gain a deferred benefit. Because of the restrictions set by the terrain and the total amount of water available, a point would ultimately be reached where additional labor invested in clearing, leveling, and canal building would yield no more productive land. With minimal maintenance labor costs, however, the developed land would continue to support far more people than were needed to farm it.

By the beginning of the third millennium B.C., virtually all the irrigable land in Egypt was being irrigated. The valley bluffs still form a sharp boundary between wonderful farmland and absolute waste; only in the eastern delta was there room for a limited expansion of irrigation, when a strong government could approach a rational maximum in water use. Within a few yards one could go from land where a family could farm an acre and feed itself and another family, land that could pay 50 percent of its product in rent, to land of which a square mile would not support a goatherd. Rather promptly the human population grew to a point where it could exploit optimally all the irrigable land. Further investment of labor would yield only diminishing returns, and within all of Egypt there was no additional place to invest this agricultural labor.

Irrigable land was an absolutely finite and almost infinitely valuable resource. Individuals, even big landholders, could not afford to fight for it because the irrigation structures were too fragile. Only an all-powerful central authority or state could allocate water and land to assure its continuing productive use. An effective government would also collect the surplus as taxes, which were really rent, and employ it for public purposes, however conceived. A weak state might protect landlords and permit them to dispose of their rents as they chose, the situation that prevailed in Egypt for several centuries prior to the present decade. Characteristically, the private landowners did little to increase productivity and sought to use their rent yields to protect their holdings or to acquire more already-productive land. Bobek identifies this system as *Rentenkapitalismus*, noting that it leads to a constricting, rather than an expanding, economy, with rising prices and rents and lowered wages as the number of landless increases.[9] Should the mass of the populace not like the way the rents were used they might revolt. The existing bureaucracy, however, supported by its great income, did its best to make this difficult, and revolts ordinarily succeeded only with external support. But the conquerors or revolutionary victors could only renew the old monolithic pattern. There was no marginal land on which a simpler system of control would serve, where the landless might eke out a bare living on even less-productive land and experiment with other political and economic systems.

Initially, the general Mesopotamian and Syrian pictures were far more complicated. Here, as in European farmlands, there were several grades of land, some of which would barely or not quite support those who farmed them, but others graded to highly productive districts. In the classic Ricardian sense, potential rents ranged from nothing through minuscule to great. There was some possibility for a small community to try to exploit marginal land. Or an individual or group could work on better and more productive land, expending its surplus or rent on protection, usually in the form of taxes to a government. Some easily irrigable land, especially in Sumer, could be made as productive as the Nile Valley. Other areas required major canal construction, which had to be maintained at greater labor cost, reducing the rent yield. Because of a limited water supply, some lands could be irrigated only briefly except in unusually wet years, so their yields fluctuated drastically. Finally, dry farming was feasible in extensive areas, but it gave lower yields for more work; in other words, it yielded little or no rent. Abraham could leave Ur of the Chaldees and try marginal dry farming and herding in the Judean hills if he did not like the government. In these terms, the kaleidoscopic political history of the region, seldom repeating earlier patterns as it seemed to do in Egypt, is expectable. Similarly, with varied societies organizing their economies in diverse fashions, one would expect a longer and more creative period of cultural inventiveness, an expectation that the historical record supports.

The 10th century may mark the last major period of cultural creativity from this Syro-Mesopotamian region, and even the preceding 1,500 years seem to have been relatively stagnant. The growing dominance of pastoralism in the steppe areas may have been a major contributory factor. The cumulative effects of overgrazing and soil erosion had made more and more of the marginal lands submarginal, so that there was ever less intermediate terrain that might support an agricultural population while yielding little or no rent. As the division between the haves and have-nots grew sharper, the power of the state had to be increased proportionately. Opportunities for technical and organizational experimentation in districts too poor to be organized and heavily taxed by a central state had virtually disappeared by the time of the Abassid Caliphate, although they were diminishing from the time of imperial Assyria.

In pre-Columbian America, environmental alteration caused by overgrazing was not a problem; but in the extreme environment of coastal Peru, which had only irrigable land and sterile waste, a rational and rigidly stable organization of society was developed early in each irrigable valley. In Mesoamerica, or more peripherally in the American Southwest, where there were many gradations in the productivity of the arid and semiarid lands, more open societies, a more fluid history, and, most of us would agree, greater cultural attainments were achieved before indigenous developments were cut off by the Conquest.

Conclusion

Today the great river valley oases that nourished the irrigation civilizations of the past continue to sustain concentrated masses of humanity but in a deplorably impoverished state. Their bordering steppe zones, terribly overgrazed, support herding groups even more miserably. For 1,000 years these lands, which once contributed so much to mankind's cultural progress, have failed either to support their inhabitants comfortably or to give evidence of cultural vitality. Despotic governments or irresponsible landlords extract high rents or taxes from overly abundant cultivators, but no advance in the general welfare is evident.

However, in the New World and in Soviet Central Asia, where labor is scarce and expensive because there are alternative demands for it, mechanization and other technological applications permit the drylands to exploit their distinctive environmental advantages. Competing with the agricultural production of humid regions within the same politicoeconomic units, they contribute to interregional trade and provide good living standards to the productive local populations. The demonstrated capacity of an irrigated oasis to support many people over long periods of time tends to make land reclamation through projects that extend irrigation a popular panacea for the ills of overcrowded lands, such as Egypt or West Pakistan. Significant benefits will be gained, however, only if per-capita production is increased. An irrigated acre may feed a family, but it will not long satisfy the rapidly increasing wants of even the humblest peoples. Unless agricultural labor is to be used sparingly and efficiently on the lands to be irrigated, as we now know it can be, the capital needed for such projects might better be directed into other, more productive channels.

The Turno in Northern Chile: An Institution for Defense against Drought

The history of the Norte Chico of Chile represents a mixture of ancient and modern irrigation practices.

The broad floodplains of the lower valley, originally used for small grains and irrigated pasture, and the little benches of the upper valley seem to have been the initial centers of development in post-Conquest times.

IN TRULY DRY REGIONS, the unusual topographic and stratigraphic features that permit water to run at the surface are almost invariably incapable of supplying a sufficient flow to meet all the demands that might reasonably be placed on it for irrigation and other purposes. This situation commonly obtains even when the water is led into the dry region from an exotic source. As a result irrigated agriculture can be developed only on a fraction of the land appropriate to it. In the Old World generally, and on the western coast of South America in the New, millennia of experience have defined the extent of the areas for which there is adequate water for irrigation. Almost without exception the presence of anciently abandoned irrigation works on valley sides indicate overly optimistic estimates of how far the water supply would go. Modern cultivation is concentrated on only a fraction of the land that once was made ready for irrigation.

Major technological developments such as high dams with large reservoirs and concrete aqueducts can enlarge an effective exotic water supply and make possible the bringing of new land under irrigated cultivation. With each development there is likely to be an overoptimistic expansion of the cultivated area and then either abandonment of some of the irrigated acreage or, if the community is rich enough, it undertakes new engineering works to make up the deficit.

The history of the Norte Chico of Chile represents a mixture of ancient and modern irrigation practices (Figure 20). The documents of the 16th-century *conquistadores* make it clear that each of the valleys that run westward from the Andes, from the Aconcagua northward, was occupied by Indians who cultivated crops with the aid of irrigation. The *encomiendas* to which these Indians were assigned, and the native villages that maintained their independence although subject to the *mita*, or labor requisition, no doubt continued to use the extant irrigation works for subsistence.

The rapidly declining native population, however, meant a reduced labor force for water management, a tendency emphasized by diversion of as much of it as possible to gold placers and to the herds of introduced European livestock. Thus when there was renewed interest in irrigated agriculture, stimulated by the developing mining industry in the 18th and 19th centuries, the developers found incompletely utilized streams and they could for some time reclaim agricultural land by the laborious but simple technique of leading long canals along the valley sides at gradients less than those of the streambed to reach relatively level terraces downstream.

The rules under which these diversions were carried out were those of the Spanish colonial government. Chilean students conclude that these rules were only slightly informed by the ancient and elaborate Roman and Moorish irrigation traditions of Valencia, Murcia, and Alicante, provinces from which few of the colonists of the New World came.[1] A considerable literature exists that treats as conflicting legal theory whether rights to specific water for irrigation in Spanish and later Chilean law belong inalienably to the *pueblo* (the people or state), or whether they can be transferred to individuals like real estate and inherited or sold. If the latter position holds, then any reduction of flow, because of new irrigation developments for example, would have to be compensated.[2] The *Siete Partidas* of Castile, formulated in the 13th century, however, make repeated reference to peculiar characteristics of water, such as its need to pass freely over lands owned by another, and the rights established by continuing beneficial use.[3] Further the *pueblo* (town or commune) in Spanish society holds great importance. It could own and distribute among its citizens, under the direction of the mayor and council, waters from springs or small rivers within its boundaries.[4] These viewpoints were retained in the Laws of the Indies and in the developing codes of the Latin American republics.

For areas such as Chile, which for most of its colonial history felt that it suffered from underpopulation and underdevelopment, a rational thread seems to run through the legal treatment of rights to water for irrigation. This is the encouragement of long-term investment by individual landholders or groups of landholders in facilities that will make water use more efficient and productive. Canals, some as much as 30 kilometers long and in places cut into sheer canyon walls, were the most important investment, but leveling, terracing, and building a water delivery net on the fields themselves, and the planting and maintaining to productivity of perennial crops such as grapes constitute a not inconsiderable addition. To induce a rational and necessarily wealthy individual, one with alternative uses for his funds, to make such an investment required firm assurance that he would have continuing access to the water he needed. Building a canal would convert a worthless terrace or bench into a valuable agricultural property.

Rapid expansion of silver- and copper-mining activities in the Norte Chico during the first half of the 19th century created both a market for agricultural products and the capital with which to construct irrigation works. One after another each of

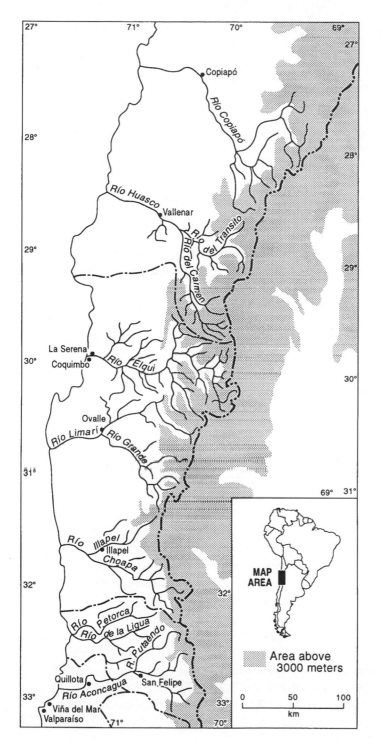

Área above
3000 meters

0   50   100
km

Figure 20.
The Chilean
Norte Chico

the valleys had committed fully their normal stream flow. For further agricultural development, water rights would have to be purchased or taken from other users by litigation. In the Spanish Empire and its successor republics, water litigation with repeated appeals is virtually interminable. Suits commonly dragged on for decades, sometimes for generations.

The other problem any irrigator must face is a sudden diminution in the stream flow because of a dry year or series of dry years, affecting its drainage basin. At such times there is likely to be an acute water deficit even in basins where the normal stream flow is not fully committed. Since such declines in flow can amount to 90 percent, limiting the irrigated area to what can be served every year would be inordinately wasteful of the flow during normal years. The technological solution to this problem, the one which has dominated water resource development programs in the western United States, is to construct major dams with large reservoirs. We may be singularly blind to the inherent disabilities in this approach, which, in addition to cost, involve evaporative water loss and the flooding of extensive land surfaces. In Chile technical capacity until recently, and fiscal stringency on a continuing basis, have effectively precluded using dams and reservoirs as a major means of assuring supplies of water for irrigation.

A Description of the *Turno*

The *turno*, a specialized social and legal mechanism for dealing with water, has enabled a remarkably high and effective exploitation of irregular flow. Basically it forces all established irrigators in a given basin to share in a shortage occasioned by drought, and in operation it recalls the meaning of its English cognate word *turn*. An important feature of the turno is that it can be called and put into effect immediately by a local governmental agency. A remarkable variety of governmental agents or agencies have at one time or another taken on themselves the responsibility of calling a turno and defining its terms. These include departmental governors or provincial intendants, local judges, municipal *cabildos*, or individual *alcaldes*. The crucial element is that the local community knew the situation and gave moral support to the action. The form of division and who may share in it is subject to appeal through the entire national legal system, a process that can continue for decades. But once a turno is called it goes into effect in the form designated and continues until the drought that induced it has ended. The hearing to call a turno is held before a local authority that knows the basin and its acute problems; monitors (called *celadores* or *Jueces de Aguas*) are appointed who are acceptable to the complaining irrigators; the fines for detected violations are prescribed in advance and are severe; they may be monetary or, more severely, involve being completely deprived of water for one or more cycles of the turno; they might be thought of as the result of contempt of a court decision and effectively are not subject to appeal even though litigation in progress may ultimately

256

change the several water rights involved and force a restructuring of future turnos.[5] The whole procedure from appeal to institution of the first cycle, depriving the upstream user of water for a fixed period of time, can be carried out in as little as one week.

The appeal to declare a turno is invariably made by irrigators in the middle and lower course of a given irrigated valley. Their claim is that so much of the stream's diminished flow is being consumed in the upper valley that none or an insufficient amount reaches them. If the complaint is found valid, the turno is called, and for a specified number of days the upstream users may take no water from the stream. Then the upstream users may take all the water for a specified number of days, and then the cycle reverts and continues until all agree that the drought has ended and there is adequate water for all. Sometimes a given valley is divided into three or even more sections, and each has its place in the cycle and a specified number of days when it gets the entire flow. In 1860 the Huasco Valley was divided into eight sections, although some were grouped to share the water for a number of days (Figure 21).[6] The number of days each section receives water is roughly proportionate to the amount of land it has under irrigation that is accepted as having a water right. The full cycle of the turno, however, should not involve more than 16 days, preferably not more than 12, since in midsummer both annual crops and vines or fruit trees could well die if they go longer without any irrigation.

Figure 21. Looking down Huasco Valley from lowest terrace on south side of valley

A section may consist of a single large estate that can manage its water flow as it sees fit, or it may involve a hundred or more small cultivators. The latter may claim their shares either by taking the whole stream for a determined time, or the flow of a single canal, or a proportionate part of a canal's flow. With the smallest holders the latter is most feasible, and even then their time to irrigate may be brief indeed. In the Putaendo Valley, some individuals get water from the canal during turno for only 2 minutes and 20 seconds.[7]

Traditionally water for certain uses has been exempted from the turno. Residents are authorized to take water for domestic use from a canal passing by or through a town, and upstream users may be required to permit a certain flow at all times in the town's canal; watering livestock also is normally permitted.[8] Water to operate flour mills, or *trapiches,* for grinding ore in theory is not consumed but is returned to the stream from which it was extracted. In the *Siete Partidas*, a mill operator could claim water and land on which to set his mill.[9] Ore processing, however, consumes and contaminates the water. By the end of the colonial period, trapiche operators were complaining that irrigation developments upstream were forcing their mills to shut down. The revenues that the crown derived from mining operations seem to have had the effect of persuading the authorities to give the mill operators continued favored access to water,[10] in effect, exemption from the turno. The economic power of the mining operators, and the political potency of their labor force, has tended to perpetuate this favored position into modern times, and their contamination of water is not strictly controlled. The turno is primarily employed in reference to irrigated agriculture.

The Landscape of the Norte Chico

The institution of the turno in the Chilean Norte Chico clearly has developed in relation to the specific physiographic and hydrologic characteristics of the region. Between latitudes 33°S and 27°S eight rivers rise in the Andes and flow through pronounced canyons to the Pacific. Their courses are from 100 to 200 kilometers in length, and their upper tributaries drain the entire western face of the Andes in this sector. The intervening streams that rise in that range's western spurs and the coastal ranges do not carry enough water to support significant irrigation. Average annual precipitation at low elevations declines from 400 millimeters in parts of the Aconcagua Valley to less than 30 millimeters at Copiapó. Significant dry farming is not possible north of Illapel, Lat. 31°40'. There is a lack of weather stations in the Andes, but it may be concluded that elevations above 3,000 meters receive several times as much precipitation as latitudinally corresponding lowland stations, the multiple increasing from 2 in the south to perhaps 10 in the north, although total values decline in the same direction. Storms are almost exclusively winter phenomena except in the basins of the upper tributaries of the Copiapó, thus the higher areas get most of their precipitation in the form of snow.

There is surprisingly little variation in average monthly flow, especially in the more northerly rivers, where the maximum is less than twice that of the minimum. The ratio for the Aconcagua is about 6:1. In all cases the heaviest flow is in late spring or summer, a fortunate circumstance since it coincides with the maximum demand for irrigation. Snowmelt must account for much of the peaking of flow in summer, but it seems likely that pervious rocks high in the Andes take up much of the melt and release it gradually in springs along the upper stream courses, permitting substantial flows to be maintained well into the fall. Because consumption of water for irrigation begins above the juncture of important tributaries in each major valley, it is not possible to state accurately the volume of the flow for each river. Allowing for variability in the sizes of the drainage basins, there is a steady increase in the flow toward the south. Average annual flow in the Río Copiapó is given as 3.7 cubic meters per second; in the Huasco it is 6.7 cubic meters per second, in the Elqui somewhat more than 10 cubic meters per second, and for the Aconcagua about 32 cubic meters per second.[11] The extent of the areas reasonably subject to irrigation if water were available, however, is approximately equal in each basin.

Although each of the major valleys in the Norte Chico has its own distinctive topographic characteristics, a number of general features are common to all. The stream profile is concave upward with a rather gentle slope near the coast, falling about 300 meters in its last 60 kilometers and perhaps three times that much in the preceding 60. The upper tributaries are extremely steep canyons used only for grazing. Major tributaries have profiles accordant with the sector where they join the main river. The breadth of each valley floodplain is variable, with narrows where it crosses more resistant rocks, alternating with broad flats up to 2 kilometers in width. The Huasco, Elqui, and Limarí valleys also have extensive high terraces of ancient alluvial materials paralleling their lower courses. At distances about 60 kilometers from the coast, where the stream course begins to steepen, the floodplain itself narrows and becomes less usable because of the mobile cobbles that form it. Small bench terraces, often associated with alluvial cones where minor tributaries join the main stream, offer the best possibility for agriculture, but even fairly steep valley sides are often pressed into the cultivation of grapes and deciduous fruit trees.

The broad floodplains of the lower valley, originally used for small grains and irrigated pasture, and the little benches of the upper valley seem to have been the initial centers of development in post-Conquest times. Pictographs and potsherd finds indicate that the upper valleys were the more intensively utilized areas in prehistoric times. A reasonable explanation is that the steeper stream gradients permitted leading water off to the area to be irrigated by short canals that could be constructed by the labor force of a village. At least some of the benches were high enough above the stream channel to be protected even during a major flood that might be completely destructive to the broad floodplain farther down valley. For the first two centuries of the colonial period, the upper and lower groups of irrigators were not in conflict. The subsistence plots (*chacras*), vineyards, and orchards of the upper valleys were on steep slopes and returned a large fraction of the irrigation water to the main channel, where it could be reused farther downstream.

The abandonment of commercial grain cultivation and the increasing market for mules and fattened cattle in the developing mines changed the water demand picture in the broader valley floodplains. Permanently irrigated pastures and alfalfa fields were rapidly expanded, and the market for animals could still not be satisfied locally. With capital derived from mining, long canals could be constructed, and lands at the higher outer edges of the floodplains and on extensive elevated terraces could be irrigated.[12] In times of low stream flow, not all the land, newly placed under canals at great expense, could be irrigated. At this point the cry to restrict upstream users, that is to institute a turno, arose.

The Norte Chico can be divided into two sectors in terms of how landholdings are arrayed along the courses of the individual rivers. The border may be placed between the Limari and Illapel river valleys. To the north the upper valleys are largely held by smallholders, descendants of the Indian occupants or Spanish and mestizo settlers who entered in the 16th and 17th centuries. Their original subsistence chacras have in many instances been converted to specialized intensively cultivated commercial plantings of tomatoes, grapes, or other fruits. The dry hillsides normally offer little or no pasture, thus the only animals the smallholders have are a few goats and fewer cattle that can be pastured along the stream course. The interfluvial slopes of the valley of the Río Grande, the main southern tributary of the Limari and at the southern, moister margin of the region, is a notable center for communal holdings of pastureland and plots that may be dry-farmed by owners of tiny irrigated plots in the valley floor.[13] Downstream there may be a few remnant smallholding areas on the edges of the floodplain, but the bulk of the land under irrigation is in middle-sized or large *fundos*, developed in the 19th century and specializing in fattening cattle.

In the southern part of the Norte Chico, a very distinct landholding pattern prevails. The head of the valley is likely to involve a few large fundos that include both irrigable valley land and extensive upland pastures. The valleys of the Putaendo, a northern tributary of the Aconcagua, and the Illapel strikingly illustrate the pattern.[14] The Hacienda Illapel, a direct descendant of a 16th-century *encomienda*, actually holds the entire upper drainage basin of the river. Downstream in this region, there are always districts of smallholders, some with properties subdivided by inheritances to plots of less than a hectare. Mixed with them are larger fundos, often with both irrigable land and interfluvial pasture, and there are some ancient communal holdings with both classes of land.[15] The size of holding is, of course, directly correlated with the owner's wealth and ability to endure drought. In less direct fashion, these factors influence the ability of one section of the valley to get more or less favorable terms in defining the turno. There are compensations; the wealth of the large landholder lets him employ the best legal talent and use court proceeding and delays to his advantage, but the political potency of a number of smallholders and their desperate needs in time of drought have forced political entities to attend to their concerns even before the populist political movements toward land reform of the past decade.

Examples of the History of the Turno in Certain Valleys

It seems probable that the turno as an emergency institution arose first in the Copiapó Valley, where there is far less water than irrigable land, and where late 18th-century mining developments placed a premium on expanding irrigated pasture. A very early formal institution of the turno, curiously, is found much further south along the Putaendo River. Although this tributary of the Aconcagua heads high in the Andes, it has a fairly limited drainage basin and a highly variable flow.[16] The upper valley had an arrangement to divide the water among several haciendas, and there were already in 1808–9 a considerable number of small proprietors downstream. During the drought of that year, the latter petitioned the governor, who appointed one Marcoleta, alcalde of the nearby town of San Felipe; he held a hearing and ultimately decreed a turno identified as the Ordenanza de Marcoleta. It seems to have been based on earlier arrangements, going back at least to 1753, that were not being followed by the owners of large properties upstream; and its general pattern, with some modifications made in 1866, is followed to the present. Even in colonial times, the privileged large holders went too far when they cut off all flow, and not even drinking water reached the smallholders. By the ordinance a small stream for that purpose was to be allowed to flow permanently.

The *Juez de Río* was always selected from among the smallholders, and while he could punish infractions by members of his community by depriving them of water for a period, he had only moral suasion to use against the large holders upstream. The *Jueces* were not afraid to speak out. Developing small fields in the riverbed has expanded the area under irrigation, and in summer when the streams's flow is highest but irrigation needs are also greatest, there is almost always need for a turno. Only in about one year in fifteen is free use of the water permitted all year (*aguas sueltas*); of similar infrequency are years so dry that the turno must continue throughout the year. The normal year now has a turno that may begin early, in December, or may not have to be called until February.[17] It is pertinent to note that with this regular application, the turno no longer serves as an institution for emergencies but merely divides the available water into aliquot parts. The emergency provision, however, has been sustained by the downstream users by setting up two arrangements. In a normal year, both the east- and the west-bank users get water for their full six days. In extreme drought the canals on one bank get only three as do those on the other.[18]

In the Huasco Valley, the necessity of declaring formal turnos does not seem to have arisen until 1832. In this case it was the large holders, with new capital derived from the mines, who were developing alfalfa fields on terraces on the valley sides near Vallenar, who complained about excessive use of water by smallholders upstream. Until 1860 it was possible to satisfy downstream needs by requiring upstream users not to irrigate at night, closing their canal intakes at sundown. At that time the stream course above Freirina was divided into sections and a characteristic turno instituted. Freirina, in the lowest part of the valley, got enough returns from upstream users so that it did not need to enter the turno until 1897.

In the Ordinance of 1880, provoked by the extreme drought of 1872, it was declared that as of the latter year, the river was serving at capacity, and any new developments had rights to water only when the river was not in turno. It also gave full rights to water only when the river was not in turno. It also gave full rights in turno to the new properties that had been developed before that date, to the considerable disadvantage of ancient irrigators in the upper valley. That there had been overcommitment is indicated by the fact that from 1850 to 1908 the river was under turno during 37 years.

A series of wet years about the turn of the century, the growing demand for local agricultural products from the nitrate industry and the great copper mines of the Norte Grande, and a plan to dam the Lagunas del Huasco high in the upper Transito Valley and run canals more than 100 kilometers to the vicinity of Vallenar induced the governor of that department to grant water to new irrigators between 1903 and 1906. Most of these failed, but some were irrigating from old canals when drought struck in 1908. They asked to be included in the turno and began a legal contest that still continues. When the dams and canals were finally constructed after 1919, the new irrigators were given preferential rights, and the smallholders upstream remain subject to the now almost constant turno.[19]

The first recorded turno in the Illapel Valley was in 1844. Evidently before that time, the hacienda upstream and the smallholders of the community of Cuz-Cuz shared water adequate to their needs. In the same year, the Chalinga, another major tributary of the Río Choapa, went under turno.[20] There seem to have been other turnos in the following years, but in 1886 there is a clear record of the governor of Illapel decreeing one after seeking permission from the intendant of the province of Coquimbo.[21] Dr. Daniel L. Stewart has examined in detail the subsequent legal contests over the terms of the turno for the Illapel Valley, and a few conclusions may be extracted. A number of middle-sized fundos were established downstream from the community of smallholders at Cuz-Cuz, and they sought to induce the Hacienda Illapel to accept a smaller fraction of the turno. Eventually this came to be 6.5 days for the upstream users and 4.5 days for downstream users, a pattern that prevailed from 1916 to 1937, and which altered the earlier 8-day to 4-day ratio in favor of the downstream users. The hacienda sold off the irrigable land in its lower section in small- to medium-sized parcels beginning in 1932, and its water rights, subject to litigation, went with them. In 1937 a lawsuit was filed to declare a permanent turno beginning 1 November and continuing until 1 April of each year. The downstream users scored an early victory, but appeals kept the suit active until 1954, and there is some suggestion that judges were suborned.[22]

Three points arise from this disputative history. The Hacienda Illapel, by selling off part of its land and water rights, improved its position in court by gaining allies who, like the downstream users, could claim desperate needs. The suit, which initially was successful in the court at Illapel, would have changed the turno from a means of resisting natural stress and near disaster to a permanent claim to aliquot parts of the stream's flow. Most important, while the suit was running in the courts, other years of drought set in. In 1945 a voluntary agreement was reached to have a 10-day turno, 4 days for the reduced hacienda, 4 for the parcels subdivided from it, and 2 for the downstream users who were also to get any runoff during the hacienda's 4 days. The city of Illapel, which, curiously, is occupied by miners' families rather than agriculturalists, was assured a continued flow for its domestic needs. With minor modification this pattern persists to the present.[23]

Pierre Denis notes that the turno, called by that name and with essentially the same characteristics as the Chilean institution, exists in the Argentine provinces of Mendoza and San Juan. He adds that a completely different approach to water scarcity obtains in the oases of northwest Argentina, where water is owned independent of the land it is used on.[24] The provinces of Mendoza, San Juan, and San Luis were part of the ancient province of Cuyo, settled from Chile in 1561 and governed from Santiago until 1776. The implication is clear that the Chilean institution of the turno was well established early in the 18th century and pervaded all the appropriate areas in that policy.

The Reforma Agraria of November, 1965, in addition to making possible the expropriation of most of the large agricultural estates in Chile, makes the positive statement that all waters are now natural resources, subject to expropriation for public use without regard to earlier established rights or to the private property on which they may arise. But the legislation specifically envisages utilizing the turno as a means of distributing waters to make them most effective for irrigation.[25] To the present time, the practice seems to be one of continuing locally established customs in operating the turno in each of the valleys of the Norte Chico. Now, however, ultimate control of the terms of a turno is in the hands of the Consejo Nacional Agraria, a national administrative agency, rather than the courts. As its administrative practice develops with experience, it will be interesting to see whether the turno is used as a basis for dividing a stream's flow among its users, or is retained as a mechanism for sharing the loss caused by a drought emergency among all irrigators and thus making it more bearable.

Summary

In evaluating the turno as an institution, it should be noted that its distinctive social and economic function can be performed only when the area under cultivation and with rights to irrigation from a given stream is such that in no more than one year in three on a long-term average does a shortage of water occur that demands its imposition. When the stream's flow is committed so heavily that the turno must be called in all or most years, it then becomes merely a means of dividing the water into aliquot parts among groups of users.

The local community, those dependent on the water of a single river, recognize their common lot and vulnerability to the forces of nature. An acceptance of the need to share misfortune, rather than for some to evade it because their lands are in favorable position to exploit the diminished stream, must be general in the community. The authority to call a turno is most effective when it is most local, and it must act on request of the first injured irrigator, not wait until a majority is in distress. The administrative authority of a municipality or a department, often acting through an elected commission, seems to have best carried out this responsibility in Chile during the last century and a half. Intervention of the courts and the normal legal processes or of governmental authority on the provincial or national basis seems, in most of the cases examined, to have led to prolonged and embittered conflicts and the splitting of communities into uncooperative if not hostile sectors. The delays such procedures engender may well have been fatal to the economically weaker entities. Such procedures in Middle Chile may have been important in expanding the lands of the large estates or fundos at the expense of the smallholders during the 19th century. In the Norte Chico, the smallholders seem to have survived much more successfully.[26]

It has been traditional to regard the governing institutions of Spain's New World empire as the epitome of centralized and arbitrary administration, leavened only by protracted legal procedures that commonly reached no decision within the lifetime of the affected parties. The institution of the turno, arising in colonial times and developing during the aristocratic period of Chile's republican history, presents a different aspect. A strong and responsible local democracy, rooted deeply in Spanish history and tradition, could make possible a viable adjustment to recurrent droughts, one of the most implacable burdens nature places on human communities.

The Head of the Colorado Delta

The rewarding area of study is to learn how [a] society and culture
reacted to a particular and knowable set of vicissitudes of history.
Such understanding both contributes to and derives from a knowledge
of the dynamic mechanisms within that culture and society.

—Review of *Warriors of the Colorado*, 1966

AFTER PASSING THROUGH an area of extreme aridity and broken terrain, but
of reasonable tectonic stability, the Colorado River enters the head of the Gulf of California in a zone of pronounced tectonic instability. An extremely heavy silt load and
widely fluctuating stream-flow have resulted in rapid alluviation and in frequent and
even sudden changes in relief and drainage pattern. The great tidal range in the Gulf
of California further deforms the southern portion of the delta. Finally, the delta lies
over a zone of active faulting associated with continuations of the great San Andreas
and San Jacinto faults. Although alluvium rapidly covers any fault displacement so

Figure 22. Desert lands west of the Imperial Valley, looking toward the Peninsular Range
(photo by M. Pasqualetti)

that there is little surface expression, such displacements have no doubt had profound effects on the drainage pattern on the generally level deltaic surface.

Since the beginning of this century major irrigation works have exploited the natural drainage pattern, bringing the Colorado River under progressively greater control. But the control remains less than perfect even now, and man-induced disasters have drastically altered the topography in unanticipated fashions. With the overwhelming natural forces locally active, the Colorado Delta will probably always remain a rich and productive, but anything but a stable and secure, part of the earth's crust. Man and nature seem to be engaged in a struggle; rich rewards are to be gained, but even the greatest of modern engineering works cannot ensure the security that is taken for granted in most of the world.

An exotic river, the lower Colorado passes through what is probably the driest and most desolate region of North America (Figure 22). Where its waters are present, life may burgeon. Away from the oasis the sparsest of vegetation supports a comparably impoverished fauna. The instability of the deltaic watercourses means that oases have been subject to sudden displacements over many miles as the result of a single season's floods.

The mountain barriers to an east-west crossing of North America are least obstructing near the latitude of the present United States–Mexican boundary (Figure 23). The extreme desert that extends for about 100 miles on either side of the lower Colorado River, however, constituted a formidable difficulty. The trail from Sonora to the river carried the name Camino del Diablo (Devil's Road), and its counterpart running westward across the delta to California, Camino del Muerto (Dead Man's Road). An easier though longer crossing from the east could be made by following Gila River, the last tributary of the Colorado, to its confluence with that river at the present site of Yuma. Until modern rail and highway facilities were built, the route across the desert to the west was always one of extreme difficulty.

This point at the head of the delta has been of strategic importance for a long time. The significance of this locality in the settlement of the Southwest by various Indian tribes in pre-Columbian times is still largely conjectural, but the first Spanish explorers who reached it in 1540 found concentrated settlements of a number of Indian tribes. This nexus at the confluence of the two rivers was visited by several expeditions over the next two centuries and changes in the names and precise location of the tribes reported along the Colorado in the vicinity suggest a continual struggle for a vital and desirable site. In 1775 the Spaniards began to establish missions and a *presidio* (military post) to secure an overland route from Mexico to California, but in 1781 all were destroyed by the Yuma Indians, and neither Spain nor, later, Mexico ever established regular land communications with the tenuously held province of California.

American and Mexican parties bound for California during the Mexican War and the Gold Rush that followed, again made the junction of the Colorado and Gila Rivers

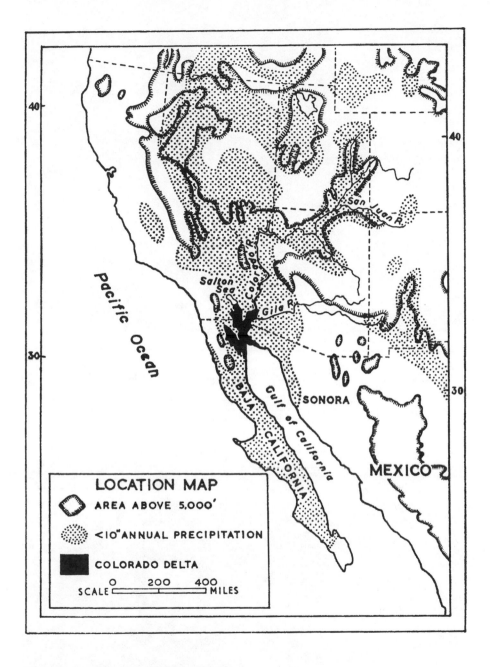

Figure 23. Location map of Colorado Delta

Figures 24 and 25. Irrigation has transformed one of the driest and least hospitable places in North America into the Imperial Valley: a fertile agricultural oasis (photos by M. Pasqualetti)

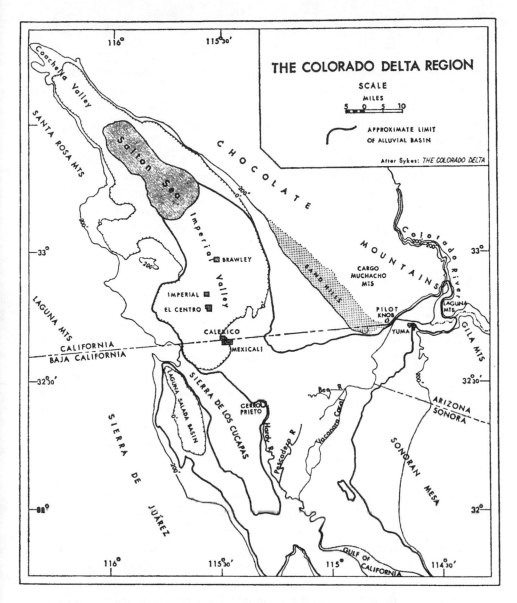

Figure 26.

a focal point. An international boundary was established, and though it was moved southward on the east side of the Colorado in 1853, the head of the delta is still a center for both effective international engineering collaboration and acrimonious, though fortunately peaceful, competitive dispute over water quantity and quality between the United States and Mexico. The irrigation works that have been expanded to make the Imperial and Mexicali Valleys, in the United States and Mexico respectively, among the most productive agricultural districts in the world began in the 1890s at this nexus, and spread out over the delta (Figures 24, 25). By now works farther up the Colorado have brought that once furious river under control, but they have also overcommitted the available water, so after a series of dry years some users, agricultural or urban, may not receive their needed supplies. Though to some degree physically controlled, the Colorado River, now in drought rather than flood, remains a political and economic threat.

This study will focus on the limited alluvial embayment that lies north of the town of Yuma (Figure 26) but below the first real canyon one enters when proceeding up the Colorado. Properly this is the head of the delta. Here there are identifiable historical landmarks, and it is possible to trace how men of several sequent cultures and technologies have perceived this unusual environment and adapted their ways of living to exploit it.

The Physical Environment

Physiographic Evolution

The last canyon through which the Colorado River has cut, separates the Chocolate and the Laguna Mountains about 10 miles northeast of the town of Yuma (Figure 26). At that locality, which may appropriately be designated the head of the delta, the river surface is about 140 ft. above sea level. From this point to the California-Mexico boundary just southwest of Pilot Knob, both the Colorado and the lower Gila Rivers used to meander on a floodplain 4 to 10 miles wide, bordered on the north and south by terrace or mesa edges that rise rather abruptly to 50 or 100 ft. above the floodplain floor. The mesas are formed of poorly consolidated sediments. West of Pilot Knob the edge of the northern bordering mesa runs almost due westward for about 25 miles, just south of the international boundary. A striking feature on this mesa surface is a belt of large, active sand dunes about 3 miles wide and 35 miles long with local relief of more than 200 ft. It runs from just south of the international boundary toward the northwest. Until this century not only the Colorado River, but irrigation canals and railways and roads had to swing far to the north, or south into Mexico, to pass this barrier. From the town of Yuma the western edge of the southerly mesa runs more or less southward to the Gulf of California. Then the delta broadens, its central axis sloping downward in a southwesterly direction for about 40 miles until it abuts against the rugged, recently upfaulted Sierra de los Cucapas and the outlying recent volcanic cone of Cerro Prieto (Figure 26). In that distance the surface falls from about 100 ft. to about 30 ft. above sea level, making the deltaic cone a rather steep one. From this central axis the drainage must diverge either to the south into the Gulf of California or to the north into the Salton Sea Depression, the bottom of which is about 270 ft. below sea level. Over the past few thousand years the river has alternated several times between these general routes, although its actual meandering course, particularly when it was flowing down the gentler slope toward the south, has varied almost continuously. The tidal range of more than 30 ft. in the upper Gulf of California has reworked the fine sediments at its head into broad, salty mudflats, cut by estuarine features, which are kept more or less scoured by the combination of river flow and tidal bore. Between the Sierra de los Cucapas and the massive granitic range of the Sierra Juárez lies the Laguna Salada Basin, the flat bottom of which is slightly below sea level. At times when spring tides meet a high river, part of the river's flow may be forced across the mudflats at the southern end of the Sierra de los Cucapas to fill the Laguna Salada with a thin sheet of water that soon evaporates.

On the occasions when the Colorado River migrated north, its flow, despite great evaporation, would eventually fill the Salton Sea Depression forming the "Blake Sea," an inland water body 85 miles long and as much as 35 miles wide.[1] Its old shoreline, at about 30 ft. above sea level, is plainly visible around most of the perimeter of the basin. The water of the Blake Sea could remain fresh because there was drainage to the Gulf across the edge of the delta at Cerro Prieto.

At one time it was believed that the construction of the delta had cut off the upper end of the Gulf of California, permitting the Salton Sea Basin to dry up, and that the old shoreline was that of the Gulf. Careful leveling, however, has shown the shoreline to be almost perfectly horizontal and appreciably above sea level. Furthermore, all the post-Tertiary shells and fishbones to be found along the old shoreline are from freshwater forms that are currently found in the Colorado River. It seems that there has been a steady subsidence of the upper Gulf, the delta, and the Salton Sea Basin since the early Pleistocene, with the alluvial deposition on the delta being just sufficient to keep it above sea level.

In the late Pleistocene, perhaps as little as 12,000 years ago, sea level was at least 100 ft. lower than it is today. The Colorado River reached the Gulf some 40 miles south of its present mouth. More significantly, the gradient of its channel toward the south was notably steepened, and the zone of heavy alluviation, which in historic times has lain both above and below Yuma, was displaced well toward the south.

It seems likely that the almost undissected mesa edges that lie on either side of the upper part of the delta have persisted since far back in the Pleistocene, and have been preserved by the almost rainless character of the region. The mesa to the north is composed of old, generally unconsolidated, alluvial material and slopes toward the Colorado at about 50 ft. to the mile. Its edges are cut by washes that carry water only a few hours per year.

Sea level rose very rapidly for two or three thousand years during the retreat of the last continental glaciers, and as the gradient of the lower Colorado was decreased, the zone of heaviest alluvial deposition was displaced upstream. By the beginning of the historic record the embayment north of the Yuma–Pilot Knob line was definitely part of the zone of deposition, over which the river, gradually building up its bottom and natural levees, was both meandering continuously and suddenly shifting its course at intervals of from a few years to a few decades (Figure 27).

Almost the entire alluvial content of the embayment then has been introduced in

271

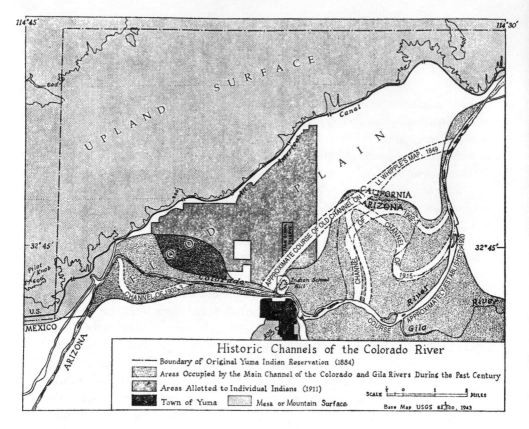

Figure 27.

geologically recent times. The Gila River has made a modest contribution on the east side, and there has been small wash of coarse sandy materials from the mesa to the north along its edge. The aggradation from the Colorado, however, has been so great as to engulf this coarse local alluvium, and there are no little alluvial cones built out from the washes that come out of the mesa edge. The Yuma Mesa to the south has contributed practically nothing. It may be assumed that the two rivers have, in the course of the last few hundred years, meandered over the entire alluvial surface. Nearly half of it has been river channel at one time or another during the past century. There is, however, one place of somewhat greater stability along the channel course. This is just north of the town of Yuma, where the channel cuts through the northern projection of the Yuma Mesa, leaving as an outlier on the north side of the river, a flat-topped hill about 60 ft. high on which the Fort Yuma Indian School is located (Figure 27). This hill, like its counterpart south of the river where the old Territorial Prison was located, is composed of a conglomerate of large granitic boulders. How the river managed to establish a channel between these two hills is not readily evident, but it must have done so a very long time ago, probably in Tertiary time, when it was flowing at about the level of the bordering mesas and before the present alluvial embayment was excavated. In any event the constriction of the flow between the hills has increased water velocity and produced a deeper channel at that point. Thus, although at times of major floods some water would pass north of the Indian School Hill and perhaps even develop a secondary channel there, as is indicated on Lt. Whipple's sketch map made in 1849, it usually returned to the constricted passage in subsequent periods of low water. This locality can be identified definitely in early explorers' accounts of journeys along the river; a strategic military position was afforded by the isolated hill on the flat north bank. It was not particularly favored as a crossing in the distant past because the river is swifter and deeper, although narrower, in this sector than immediately upstream or downstream. When bridge construction was considered, however, the relative ease of access to both banks, the firm base, stable channel position, and short crossing virtually determined that Yuma would become the regional communication center.

Climate

The entire zone of the Colorado Delta qualifies as a hot desert. The major uplands to the east and west of the Colorado Valley act as barriers to the flow of moist air from both Pacific and Gulf of Mexico sources; the Gulf of California only rarely is a source of unstable, rain-producing air masses, though in its immediate proximity relative humidity of the surface air is raised enough to make the summer heat especially oppressive. Local topographic features around the delta are not big enough to bring about significant variations in temperature and precipitation. All stations on the delta show mean annual rainfall values of between 2 and 4 in. At Yuma, where there is more than an 80-year record, the average annual rainfall is about 3.25 in. In 5 of 80 years more than 6 in. of rain were recorded, with a maximum in 1905 of 11.41 in. In 9 years less than 1 in. fell, with a minimum in 1928 of 0.47 in. There are two highly uncertain seasons when rains may occur. Cyclonic storms from the Pacific Ocean may bring rain at any time during the winter. The area is near the southern edge of the storm tracks from the North Pacific, and much of the moisture borne in these storms falls over the Peninsular Ranges to the west in California and Baja California, so the rains are usually brief and light. Occasionally they will persist for days and at times be vigorous. The summer storms may occur from July to September. They involve intense convection in a tropical air mass, are likely to be quite localized, and can be extremely violent. In many years none at all occur, but about one September in five a tropical cyclone, locally called a Chubasco, moves up to the head of the Gulf of California. The instability associated with these storms typically brings widespread downpours throughout the delta region.

During the months from June to September the mean daily maximum temperature exceeds 100°F; 120°F is not infrequently recorded. Nights, especially in midsummer, are comparably warm, so the average temperatures for the months of July and August exceed 90°F. The mean January temperature at Yuma is about 54°F, but, under clear skies and with very low humidity, there is a great diurnal temperature range on almost all winter days. Frosts can occur from late November until February, but they are almost always mild and in many years the temperature does not fall below 32°F. Where water can be introduced, plant growth can proceed vigorously throughout the year, though it has been discovered that for certain crop plants, notably alfalfa, the midsummer heat is so great as to exceed the physiological optimum of the plant, and there is a considerable slowing of growth at that season.

Wild Vegetation and Fauna

The upland and mesa surfaces around the alluvial floodplain or delta carry only the sparsest xerophytic vegetation. Small bushes of ocotillo (*Fouquieria splendens*), creosote bush (*Larrea tridentata*), and *Franseria dumosa* are spaced no more densely than one per square rod on the uplands, though palo verde (*Cercidium microphyllum*) and ironwood (*Olneya tesota*) occur in the washes somewhat more densely; these form the perennial vegetation. Following the occasional heavy rain an ephemeral carpet of wildflowers may spring up to cover the ground for a few weeks; then it dries up and blows away leaving a bare reg surface. The animal life is correspondingly impoverished, and other than insects that drift out from the riverine oasis is largely limited to a few lizards and an occasional jackrabbit.

Life is far more abundant in the floodplain. In addition to quite regular and extensive though seldom universal inundation from the seasonally flooding Colorado and Gila Rivers, the water table within the embayment is quite high. Once established, deep-rooted trees like mesquite (*Prosopis* spp.) and cottonwood (*Populus fremontii*) could probably sustain themselves permanently, or at least until they were washed out by a shift in the river's main channel.

Levee and canal building, leveling, drainage, and cultivation during the past 70 years have converted all of the upper delta save for a few saline and sandy spots, and narrow strips between the artificial levees and the rivers, into a completely artificial landscape. Only by looking at the southerly part of the delta in Mexico, where the extension of irrigation and cultivation is much more recent and still proceeding, is it possible to gain an impression of the variegated detail of the microtopography and related vegetation patterns that existed on the floodplain prior to modern irrigation developments.

In a four-mile transect from the river to the base of the bordering mesas the local relief amounted to 10 or 15 ft. A typical profile would show a zone rising for a few hundred yards from the low-water river to the crest of the natural levee, formed and maintained by deposition of silt in the shallows at the river's edge in times of higher water. When the Colorado's flow is impeded by vegetation, the silt-laden river, 10 percent of whose mass is sometimes suspended solids, can deposit tremendous volumes of sediment in a single flood. Godfrey Sykes estimated that from summer 1931 to summer 1933 an area of 40 to 50 square miles at the lower end of the Vacanora Channel farther down the delta was raised an average of 8 ft.[2] Of course great floods such as occurred in those two years probably removed as much as they added to the head of the delta, and the river levees there are likely to have been developed during less extreme periods of high water. Since 1935 the major dams upstream have virtually eliminated floods and impounded most of the Colorado's silt burden, so growth of the delta has nearly ceased.

A thick and nearly pure stand of arrowweed (*Pluchea sericea*) is the pioneer vegetation that develops on new surfaces close to the river when they are exposed after a flood. Such stands exist today on the flats between the normal low-water river and the protective levees a few hundred yards back of it. Willows tend to replace the arrowweed after a few years if no new deposition occurs. Along the bank or crest of the natural levee and on its blackslope isolated cottonwoods and clumps of mesquite are likely to be present.

Away from the river the levees slope off into broad swales, perhaps a mile across. This lower surface may have small ridges and gouges in it, formed by floods that overtopped the main levee. Clumps of mesquite, and screwbean (*Prosopis odorata*) in the lower wetter spots, mixed with grasses and weedy herbs such as amaranths, cover these swales. The tangle of trees and bushes becomes a veritable jungle following a flood, when spillover stands on the surface to evaporate or soak in. After a year or two without flooding, the annual plants and those without especially deep roots will die and little but the mesquites whose long roots have penetrated to groundwater will remain.

Almost any transect will cross the oxbow of one or more old meanders and their bordering natural levees. Although deposition has filled the channel of the former meander at the points of cutoff and perhaps elsewhere, parts of the channel may remain as depressions several feet deep. Following a flood, water will remain impounded in these depressions for some months, and at nonflooding times of high water, groundwater may appear at the surface, forming swamps or even shallow lakes such as exist just south of Bard on the great oxbow to the northeast of Yuma that was cut off in 1915 (Figure 27). Around the edges of these marshy localities an occasional cottonwood is to be found along with dense thickets of willow and screwbean, cane brakes (*Phragmites communis*), and such herbs as *Rumex* spp. and sedges. Where water stands more or less permanently cattail rushes appear. At present the introduced tamarisk is developing into a virtual forest wherever its roots can find permanent water. Finally, at the mouths of washes coming down from the edge of the mesa to the north there is sandy soil and slightly greater elevation but some underground water percolation. Mesquite, ironwood, and palo verde trees or bushes may be scattered over the otherwise bare sandy surface. Following a heavy rain or general flood a heavy cover of grasses and herbs will spring up on all surfaces where it is not excluded by a dense phreatophytic perennial vegetation.

At the southern end of the point lying east of the old meander that was cut off

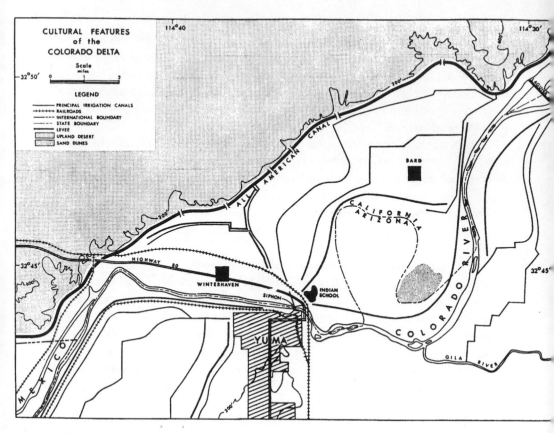

Figure 28.

in 1920 there is an area of two or three square miles of small but active sand dunes (Figure 28). The characteristically irregular relief, unstable as to position, amounts to about 5 ft., and low perennial desert shrubs are scattered over the generally barren surface. *Atriplex* spp. and *Allenrolfea* are dominant among the shrubs. Similar dune areas adjacent to old meander scars are to be found at other localities farther down the delta.

Soil Development

The river-laid sediments that form the floodplain and delta were somewhat sorted as to size by their relations to the river flow at the time of deposition. Old banks between channels of the former main stream are likely to be sandy. Swales away from the stream where slow-moving water during floods was ponded to evaporate later have sediments with a high clay content. The most characteristic soil texture is a silty loam, but the soils of the whole floodplain are placed in the Gila series, broken into some 20 phases on the basis of the texture of both the surface material and the subsoil.[3] The latter is likely to vary independently of what is on the surface, for when it was laid down on the aggrading surface, local topographic conditions are likely to have been quite different.

Above all, these features of the microtopography of the floodplain that forms the head of the delta and their associated vegetative cover should be recognized as extremely unstable as to position under conditions that prevailed prior to the construction of the great dams. All the features mentioned are likely to have been present within a quite limited area, but any could be transposed during a flood when the river constructed a new main or subsidiary channel. At such times heavy local deposition might completely cover all the existing vegetation or a swiftly running side channel scour a long narrow depression into the underlying alluvium. Thus transient vegetative associations such as the pioneering arrowweed were always being formed anew, and then succeeded by other forms.

The deeper depressions in old oxbows that held ponds and marshes tended to accumulate high concentrations of soluble salts. In addition to their heavy silt burden both the Colorado and Gila Rivers carry large quantities of dissolved solids, as might be expected from the arid character of their lower drainage basins. More than a thousand parts per million are often present, though proportions do go down at high-water stages, and the dangerous sodium radical forms an important fraction of the solutes. Marshy depressions where water enters, stands, and evaporates tend to become alkaline. A map showing soils affected by alkalinity shows them in strips that follow the paths of known former channels,[4] and it may be postulated that similar alkali concentrations mark the courses of ancient oxbows which now have no surface expression.

Animal Life

The nutrient content of much of the floodplain vegetation is high; many of the shrubs are leguminous, contributing as well as concentrating nitrates, and the regularly replenished alluvium from a diversity of upriver sources provided all the minerals needed for animal life. The thickets of several sorts provided a diversity of cover. From rabbits and game birds to beaver and deer, as well as the predators, the concentration of game in whatever parts of the delta area were currently receiving water flow was tremendous. The undeveloped lower delta is still a hunter's paradise, and in 1922 when Aldo Leopold described hunting in areas, then wilderness but now developed, just south of the international boundary, he knew no match for it.[5] Most of the lands in the head of the delta are now under cultivation, but an observer must be struck by the abundance of cottontail rabbits, quail, and doves wherever a little patch of cover is afforded around a marsh or along a levee or fence line.

The muddy lower Colorado is rich in coarse fish. The humpbacked sucker, a minnow that grows up to three ft. long, and the bonytail are the principal species. These were exploited enthusiastically by the historically known Indians.

Significant Aspects of the Natural Environment

The oasis created by the Colorado, which provided both water and alluvium as it approached its mouth, afforded support for a concentration of plant and animal life scarcely equaled in the world. Solar energy for plant growth was always abundant; cold was scarcely ever a problem. Isolation by the surrounding barren deserts is extreme; even the Colorado itself can only be followed a modest distance upstream before its canyons make it a complete barrier even to travel on foot. Finally the oasis was actually mobile in its lowest portions. When the Colorado River flowed into the Salton Sea Depression and abandoned the southeastern portion of the delta an area of several hundred square miles became absolutely desolate waste. A shift back to the south would render an even larger area, the entire northern arm of the delta, barren. These drastic changes could occur in a single season, or at most in a few decades while the Blake Sea dried up and became too salty to be used. Capital accumulation in permanent settlements on the lower delta was patently impossible until very recent times. Shifts in the river course at the delta's head in the Yuma embayment were over shorter distances, permitting more stable occupation, but any point on the floodplain surface might in a single flood season be changed from a levee bank or a protected swale to part of the main river channel.

Indian Occupance

Archaeological Data

The earliest occupation of the Colorado Delta is likely long to remain obscure. We may postulate that this oasis with its concentration of water, wild plant foods, game, and fish in a locality of relatively mild climate was attractive to man well back into Pleistocene times. But the living sites were on the delta or floodplain surface, and the continuing alluvial buildup of that surface has undoubtedly covered with many feet of alluvium all camp sites and artifacts that are more than a few hundred years old. Further, because of the mobility of the river no spot is likely to have remained attractive for settlement for more than a few decades or a century at most. The stratified site of long-continued occupation where changes in the culture pattern of the occupants can be readily arranged into a temporal sequence probably does not exist. When the excavation of a canal or irrigation ditch fortuitously brings to light some nonperishable artifact such as a potsherd or stone tool, there is no cultural sequence in which to place it. The historic Indian inhabitants of the floodplain did not have a complex lithic technology; most of their artifacts were of organic materials, a situation to be expected, since stone is not present in the alluvium on which they spent most of their lives. With recurrent flooding and drying out of open sites, such organic artifacts would almost certainly disappear within a few decades.

Any reconstruction of prehistory must rely heavily on ceramic types. On the basis of datable intrusive potsherds from the Hohokam culture of south-central Arizona, Schroeder estimated that the indigenous pottery-making tradition along the lower Colorado began about A.D. 900.[6] Farther back in time we can only speculate. One question of interest is when the distinctive pattern of farming in the mud left behind after a flood, utilizing specialized, quick-maturing crops adapted to develop after a single irrigation, developed. It may be conjectured that this successful farming technique was developed in or introduced to the upper delta at the same time that indigenous pottery appears.

Archaeological data from a locality quite distant from the head of the delta appear to throw light on the notably fluid pre- and protohistoric tribal movements and the war pattern that characterize the lower Colorado as we know it in history. Along the more than 200 miles of shoreline of the Blake Sea that formerly occupied Imperial Valley there is an almost continuous though fairly thin midden of fish bones and potsherds. The fish bones are all of freshwater species which currently live in the lower Colorado, and by appearance are certainly not very old. Radiocarbon dates for charcoal from this long midden indicate an age of only 200 years. Radiocarbon is not a precise instrument for dating such recent materials, but it is probable that a vast freshwater lake occupied Imperial Valley no more than 500 years ago. When the Colorado River shifted its course to flow directly into the Gulf of California the Blake Sea began to dry up, probably falling at the rate of 10 ft. per year. Within a few years all the fish died and the water became too salty to drink. The Indian population then resident along the shore, numbering many thousands, had to find new, habitable land or starve.

Thus, at the beginning of the historic record major tribal migrations and dislocations had taken or were taking place. There were two general areas to which the shore dwellers could go. One was coastal California; the other was to the Colorado and its delta. The sudden influx of displaced tribes into the already occupied river valley and delta and the economic and social tensions thereby produced are a likely cause for the development of the violent war pattern that has characterized the several tribes of the latter area since they have been known to history. Curiously, however, the nonfarming coastal peoples to the west, though they must have experienced a comparable influx of immigrants, are reported by the first explorers to have maintained essentially peaceful intervillage and intertribal relations.

Tribal and Linguistic Patterns

All the Indians known to have occupied the delta and lower valley of the Colorado spoke languages belonging to the Yuman family. The several languages of the area are closely related, probably closely enough to permit some mutual intelligibility. The Yuman linguistic family is classified as part of the great Hokan-Siouan linguistic stock, representatives of which are found widely scattered through California and in other parts of the United States and Mexico. The several linguistic families within this stock have widely divergent languages, indicating that a very long time has elapsed since they all formed part of a single speech community. This and their geographic distribution convince most linguists that the Hokan-speaking tribes represent a very ancient wave of Indian occupation in North America. The several families were split by intrusions of peoples with other linguistic affiliations.

Other Yuman-speaking tribes occupied the plateau country of western Arizona and the Gila Valley west of Phoenix; they also extended westward to the California coast in the San Diego area, and southward to occupy at least the northern three-fourths of the peninsula of Baja California. Shoshonean tribes, representatives of the great Uto-Aztecan linguistic family, occupied both the desert and the coastal valleys of California. Historically the Salton Sea Depression was divided just about equally between Yuman speakers in the south and Shoshonean speakers in the north. The boundary between the two groups ran almost due westward across the mountains to the Pacific Coast, and in doing so it passed through a far richer and more thickly set-tled area. This division probably antedates the last occasion when the Blake Sea was filled with freshwater. When the Blake Sea dried up, most of the Shoshoneans had to join their linguistic congeners to the west and north. The Yuman speakers who went west were probably Diegueños or were quickly assimilated by them, as that is the only group to be found to the west and south. Those who moved south and east into the Colorado Valley parts of the delta where water was present may have been from various tribes, since in historic times at least six and probably more distinct tribal groups have been reported from the region.

There is a striking difference in sociopolitical organization between the tribes of the delta and the lower Colorado as far north as the Mojave and the myriad Indian groups that occupied the rest of California, even between the Colorado tribes and the Diegueño, their fellow Yuman speakers to the west. The characteristic California pattern of social organization involved politically independent communities of from 200 to 500 people, with fewer in extremely poor and more in extremely favored lo-calities. The terms *ranchería* or tribelet may identify them better than village, since only in favored places could all gain a living from a single center of residence. An in-dividual's loyalty was to his own ranchería; most marriages took place within it, and it was uncommon for a person to move from one such community to another. Neigh-boring rancherías might speak mutually intelligible dialects and visit one another or there might be feuds between them; rarely was there real warfare, and permanent al-liances between rancherías were almost unheard of. Inter-ranchería intercourse was almost as likely to cross major linguistic barriers as it was to cross minor dialectic ones.

The groups that lived along the lower Colorado, on the other hand, were real tribes involving as many as 5,000 people. They lived in small settlements of half a dozen families each, dispersed along the levees and channels where flooded land that might be planted to crops could be found. New settlements could be established as changes in the river's course offered new localities for farming, and an individual might change his local associations several times in a lifetime. The whole tribe was his community and he was loyal to it.

Warfare was an important social value, and a man's normal route to social prestige was by demonstrating his fighting prowess. Bows and arrows were for hunting and for boys to skirmish with. Men fought with short potato masher clubs, made of the hard wood of mesquite branches; they attacked en masse, seizing their opponents by the neck and smashing their faces, and leaving them for the boys and women to beat to death with longer clubs.[7] The riverine Yumas have always been big powerful men physically; six-footers weighing over 200 lb. are common. Warfare was bloody in contrast to the rest of California, where one casualty might end a fight; a tribe might be almost exterminated in a losing battle and the survivors might have to flee hundreds of miles. Leslie Spier has presented convincing evidence that in sequence the Maricopa, the Kaveltcadom, the Halchidoma, the Kohuana, and the Halyikwamai were forced to flee the Colorado Valley and settle on the less attractive lands of the Gila River. The Maricopa had been driven eastward in prehistoric times; the last groups fled in the late 1830s.[8] It is noteworthy that in 1781 the Quechans or Yumas were able to overwhelm and wipe out two well-garrisoned Spanish missions which had been established at the head of the Colorado Delta. The Spaniards suffered no other defeat so crushing and decisive in any of the northwestern provinces of New Spain.[9]

Ethnographers who have worked with them consistently comment that the riverine Yuman individuals, for all their violent militarism, are strikingly direct and frank, friendly and demonstrative in their social relations. They are quick to anger and quick to forgive. In these characteristics they stand in striking contrast to the secretive, grudge-holding, though overtly peaceful sorts of people that are found in the typical California Indian communities.

Economic pressures are not likely cause for the intense war pattern of the Colorado Valley and Delta. If the river failed to flood in a given year, crops could not be planted and all would go hungry, but if the river did flood there was far more productive land on which to raise food than could possibly be used. It is more likely that social tensions produced by several tribes, each composed of individuals closely loyal to each other, living in close proximity to one another along the river and its distributaries, provoked fights and feuds. Tribal loyalty extended the feuds to full-fledged warfare. A reasonable hypothesis is that the social, not the economic, crowding on the river and delta that resulted when the shoreline of the Blake Sea had to be abandoned precipitated the development of the violent war pattern that is known from historic times.

Historic Records

In 1540, as part of the Coronado expeditions to the north, Hernando de Alarcón reached the river mouth and with his party traveled in small boats up the Colorado River. He reckoned that they traveled 85 leagues, which would place him well above the mouth of the Gila and the head of the delta. He found a large Indian population farming along the river, and mentions, probably as tribes, the Quiquima (possibly Halyikwamai) and the Coana (possibly the Kohuana of later reports).

Juan de Oñate's 1604–5 expedition from New Mexico to the Colorado and down the river to its mouth mentions several tribes in sequence: the Amacavas (who are the Mojaves), the Bahacechas (possibly the ancestors of the Quechans or Yumas who up to this time are never mentioned), the Halchidoma, living below the confluence of the Gila River, the Coguanas (Kohuana), the Agalequamaya (Halyikwamai), and finally the Cocopa, living closest to the Gulf of California as they do today.

Father Eusebio Francisco Kino, between 1698 and 1702, made a series of well-reported visits to the lower Gila and Colorado Valleys, traveling overland from the Pimería Alta in what is now northern Sonora and southern Arizona. The most important of the several tribes he identifies were the Quechans, or Yumas, who lived at the strategic head of the delta, about the confluence of the Gila and Colorado Rivers. The Halchidoma had by then moved north of the confluence. Quiquimas and Hagiopas (Cocopas) are identified on the delta farther south. Kino's resources did not permit him to establish permanent missions along the Colorado although he indicates that the Indians were very receptive to the idea. Nor was Father Jacobo Sedelmayr, a later Jesuit who visited the Colorado three times in the 1740s, able to establish a mission there.

After the Jesuit expulsion in 1767 and the establishment of the Franciscans in the missions of the Pimería, there was an intensification of Spanish interest in extending the northwest frontier of New Spain and supporting its several outposts, particularly California, by secure overland communication routes. Missionary journeys by Father Francisco Garcés reached the strategic point of the confluence of the Gila and Colorado in 1771, and Juan Bautista de Anza's series of overland expeditions to California in the 1770s all passed through that critical crossing. The Quechan, or Yuma, tribe, which occupied the head of the delta, quickly became the most important of the riverine Yuma groups, receiving the most intense acculturative influences from the Spaniards in terms of horses, new crops, and ideas about European civilization.

Two garrisoned missions were finally established in 1779, one near the Fort Yuma hill on the north bank of the Colorado and the other two or three miles south of Pilot Knob in what is now Mexico. In the summer of 1781 both were destroyed completely in the successful revolt. All men were killed, and women and children held as captives. A punitive expedition managed to ransom the captives, but in the thickets and sloughs of the floodplain, Spanish horses were of no military value, and at close quarters the Yuma fighter was about as good as the fine Spanish soldier. Pressed by Apache marauders in Sonora and New Mexico neither Spain nor its successor, Mexico, ever again seriously attempted to control the strategic overland passage to California. A few groups were permitted by the Indians to pass through, and a famous American party which included James O. Pattie trapped beaver on the lower Gila and Colorado and in the sloughs of the delta in the 1820s.

During the first half of the 19th century the history of the area is one of changing military relations among the several Indian tribes along the river. The Halchidoma, Kohuana, and Halyikwamai had moved up the Colorado north of the Quechan, or Yuma, tribe at the confluence of the Gila. In alliance with the Mojaves the Quechans crushed these tribes, and the remnants fled up the Gila where they were supported by the Maricopas and Pimas, the former a Yuman-speaking group, who probably had earlier been driven from the Colorado. The allied Quechans and Mojaves exchanged bloody raids with the Pimas and Maricopas across 150 miles of unpopulated desert of the lower Gila Valley, the latter probably enjoyed the best of the fighting until 1857 when a Quechan-Mojave attacking party was crushingly defeated near the junction of the Salt and Gila Rivers, losing a hundred men. The Quechans also fought with the Cocopas who lived farther south on the delta, but they permitted a band of Diegueños, the Kamia, to move into and learn to farm in the delta south of Pilot Knob.[10] Clearly, getting more territory to exploit was not their aim in fighting other riverine tribes.

Quechan independence was not threatened until after the Mexican War, when in 1848 the United States took over California and the lands north of the Gila River in Arizona. During 1849 large numbers of emigrants, both from the United States and from Sonora, when participating in the Gold Rush to California followed the Gila-Colorado route. The river crossing at the head of the delta was again a strategic point and the Quechans effectively levied formal and informal tolls on the emigrants. The Americans decided to secure the passage by establishing, on the California side of the river, what later became Fort Yuma. A garrison was established in 1850, but the molesting of travelers continued. Major Heintzelman found the Indians able to flee to the sloughs and floodplain thickets as they had from the Spanish punitive expedition of 1781 and to continue to assert their independence.

Once navigation up the Colorado River had been established and the supplies for the American garrison thus assured, Indian encampments near Fort Yuma were raided continuously and food stores destroyed. The Quechans moved south into the delta in Mexico or upriver away from the fort. By the late summer of 1852 the American army had found and destroyed the fields of the Quechans and thus brought them to terms. Threatened with starvation they were effectively placed on relief at Fort Yuma. It is striking, however, that the Quechans as late as 1857 still felt able to engage in what proved to be a disastrous war with another Indian group, the Maricopas.

The Indian Economy

All the riverine Yumas practiced a specialized though simple flood farming as the principal basis of their subsistence economy. Two quotations from observers who saw the Indian economy while it functioned describe the system quite well. Major S. P. Heintzelman, who commanded the U.S. garrison at Fort Yuma in the early 1850s, wrote, "Their agriculture is simple. With an old axe (if they are so fortunate as to possess one), knives and fire, a spot likely to overflow is cleared. After the waters subside, small holes are dug at proper intervals a few inches deep with a sharpened stick, having first removed the surface for an inch or two as it is apt to cake. The ground is tasted, and if salty, the place is rejected; if not, the seeds are then planted. No further care is required but to remove the weeds which grow most luxuriantly wherever the water has been."[11] Father Thomas Eixarch, who was in the area in 1775–76, adds the following relevant notes: "[T]he land is so good that only with the bathing given it by the river during the time of its flood it conserves enough moisture so that it produces wheat and also maize, beans, watermelons, calabashes, etc."[12]

Since the Colorado is fed primarily by snowmelt from the Colorado Rockies, the normal time for it to flood is in the latter part of June. Typically, high water lasts for about three weeks and then subsides gradually. Occasionally there is high water in winter, and the summer flood may occur earlier or later and sometimes may show more than one peak. The intensity of the flood too can vary enormously. A flow of as much as 186,000 cfs has been recorded at Yuma, whereas the normal annual maximum flow was about 81,000 cfs. At the lowest recorded water, during late summer, in September 1924, the flow was 1,200 cfs.[13] From the Indian viewpoint the most serious difficulties arose in years when the river failed to rise and temporarily inundate prospective planting grounds, or when a late second flood washed out half-grown plantings. At such times severe scarcity of food might afflict entire tribes, and sustenance had to be eked out largely from wild plant foods, of which mesquite beans were most important. Occasionally, however, double-peaked floods permitted the raising of two crops.

The principal aboriginally cultivated crops were the standard ones for North America: maize, beans, and squash. The varieties, and in the case of beans the species, were adapted to the peculiar cultivation pattern. They matured so rapidly that the river mud in which they were planted did not dry out until adequate growth was attained. Maize reached the roasting ear stage in as little as 60 days and was fully mature and dry for harvest and storage in 80 to 90 days. The bean of the region is the tepary (*Phaseolus acutifolius*), noted for its quick germination, rapid growth, and resistance to extreme heat. It too matured in 60 to 80 days. The several varieties of *Cucurbita* grown are also all rapidly maturing vegetables. The harvest season tended to occur in September, but could vary with different dates of onset and termination of the flood. Maize and beans were shelled and stored in pottery ollas or baskets. Squash or pumpkins could be peeled, cut into strips, and dried on house roofs either before or after cooking. There were other minor native crops, including gourds (*Lagenaria*), tobacco, and possibly cotton.

Wheat and barley were early introduced or diffused into the area from Spanish sources. Father Kino found the Indians growing those grains in 1700. They were not preferred as foods, but their short growing season permitted them to mature following off-season flooding, thus enhancing the security of food supplies. Watermelons and cantaloupes similarly had reached the area by Kino's time. Here, however, the taste was much appreciated, and was influential in their adoption as crops. Black-eyed beans (*Vigna sinensis*) were another significant borrowing from Spanish sources. They were eaten both green and dry, but the former possibility, which could be utilized before most crops were ripe, was an important reason for their adoption.

A typical family cultivated one or two acres. If floods occurred at the proper time, fields in the rich delta soil would yield far more than could be eaten by a family. The Quechans had at their disposal at least 30,000 acres, which were sometimes subject to flood. It is doubtful whether as much as 1,000 acres were ever cultivated.

Wild plant foods were gathered by women, especially in the period prior to the harvest of cultivated crops. The most important were mesquite and screwbean pods and both the seeds and the leafy parts of the amaranth, a weedy plant that grows especially abundantly in the delta. All these crops were stored just like the cultivated ones. A wide variety of grass seeds and other seeds of desert bushes were collected, pounded into meal on metates or in mortars, and made into a gruel to be eaten, but the latter group of wild foods was of importance only in food-short seasons or in famine years.[14]

Men hunted and fished, generally on a casual basis, when they were not busy with clearing, planting, and harvesting. The gathering of wild plants and the rather laborious task of food preparation, especially grinding, kept women busy more or less constantly. Man's work, however, was sporadic. The activities associated with crop cultivation would not involve much more than four months of any year. But work had to be intense and rapid, especially planting, since the drying out of soil would result in crop failure. The other principal male activity was warfare, classically a combination of long waits and short violent bursts of activity.

Housing was simple; for most of the year the Indians lived under shades formed by placing arrowweed stalks on a pole framework. A half-pithouse dug into a sandy ridge and roofed with arrowweeds formed a warmer dwelling for several families in the coldest part of winter. Though these shelters might be used for several years in a row, they were easy to construct. If a house was destroyed in a flood the principal loss was any stored food that could not be carried to safety. A family would also readily abandon its house and move to a new planting site. Clothing was simple and, like the tools used, required little effort on the part of native society.

The sources of Indian subsistence in terms of gross caloric intake may be estimated as follows:

Cultivated crops	40%
Semicultivated grasses and herbs	10%
Wild plants	40%
Fish	7%
Game	3%
	100%

In years of crop failure the proportion of food derived from cultivated plants would decline, perhaps almost to nothing, and in good crop years the wild plant foods and game would be of less importance.

Population

Oñate estimated in 1605 that there were 22,000 Indians living along the Colorado River between the mouth of the Gila and the Gulf. About half of them lived at the head of the delta, and may be identified with the powerful Quechan tribe. Later reports suggest reduced numbers, probably a result of really bloody intertribal warfare and the diffusion of European diseases such as smallpox, measles, and perhaps syphilis from the Sonora settlements. These highly infectious diseases could easily pass from tribe to tribe, even those having no direct contact with Europeans. In 1776 Garcés estimated that there were 3,000 Quechans, or Yumas. Data available from U.S. Army reports in the 1855 indicate that the total Quechan population was about 1,700, and it was significantly reduced by the disastrous battle with the Pimas and Maricopas in 1857. The Indian Service censuses of 1905 and 1910 recorded populations of 900 and 834 respectively.

Since that time the decline in population, the latter part of which is clearly the result of disease and dissolution associated with contact with the white community, has been stemmed. The number of Indians on the "Reservation" rolls is a little over 1,200. Some individuals are of largely white ancestry, but these may be balanced against those who have moved to large cities and severed their tribal connections.

Even for the largest numbers cited above there was no shortage of arable land or food in seasons of normal flooding. One cultivated acre would readily feed a family, and perhaps 5,000 acres were flooded and subject to cultivation in the head of the delta district in a normal flood season.

The Development of the Modern Landscape

Once an American garrison was firmly established at Fort Yuma a slow but steady acculturative pressure developed, involving the Indians ever more deeply in the American economy. A steamship service up the Colorado from the Gulf of California was developed, first to supply the garrison and later to bring provisions to mining camps served by such river ports as La Paz and Ehrenburg farther upriver. The shifting channels and great variability of flow made navigation on the Colorado River extraordinarily difficult and dangerous, so ships were often grounded and lost. Remnants of their hulks and boilers are still to be found in long-abandoned channels on the lower delta.[15] But the desert barrier to overland transit to the east and west continued to make the river route from the Gulf useful until after the Southern Pacific Railroad reached the river from the west in 1877. In 1880 rail connections with the east across southern Arizona were completed. As late as 1895 freight was still hauled by steamboat to points along the Colorado not served by rail.

The ferry crossing, river port, Territorial Prison, and some other frontier government services came to be centered at Colorado City, later called Yuma. The bulk of the Quechan Indians continued to practice flood farming, residing in villages on the west or California side of the river from 10 miles above to 10 miles below Fort Yuma, the latter locality extending into Mexican territory. Some income with which to purchase manufactured goods could be obtained from selling cordwood used as fuel by the river steamers, cutting wild hay, selling vegetables and melons, and performing casual labor in the town of Yuma. The Indians did not give up farming, but the activity certainly became less important. At the time of railroad construction Indian labor was in demand, as were the personal services of uninhibited Indian women.

Contemporary observers speak of a degraded Indian population living in a marginal relation to the town of Yuma, racked by indolence, alcoholism, and venereal disease. Subsequent events showed the picture to be somewhat less than accurate. Though part of the Indian community did come to rely on scraps and handouts from the town, a substantial though less visible part continued to seek its own support through the traditional means, aided by a few metal tools they had acquired.

There were some difficulties with American squatters on the floodplain, but there were too many Indians present for ranchers to be able to drive them out and use the floodplain for grazing cattle without military aid. Right at a military headquarters such aid was not to be afforded, and the Indians continued to live and farm on the Fort Yuma Military Reservation. In 1883 the area was converted into an Indian Reservation by Executive Order of the U.S. President except for a few spots to which squatters had established title, and in 1884 another Executive Order expanded the Reservation into an area of approximately 40,000 acres (Figure 27). About half the reservation lay in the floodplain north and west of the Colorado, and the remainder consisted of wasteland on the mesa to the north.

In Arizona south of the Colorado and west and south of Yuma the land was surveyed and sold or conveyed to agricultural homesteaders who, by diverting the river, developed canal irrigation. The Quechan Indians had neither the organizational and technological skill nor the capital for such activity. But by continuing flood farming, hunting, and collecting, and with casual work in the town or on the railroad they maintained a degree of economic independence, associated with but not completely dependent on the town of Yuma.

A Roman Catholic Mission of the Sisters of St. Joseph, under contract to the U.S. Government, also operated an Indian School on the site of Fort Yuma. The town of Yuma developed as a strategic transportation center serving the miners, ranchers, and incipient agriculturalists in a large area of southwestern Arizona and southeastern California. In the late 19th century its population was composed of about equal part of Anglos (English-speaking immigrants) and Spanish-speaking immigrants from New Mexico, southern Arizona, and Mexico. The latter, in general totally dependent on wages, were slightly preferred as unskilled laborers, since the Indians always had some option of quitting and returning to the Reservation.

The potentialities of much greater, controlled irrigation projects on the Colorado, however, were recognized. In 1893 the U.S. Indian Commissioners worked out an agreement with the Yuma Indians in California (perhaps better identified as the Quechans), which was signed by most adult male members of the tribe, whereby the Indians would relinquish their rights to the Reservation and would receive in exchange a five-acre tract of irrigable land for each man, woman, and child enrolled in the tribe. Although individually owned and subject to inheritance, these tracts could not be sold. The money derived from sale of the remainder of the Reservation would be held in trust by the U.S. government for the purpose of paying the costs of irrigating the Indian lands once the major works had been completed by private investors who were expected to buy the irrigable land of the rest of the Reservation.

Nothing happened for more than a decade. The task of controlling the Colorado River at the head of its delta was too big a project for private investors to undertake. Only after Reclamation Act of 1902, amended in 1904, which authorized expenditure of federal funds for irrigating arid lands with the costs gradually to be repaid by the beneficiaries, was serious work initiated. Then the Laguna Dam, right at the head of the delta at the northeast corner of the former Indian Reservation, was started in 1905 and completed in 1909 (Figure 28). It should be noted that this first dam on the Colorado River enabled the river level to be raised a few feet so that irrigation water could be diverted at times of low water. But the dam had virtually no storage reservoir, so it afforded no protection against floods. Canals were dug immediately afterward so that some water could be delivered by 1910, and the basic system of lateral distributaries was essentially complete by 1915. In 1911 the allotments of land to each Quechan Indian were increased to 10 acres.

In the meantime far bigger and more exciting developments were taking place farther down on the delta. On various occasions in the 19th century floodwaters of the Colorado had spilled northward toward the empty basin of the Salton Sink. Attractive pasturelands for cattle in formerly barren wastes were thereby created, and the idea arose of diverting part of the Colorado's water to irrigate fully the vast alluvial slope that was to become the Imperial and Mexicali Valleys. Between 1900 and 1904 some 600 miles of canals were constructed and water was diverted just north of the international boundary, was passed into Mexico to get around the Sandhills, and was then delivered to settlers who were establishing intensive agriculture in Imperial Valley. When the summer floods of 1904 proved to be low, it was feared that the somewhat silted Imperial Canal could not supply all the water that was needed, so a direct opening of the river toward the west was made four miles south of the boundary. In 1905 the entire Colorado River broke through the new opening and flowed

Figure 29. Salton Sea, looking northwest toward Mt. San Jacinto (left) and Mt. San Gorgonio (photo by M. Pasqualetti)

into the Salton Sink. A major disaster followed; the Salton Sea was formed and when in 1907 the Colorado River was finally controlled and turned back to the Gulf its surface had risen from −270 ft. to −197.5 ft.; the Southern Pacific Railroad's main line had to be moved several times to keep it out of the water, and extensive newly irrigated farmlands were drowned. The surface of the Salton Sea then was lowered by evaporation to −249.5 ft. in 1921, but increasingly extensive irrigation, especially in Mexico, has brought its level up to −233 ft. at the present time (Figure 29).

At the head of the delta the Indian allotments took up about half the irrigable floodplain lands of the former reservation. They were selected in the area north and west of Fort Yuma Hill, leaving the eastern sector for sale to white agriculturalists (Figure 27). Laguna Dam also permitted the diversion of water to the east side of the Colorado River on both sides of the lower Gila and west and south of Yuma, and there also land was sold to white immigrants who undertook irrigated agriculture.

Levees were constructed along the still uncontrolled Colorado River to protect the newly irrigated and valuable floodplain lands (Figure 28). In a flood in 1920, however, the big meander that lies to the northeast of Yuma was suddenly cut off and the Colorado and Gila Rivers assumed approximately their present configuration. Nearly

10 square miles of the State of Arizona suddenly appeared on the west, or California, side of the river. A jungle of willows, arrowweed, and mesquite, bordered by old artificial levees and in part by oxbow lake swamps, lay adjacent to the former reservation lands, now irrigated and developed. Cloudy land titles, in an area where it was not certain which state held jurisdiction, and for a time, fear that the river might return to something approximating its old course, have even to the present time impeded full development of what is generally first rate floodplain alluvium.

The erection of Hoover Dam far up the Colorado River in 1934, with the great storage capacity of Lake Mead behind it, has finally brought the river under control, and the Parker, Davis, and Imperial Dams downstream, but still above the delta head, have further regulated the flow. The flow still varies substantially, but it is now controlled in response to needs for power generation and water for irrigation or urban consumption. Major floods are no longer to be feared, and the principal concern now is that there is more demand for Colorado River water than the river can supply in a series of years of subnormal flow. Also the bulk of the detrital load of the river is being impounded in the reservoirs upstream. Deposition of silt during floods is no longer a problem, but the failure of such silt to arrive on the delta may well be creating a new physiographic instability at the Gulf of California margin of the delta.

One other major engineering project has modified the landscape at the head of the Colorado Delta. The All American Canal was begun in 1934 and effectively completed in 1940, though distribution lines in Imperial and Coachella Valleys were not finished until after the Second World War. It takes water from the Colorado at Imperial Dam, just above the old Laguna diversion dam, desilts it, runs it along the edge of the mesa north of the floodplain, and cuts through the sandhills north of the international boundary, conserving elevation to be able to pass water right around to the north end of the Salton Sea. A good deal of cut and fill was involved in setting a 200-ft. wide canal into the edge of the mesa, but economies of earth movement made it appropriate to follow the mesa edge as closely as possible. Bridges have been constructed across the canal to carry over it the detritus-laden water that occasionally flows off the mesa during flash floods.

Modern Agricultural Developments

After the Fort Yuma Indian Reservation was abandoned and part of its good land allotted to the Indians, the remainder of the irrigable land was sold to American settlers in 40-acre plots. Similar sales were effected on the east side of the Colorado River both north and south of the Gila east of Yuma, and in the area west and south of Yuma. As the canal system and its laterals were completed, land was cleared, leveled, and taken under cultivation. Very sandy localities and former marshes and lakes where drainage was too poor and soil alkalinity was too high have been left as little jungles. These are still quite extensive in the cutoff meander on the north side of the river.

Water charges are imposed by the Bureau of Reclamation at such a level as to permit the slow amortization of the costs, including maintenance, of the facilities for providing and distributing water. Accordingly the cost of water is lower at the head of the Colorado Delta than in most places to which water from that river is taken for irrigation. At present, costs for the major dams and canals have been repaid. Water can be used as freely as the crop can use it, and since pump maintenance is the only cost, the more that is used the lower is the cost per acre foot. Alfalfa uses more than 4 ft. per year. Limitations on water use are enforced largely by drainage difficulties at certain points.

The white settlers from the first practiced commercial agriculture with cotton, alfalfa for hay and seed, barley, watermelons, and cantaloupes, as the principal crops. Specialized out-of-season production of such vegetables as lettuce and peas, taking advantage of an almost complete freedom from frost, has not been so important as in the Salt River Valley to the east and the Imperial Valley to the west. The isolation and relatively small size of the Yuma district seems to put the area at some disadvantage for the marketing of perishable vegetable crops. In the past 15 years, increased mechanization has tended to increase the minimum size of an efficient farming unit to about 80 acres, and the operating farms have been expanded by purchases or leases of land to bring them to at least that size. The typical farm operator lives in Yuma, keeps his mechanized farm equipment there, or on one of the scattered 10- to 40-acre plots that he works, and moves it from plot to plot as he needs it. Extra labor will be needed on such farms seasonally, but the principal crops make only relatively small demands on labor, by contrast with the fresh vegetables grown in the Imperial Valley or the Salt River Valley.

Cotton is regarded as the crop yielding the greatest net return per acre, but there is a government limitation to the number of acres that can be planted. This amount nowadays is only a fraction of what any farmer planted 10 years earlier. Alfalfa, which gives several cuttings per year, and barley are grown in the rotation, so that a given piece of land is seldom fallow. Twenty-five years ago a considerable amount of land was devoted to Bermuda grass, which was topped for seed once a year and at other times used as permanent pasture for cattle, but by now more valuable crops have displaced almost all of it. Fertilization is heavy and yields have been pushed to very high levels. Cotton, for example, averages about two bales of lint (over 1,000 lb.) per acre, with the seed weighing about twice that much. Floodplain agriculture in the Yuma region can be characterized as capital-intensive and highly productive.

Development of the lands allotted to Indians, although they were of excellent quality, tended to be somewhat slower, primarily because the Indians lacked the capital needed for clearing and leveling their lands. An improvement lease program was worked out by the Indian Service whereby the allotments were developed by white lessees. Of the 8,110 acres held by Indians, only 136 were cultivated in 1913, but by 1920 the cultivated area had increased to 7,820 acres. Of these, 6,000 acres were held by lessees. As the leases expired the Indians resumed possession of the land and by 1924 only 1,587 acres were being worked by lessees.

The next 15 years saw a steady decline in the agricultural use of the allotted Indian lands, a decline accentuated by the onset of the Depression. By 1932 only 2,991 acres were being cultivated, 1,106 by lessees and the remainder by Indians. By 1940 the area under cultivation had fallen to 2,246 acres, of which 300 were double-cropped. At that date Indians cultivated 1,243 acres and lessees 1,003 acres. Commercial crops, alfalfa and cotton, accounted for nearly all the leased acreage, but about half of the land worked by the Indians was devoted either to subsistence crops such as maize, beans, and garden crops or to pasture.

When the allotments were made to the Indians it was expected that a four-member family would pool its land into a farm of 40 acres, which at the time was at least as much as it could cultivate. Such a farm would then have been a viable economic unit. Only a declining fraction of the Indian families chose to make farming their primary occupation. Also, in the more than 50 years since the allotments were made, equal share inheritance to the descendants of allottees has complicated ownership patterns. Records are carefully kept by the Indian Agency at Fort Yuma and the ownership of some plots is shared by as many as 32 individuals, not all of them holding equal shares. Many individuals own some share in a large number of separate plots. This ownership complexity is destined to increase in geometric progression as time goes on, even though the population is increasing only very slowly, since there is no provision for sale of shares to simplify ownership patterns. Thus the inheritance rules and prohibition of land sales enforced by the Indian Agency are rapidly destroying any sensible pattern of individual land ownership and rapidly returning the allotted plots to an ownership pattern in which they can only be dealt with if treated as subject to common tribal ownership.

Stimulated by the demand for crops in the Second World War, and even more so by the Korean War and subsequent price supports, almost all of the Indian-owned land has been taken back into cultivation. By 1955, 4,085 acres were cultivated, and at present almost all the allotted lands are being farmed. A major change is that the Indian owners have effectively given up farming. Only 466 acres were farmed by Indians in 1955, and at present no Indian is attempting to support himself by farming, though a few have kitchen gardens at their homesites.[16] The land is rented to commercial farm operators who seek to work at least 80 acres. The bulk of the Quechan tribe members who remain in the Yuma area have constructed their scattered houses in the area immediately north of the Indian School. The good allotted land is rented for as much as $50 per acre per annum. The Indian Service arranges the leases and undertakes the complex bookkeeping to see that payments are divided among the multiple owners of each plot. The multiplicity of inherited ownerships results in a fairly even distribution of the rental income.

Relatively few of the resident Indians work as farm laborers for the farmers renting their land. Mexican immigrants or braceros (contract laborers from Mexico) provide the needed extra harvest and planting labor on the Indian-owned lands as well as in other irrigated districts at the head of the delta. With their powerful physiques and high energy output, Quechan Indians are regarded as excellent workmen, but they are also considered unreliable, being willing to quit or lay off with the crop half planted or harvested as soon as they get paid. The discipline of a town job or day labor fits them better, and many individuals have semiskilled jobs in Yuma. In addition to a modest income from rental of land each family has an inalienable homesite. In a climate which does not require elaborate housing, each of the approximately 1,000 Indians has at least minimal economic security. Wage labor, if available and obtained, affords comforts, luxuries, and liquor.

Urban Developments

The city of Yuma on its mesa on the Arizona side of the river, has recently experienced vigorous growth. From 1950 to 1960 it grew from about 9,000 to 24,000 people. It serves as a market town for all the agricultural districts at the head of the delta and for the newly cultivated areas some distance up the Gila River near Wellton. The city also has some resort, retirement, and recreational attractiveness, especially in winter. While Yuma is still an important point on the Southern Pacific Railway's main line to southern California, Highway 80 from the east to southern California has become the most important element in the still vital transportation nexus of the Yuma crossing. On the most southerly and one of the lowest transcontinental routes to the Pacific Coast, Highway 80 is the preferred route for both truck and automobile traffic in winter as there is no problem of the roads over the continental divide to the east being blocked by snow. A reasonable day's drive from the West Coast cities, Yuma supports a vast number of motels, restaurants, gas stations, and other services to both the motorist and the truck driver.

Most of its population is composed of recent immigrants. Many are sunseeking settlers from the east, but nearly half are Mexican and Negro workers who can get employment in agriculture or the service industries in town. The Quechan Indians, while maintaining their numbers and identity on their allotted lands on the California side of the river, have thus become a small minority in the lands at the head of the delta, where they were once the exclusive occupants.

Current Instabilities and Prospects

Physical Changes

The dams on the Colorado River have for 30 years almost completely eliminated the deposition of silt on the Colorado Delta. There is good evidence for isostatic subsidence of the whole head of the Gulf of California while the silt was being deposited, just about balanced by alluvial buildup of the surface. Only an elevation of about 30 ft. divides the Gulf with its great tidal range from the extensive valuable areas below sea level in Imperial Valley. Further subsidence, erosion of the southern margin of the delta by the tidal bore, and the possibility of major movement along the great and active fault lines that underlie the delta pose imponderable threats to Imperial Valley. Nothing may happen, but there is no assurance that there exist engineering means which could prevent the waters of the Gulf of California from invading the whole region should major earth movements occur fairly suddenly. The head of the delta near Yuma is more than 100 ft. above sea level and is reasonably safe from such physical disturbances.

Economic Problems

Those who currently use the water of the Colorado have already laid claim to more water for irrigation and industrial and domestic use than the river can supply in a normal series of dry years. Especially critical is the situation in southern California, where about 10 million people are dependent on the Colorado River for part of their water supply. By Supreme Court decision California's right to all the water needed in time of shortage has been denied. Vast plans are under way to develop other sources of water for the Southwest. The Feather River [California Water] Project is under construction and in six or seven years will carry water to southern California from the northern part of the state, increasing costs but relieving the pressure on the Colorado. But population and water consumption in the southwestern United States continue to increase in explosive fashion, and there is talk of bringing in water from as far away as the Columbia and Snake Rivers. Such a program will require decades for its execution. The important question is whether, in the event of a succession of years of low flow in the Colorado, it would be politically feasible to deny domestic water to millions of people in southern California in order to honor the water rights of a few thousand farmers irrigating crops along the Colorado River and its delta. The position of the farmers, in this eventuality, is weakened by the fact that, at least at present, the commodities they produce are in oversupply and are contributing to the extremely costly farm surplus.

By treaty the United States has guaranteed that 1.5 million acre feet of Colorado River water will be allowed to reach the Mexican part of the delta every year. There has been a steady expansion of cultivation in this area, and it is clear that there is more first-class land subject to irrigation than the 1.5 million acre feet will adequately irrigate. In contrast to the United States, Mexico, with its enormously growing population, has no agricultural surplus problem, and the produce of its part of the Colorado Delta is of critical importance to the national economy. The city of Mexicali itself has rapidly grown to nearly 200,000 people. In wet years, when more than the minimal 1.5 million acre feet of Colorado River water reaches the border, lands are taken under cultivation that could not be irrigated if Mexico received only its treaty allocation. Further, minimal irrigation is carried on for such crops as cotton which have only modest water demands. This maximizes the area that can be irrigated. Unfortunately, however, there is little flushing of the soluble salts introduced by salt-rich Colorado River water, which therefore remain at the surface when the water evaporates.

In the last few years an acute problem of crop failure has arisen due to the excessive salinity of irrigation water. Its immediate cause is an expansion of cultivation in the Wellton area of the Gila Valley, based on the exploitation of almost brackish well water. To produce crops with this water it is necessary to overirrigate, thereby preventing salt accumulation in the fields where it is applied. The runoff enters the Gila River and thence the Colorado just before it enters Mexico. There seems little question but that the quality of the water reaching Mexico has deteriorated markedly in the last few years. To dilute the salt solution entering the Colorado from the Gila to a level characteristic of natural conditions would require more than a doubling of the amount of water allowed to reach Mexico. But the flow of the Colorado is already overcommitted upstream. The prohibition of irrigation by well water along the Gila is probably the most reasonable means of restoring the quality of the Colorado River water received by Mexico, but the political effects of telling farmers that they must abandon their land, and the towns that serve them that their business will be destroyed, is a problem that the United States will not yet face. There is even talk of building a 75-mile-long cement-lined conduit from the mouth of the Gila River to the Gulf of California thereby diverting the saline Gila waters from the Colorado. The economic absurdity of such a program is emphasized by the fact that the cotton produced in the Wellton district is at present a surplus crop requiring price support. Further, the groundwater which causes the trouble is being exploited at a rate far in excess of the possibilities of replenishment, so the whole agricultural development based on this groundwater can continue only for a few years.

In the 20th century the dominant element in the use of the head of the Colorado Delta has been the fuller exploitation of Colorado River water for irrigated agriculture. As compared with aboriginal times a more than hundredfold increase in production has been achieved, and the wild floodplain landscape transformed into an almost completely cultivated one. It seems clear that locally this development has reached, if not exceeded, rational limits. Further developments at the Yuma Crossing are likely to be associated with its strategic role in transcontinental and interregional transportation, with the recreational opportunities afforded by the controlled and dammed waters of the Colorado River, and with the desert climate, which is generally attractive in winter and easily ameliorated by air-conditioning in summer. Such developments will depend on sustained prosperity and high-level energy consumption in the American economy.

NATIVE

ENVIRONMENTS

Introduction

George F. Carter

Hᴏᴡ ᴏɴᴇ ᴍᴇᴀsᴜʀᴇs a fellow geographer must depend in part on one's definition of geography. For me this is very simple. Geography is that subject that asks questions about the relationship of human beings to the land they live on. It is distinguished from history with its focus on human actions at specific times, from sociology with its emphasis on human organization, and from anthropology with its emphasis on human learned behavior, or culture. It shares a common body of knowledge with these allied humanities, but it differs in its emphasis on the human-land relationship. It also differs in that this emphasis on the land requires the master geographer to be at home not only in the humanities but also in the earth sciences. It is a relatively rare geographer who is truly masterful in so broad a discipline. It is from this background that a few of Homer Aschmann's papers on native environments are considered.

Aschmann was particularly interested in the relationships of native peoples with their environments—as we have earlier seen in his publications on Baja California, the Guajira Peninsula, the Canaries, and Chile—but much of his work on the subject concentrated in the American Southwest. His work there reflects the depth that sustained interest develops. It is axiomatic that human occupance of the land must be seen in some time-depth, for every people inherits a landscape long occupied and to some extent modified by the prior inhabitants. Equally, each new occupant group to some extent learns from its predecessors. Aschmann's work contains not only this attention to depth and continuity of cultural time but deep knowledge, both of his chosen area and of similarities and differences on a worldwide scale.

In discussing human use of fire and its impact on the evolution of landscapes, for instance, Aschmann moves knowledgeably through the time period spanning the Paleolithic to the Historic, and he moves with equal surefootedness throughout the world, often focusing on the lands of the Mediterranean climates. This allows him to compare the impacts of differing cultures in somewhat similar environments over the course of a vast amount of time. His conclusion that human beings have greatly modified their environment through the use of fire is very much in the tradition of George Perkins Marsh, Carl Sauer, and Omer Stewart, scholars who were among the most insistent on the importance of fire. Nor does he fail to comment on practical modern applications of this knowledge, noting gently that Southern Californians would do well to fire-manage their environment to avoid the enormously destructive fires that occur with dismal frequency due to the accumulation of combustible plant growth. Native Australians did better.

Penetrating the thinking of those who precede us often requires that we are able to read the original languages. For study of the Southwest this requires ability to read Spanish, even colonial Spanish. Aschmann was at home with Spanish and not only read in this language but went so far as to translate an original Spanish document, supplementing the translation with notes from his own experience in the region under discussion. Found in the first section of this volume, this translation of the Vizcaíno expedition up the western coast of America is rich in insights into the missionary's view of Baja California. Those who love the desert love Baja California, but in the main it is a desolate and forbidding land, a quality illustrated in the other selected examples of Baja California papers. One "fertile plain," so described by the missionary, Aschmann characterized as one of the most desolate parts of the earth. However, to the missionary, all land was good land that had souls to be saved, and this zeal led eventually to an enormous effort to do something with this barren land. It was in spite of the physical geography, not because of it, that this effort occurred, and this is beautifully conveyed in the translation.

Little now remains of the missionary effort in Baja. Some churches are in ruins. How would one understand how this occurred without looking into human motivations? And how would one get at an understanding of human motivations without facility in the native language and the drive to examine the relationship of people with the land? One could well say from our momentary view of things that neither minerals nor good arable land led the Spanish to pioneer this land. Nevertheless, a motivation existed and can be determined and understood, but only after study in depth of time, language, and culture.

Aschmann's command of facts and breadth of training led him at times to be impatient with the work of others. In a review of a study of the Yumas of the lower Colorado River area, he showed skepticism at an author's attempt to see things from the Indians' side. At one point the author uses his understanding of the oral histories to locate places with precision on the featureless and shifting Colorado River delta, an improbable thing to do successfully. Unquestioning acceptance of Indian tribal legends, especially of origin legends, receives Aschmann's stringent criticism, as well it might. The recently arrived Athapaskans in the Southwest all claim to have originated there, as Aschmann points out in his paper "Athapaskan Expansion in the Southwest," reprinted elsewhere in this volume. The author, Aschmann finds, relied too much on accounts of informants from outside the cultures in question.

He also admonishes the author for simply assuming that the Indians are like him and would "react as does he to comparable stimuli—the very mark of Indianness is that the stimuli are perceived through a distinctive cultural screen the appreciation of which will involve long study and subtle appreciation."[1]

That final sentence contains the essence of the geographers' view. People everywhere have to be understood in their own terms. It is their value system, not the observer's, that determines what use they will make of their land. The Yumans in question lived in a potential Egypt: an annually flooded riverine floodplain environment in a subtropical climate with every needed resource to replicate Egypt. They did not choose to go the route of Pharaohs and pyramids.

With his background of long experience in the Southwest and extensive reading of its history, ethnology, and archaeology, combined with sound training in the earth sciences, Aschmann was drawn into the Indian Land Claims that make up the bulk of the present section of this book. The federal government waived all statutes of limitations and invited the American Indians to present their claims to their lost lands. The government hired scholars to examine the records to determine who was where at the time the federal government took over the land, and Aschmann participated in the task, writing what amounted to two short regional geographies, "Environment and Ecology in the 'Northern Tonto' Claim Area," and "Terrain and Ecological Conditions in the Western Apache Range."

The specialty of regional geography focuses on developing a comprehensive understanding of the many factors at work in shaping landscapes, and it sometimes receives the facile condemnation of being a mere recitation of an outline, beginning with the very earliest geological epoch. The implication is that regional studies amount to little more than filling in a checklist with the headings geology, geomorphology, soils, vegetation, and human inhabitants. Aschmann's two reports suggest the exact opposite. They demonstrate that a good regional study deals with what is significant in the environment for the understanding of the people occupying that environment and skips all else. It is of interest then to see how Aschmann proceeds in his description of the landscape as occupied by various American Indians.

Aschmann deals knowledgeably with the Yavapai (old occupants) and the Apache (recent intruders). He sorts through the complex of Apache, Tontos, Apache-Mojaves, Yavapai, and others, using Spanish documents and early army accounts (though keeping in mind that the army tended to call all these groups Apache). Aschmann's description of the environment concentrates on the significant features: the pertinent climatic statistics, and vegetation. Where geology is significant for explanation of land forms and water distribution, it is introduced. This is no checklist, but a meaningful presentation—meaningful for understanding the human occupance that is the goal of the report.

Concerning the Western Apache, he comments: "The Mogollon Rim is the dominant terrain feature in relation to which the settlements of the Western Apache Indian bands were located." He does not bother to start with the Precambrian era, but sets the stage in a more recent and relative landscape. After a review of this difficult terrain, Aschmann concludes that the Western Apache were infiltrating into a refuge area. He quickly adds that the area had once been occupied by the sedentary agricultural Mogollon culture and to the south by the Sobaipuri. The Mogollon, he thinks, were long gone, leaving an empty area into which the Apache could move.

The occupation of this land by the Apache, of course, rests on their subsistence economy. The role of agriculture was estimated as supplying 25 percent of the subsistence. Hunting was important, though deer were scarce and most of the meat was supplied by rats, hares, and rabbits. There was a notable food taboo: not only fish but, oddly, all water-related animals were avoided, even though there was a rich fish resource in the Verde River and protein was in short supply. Further, it is hard to know where this taboo originated. The Athapaskans of the northern forests ate fish, but these Southwestern Athapaskans avoided them. Clearly, the relationship of the people to the land is anything but simple and functional; rather, it is heavily determined by culture.

Raiding was less important for the Western Apache than for the Eastern Apache due to the greater distance of the former from the Spanish and Mexican ranches. Warfare, though extensive, was uneconomic and cost heavily in lives. Population density was low: one person per three square miles overall, and only about one per square mile in the areas considered valuable by the Apache.In this case a land rich in pack rats (often as big as squirrels) marked a good land. Today we are interested in pack rats because their nests have proved to be at times of great antiquity and to contain a record of past climates. Despite limited resources, the various bands held loose gathering rights in their territories and normally allowed other bands to gather there also, provided that they asked permission. These insights into some of the essential features that guided occupation of the land by the Apache provide a sampling of Aschmann's approach. Together, such insights comprise a cultural assessment of the land and its resources as determining its occupance.

What emerged from Aschmann's investigations into the Land Claims are little classics, clear descriptions of landscapes and their inhabitants before they started to give way to the Anglo settlers who continue coming to this day. They demonstrate "Man and the Land" in the classic sense, rich in understanding of both the land and the people. The set of papers shows Aschmann as a master geographer, balancing the human and the physical considerations and understanding fully that it is the human outlook that determines the use of the land. Whether dealing with early Spanish contact with the landscape, or with that of the late-arriving Athapaskans, the outlook is firmly fixed on a cultural-historical understanding of the geography.

Aboriginal Use of Fire in Areas of Mediterranean Climate

> Man as a burning agent . . . has affected . . . vegetation through at least
> one full glacial cycle of advance and retreat, some hundreds of thou-
> sands of years. . . . [The] intent was to change completely the vegeta-
> tion of the lands they used.

CONCERN FOR whether, why, and to what extent aboriginal or prehistoric occupants of the several Mediterranean-climate regions burned the wild vegetation has
relevance . . . in two ways. While naturally caused fires do occur in all continental
Mediterranean regions the frequency of burning and thus its intensity based on accumulated fuels would be strongly affected. Secondly the extended time periods of
aboriginal burning would allow plant communities time to evolve as adaptations to
environments in which fire was regular and frequent.

Omer C. Stewart has taken perhaps the strongest position on the universality of
burning as a human culture trait, going back to the earliest times that man controlled
fire and practiced wherever the vegetation was not too wet or too sparse to burn.[1]
The vegetation of Mediterranean-climate regions is consistently inflammable at least
during parts of the year. Stewart's preoccupation with the subject seems to have arisen
from his own fieldwork among the Pomo Indians, who occupied both coastal areas
and interior valleys of the Coast Ranges north of San Francisco Bay.[2] His informants'
allusions to places they burned annually or less frequently suggested to him that the
importance of widespread burning as a cultural technology and a modifier of the society's environment had received inadequate ethnographic attention.[3]

Examining the documentation Stewart collected from all parts of the world, from
classical literature through travelers' accounts to modern ethnographies, makes it
hard to deny his conclusion that deliberate burning of vegetation as well as starting
fires accidentally has been characteristic of human societies.[4] Until extremely recent
times proscribing such action was unusual.

The Temporal Limits of Aboriginal Burning

Termination

[T]he aboriginal period for the lands at the eastern end of the Mediterranean Sea
ended before 2000 B.C. By 1500 B.C. it was over in Greece and its associated islands.
Early in the first millennium B.C. Phoenician and Greek colonists were bringing commerce and cities to Sicily and Southern Italy, Tunisia, Southern France, and the south-

ern and eastern coasts of Iberia. In these western lands, however, the transformation was not complete, though in Italy it shortly became so. In Iberia and the Maghreb, tribes in the interior continued their old patterns of land utilization for many centuries, and not until the full establishment of the Roman Empire at the beginning of the Christian Era could they be considered as other than aboriginal, in parts of Algeria and Morocco not even then. Extending over many centuries, rather than a single one, the spread of the more advanced culture into the western Mediterranean Basin resembles the entry of Europeans into California, beginning in 1769 and continuing into the 1850s. At points of settlement aboriginal behavior was modified drastically but elsewhere it persisted with only a little change. The final replacement of aboriginal behavior patterns, in the Canary Islands, did not occur until the 15th century A.D.

The conquest of most of the Mediterranean-climate areas of Chile late in the 15th century by the Inca Empire brought that region into the realm of Andean Civilization prior to the arrival of the Spaniards in the 1540s.[5] Except in the irrigated valleys of the Norte Chico, however, the native way of life, essentially a neolithic one, does not seem to have been significantly modified by the Inca overlords at the time of the Spanish conquest, about 1545, which can be taken as the end of aboriginal conditions.[6]

In the Cape District of South Africa, there had been a major cultural change in the native occupation in the century or so prior to European contact. Hottentot pastoralists had almost completely displaced the Bushmen hunters and gatherers from the limited area of Mediterranean climate, though the groups were thoroughly mixed in the extensive dry regions of the southwest part of the continent.[7] To the extent that aboriginal burning was deliberate there may have been significantly different patterns practiced by the pastoralists for a century or two as opposed to those of the Bushmen, which were followed for many earlier millennia. Though reached by the Portuguese in the late 15th century effective European contact began with the Dutch provisioning station founded at Table Bay in 1652. Although the Dutch East India Company tried to restrict its territorial outreach the area of Mediterranean climate is so small that within a few decades it was fully under European domination.[8]

European entry into California found an aboriginal population of hunters and gatherers but one which was remarkably large and had to be extremely adept at exploiting the wild vegetation in its environment. After rapid expansion from San Diego in 1769 to San Francisco in 1776 the mission system stabilized in the near-coastal region where, by plan, it obliterated the aboriginal culture. Well over half of the Mediterranean-climate portion of California, however, continued to be occupied by relatively undisturbed Indians until the Gold Rush of the mid-19th century. As a result we have fuller and better records of aboriginal burning practices from California than from any other Mediterranean-climate region of the world.[9] . . .

Even for Australia the effective entry of Europeans into the two regions of Mediterranean climate on the continent was remarkably late, 1827 in Western Australia and 1837 in South Australia. In those areas the land-hungry European pastoralists were in direct competition with the aborigines and had effectively destroyed them culturally before much interest in the aboriginal ecology and way of life arose. The abundant ethnographic data on Australian Aborigines that we have, in which extensive burning for several purposes figures prominently, come largely from the Center and North, areas with strikingly different climates.

Beginnings

Abbevillian and Acheulian archaeological materials found at many places around the Mediterranean Basin demonstrate that man has inhabited the region at least since mid-Pleistocene times. His experience included several glacial advances and retreats in areas farther north with parallel displacements of weather and vegetation belts around the Basin. Protected by its northern mountain rim the area probably experienced only moderate temperature depression, but more southerly storm tracks notably increased precipitation and reduced its seasonality with expectable effects on vegetation distribution.[10] Whether man's ability to enter and occupy the Mediterranean Basin at all depended on his control of fire is less than certain, but there is scarcely any doubt that he controlled it during the last interglacial. Man as a burning agent then has affected the region's vegetation through at least one full glacial cycle of advance and retreat, some hundreds of thousands of years.

The southern tip of Africa may well have had a human population even longer than the Mediterranean Basin. The question of whether he controlled fire for so long a time is more vexed. Nonhuman hominid populations that probably did not control fire evidently survived into fairly recent times, and except in high areas fire is not needed for survival. Elaborate fire-making traditions among the Bushmen, with their generally simple culture, suggest that the culture trait is ancient among them.[11]

Recent archaeological work in Australia seems to confirm the presence of man in southern parts of the continent 35,000 years ago, full glacial times when, because of lowered sea level, he could walk from New Guinea. There is little doubt that he was in control of fire from the beginning of this period.

The hotly debated question of whether man's entry into the New World was immediately postglacial, perhaps 12,000 years ago, or much older is relevant but too complex to confront here fairly. My own inclination is to regard the arrival as much earlier, at least 35,000 years ago.

Firm radiocarbon dates from a definite archaeological site at San Vicente Tagua-Tagua in the Mediterranean-climate region south of Santiago, Chile make it clear that he had occupied both continents by more than 10,000 years ago.[12] Further, since passage through Northeast Asia and Alaska would be impossible without control of fire, he is certain to have been able to burn through his entire sojourn in the Americas.

Madeira

There is one area of Mediterranean climate in the world that has no prehistory. The island of Madeira was discovered by Portuguese explorers early in the 15th century. Both its name and early descriptions indicate that it was enormously heavily wooded with great evergreen trees similar to but larger than those in the surviving patches of *laurisilva* or *monteverde* of the higher Canary Islands. Shortly after the first settlement in 1420 fires were started to clear lands for planting. They got away and burned uncontrollably, forcing settlers into the ocean for their safety. Some early sources say the fire burned for seven years; others offer a more probable six months, and it is possible that fire clearing of most of the island went on another six years.[13] In any event a permanent change in the vegetation of the island was effected.

In its oceanic position Madeira experiences lightning infrequently and only in connection with winter frontal storms when it is thoroughly soaked. It is likely that, as the chroniclers report, millennial trees and an enormous amount of deadwood offered fuel for an almost unheard-of conflagration.

The discoverers' accounts seem to indicate clearly that the forests came to the water's edge. Today the lowest 200 meters of the island carry a thorny or succulent vegetation of desertic aspect. Extensive relatively level lands at the crest of the island above 1,000 meters are a grassland, grazed almost to a lawn by sheep. Excavations directed toward developing an airport on the high marshy flats of Paul da Serra exposed charcoaled fragments of tree trunks half a meter in diameter, representing a forest matched nowhere on the island today.

Because lightning-caused fires do occur in most continental Mediterranean-climate areas one would not expect to match Madeira's pristine forests in many places were humans not present. But their survival into historic times does suggest that without man vegetative patterns quite different from those we know could exist.

Ethnographically Recorded Aboriginal Burning Patterns

Where inflammable materials are present no modern society and probably no aboriginal society that used fire could avoid accidental fires started by campfire embers, torches, or slow matches being carried between campsites. The long dry summers, capped in all Mediterranean-climate regions except Chile by strong and unusually warm easterly winds, Santa Anas in California, make those regions especially vulnerable. An accidentally escaping fire per camp per year might be a reasonable estimate, not a particularly great incidence compared to the scores of lightning-started fires in a single California mountain range on a day of scattered thundershowers. Lightning, however, occurs only under conditions of high humidity and is generally accompanied by some rain, and most fires caused by it smoulder out. A camp or cooking fire is most likely to escape under the worst conditions with tinder-dry vegetation and strong winds. Because of the palpable danger of brush or chaparral fire under those conditions one would expect special caution at camps in brushy areas, less in grassy ones.

Accidental fires would then be proportionate in frequency to the population density or the number of villages or camps. Most would not spread significantly but a few would travel over vast areas. My guess, and it could not be other than that, is that in typical areas of Mediterranean climate accidental fires caused by aboriginal man would double the frequency of burning of a given area over what would result from naturally caused fires.

There are abundant reports of the vegetation being ignited deliberately for a variety of reasons. Perhaps dividing those reasons into immediate effect and deferred effect sets is useful. In the former set of fire drives for hunting both large and small animals are the most widely reported. Ethnographic references to 33 tribes of Indians in Central and Northern California using fire to drive game have been uncovered, and they may be incomplete.[14] There are similar references in other parts of the world, though they are less abundant. A fire drive in dry grass set in late summer or fall would be easiest to start and less risky to the burners than one in chaparral. In areas with adequate precipitation such drives could take place annually; in drier areas a fire might carry only after a wet season. Burning a definable, limited area would be efficient, perhaps a grass opening of at most a few hundred acres that could effectively be harvested annually. A wildfire that burned many square miles would tend to denude a tribe's territory of game for many years.

There is a circumstantial account of the Australian aborigines using fire to aid in attacks on John McDouall Stuart, the first European to enter their part of Central Australia,[15] and the general uniformity of material culture in Australia would suggest that the technique was known throughout the continent. The use of fire for both attack and defense is known around the Mediterranean Basin from early historic times. Generally speaking such use was a desperation measure, and only occasionally would both the physical and strategic situation be suitable. An incident might be remembered for generations.

Recreational burning is less well documented, and there were undoubtedly tribal taboos against it, though I doubt that they were universal. My own experiences with guides in Northern Colombia and Baja California involved backwoodsmen but not aborigines. In Colombia lighting fires in brush country was considered the work of a good citizen. In Baja California the shag of a *Washingtonia* palm was ignited for fun, and on the windy evening it was only by chance that the whole grove did not burn.

The number of instances in which California Indian informants said that they burned grass or brush to manipulate the environment is remarkable, particularly since most ethnographers were not trying to elicit that sort of information. Lewis mentions 35 tribes in Central and Northern California who said they burned after harvesting seeds to increase the next year's yield of grass seeds and edible tubers, i.e., in late summer or fall.[16] Planting tobacco on burned-over ground is reported by 22 otherwise nonagricultural tribes. Bean and Lawton speak of this regular burning of specific valleys and meadows as semiagriculture, supporting their position by noting the unusually large populations maintained by the nonfarming California Indians.[17] The Australians for whom tubers played an important dietary role could well have had a like practice, since burning normally favors such plants, but there is little ethnographic documentation.

In two ways burning may increase the future hunting potential of a locality. Maintaining edge features between grassland and brush or woodland produces maximally favored habitats for game. This moderately sophisticated wildlife manager's concept is not often reported. Perhaps the aborigines did not have it or the ethnographers did not recognize it when told about it. Resprouting can occur in brush and bunch grasses even before the beginning of the rainy season, and the tender shoots are attractive forage especially late in the dry season when other leaves are dry and hard.

Finally there are reports of burning brush in forests and woodland, particularly in Northern California, to make passage easier and to increase the visibility of game for hunting purposes. In deliberate burning for environmental manipulation the frequency could easily rise to an annual level. Not enough fuel would be accumulated to produce a wild conflagration except under dry windy conditions and there are indications that the California Indians avoided setting fires at those times. Thus a grassland would burn to the woodland edge but not beyond, or a floor fire would run through a forest without crowning.

Burning grass to obtain better pasture for their livestock is widely reported for the Hottentots.[18] It probably was also done around the Mediterranean Basin from Neolithic times to the era of the *Mesta* in Medieval and Early Modern Spain. In areas of Mediterranean climate the general timing of such burns would not be much different from that of hunters and gatherers, but the pastoralists would have reason to burn more extensively. Selecting weather conditions conducive to a maximum conflagration would seem profitable and a wildfire that burned brushland or grassland would be beneficial, at least for a few seasons.[19]

Finally there is the burning for clearing and planting carried on by Neolithic cultivators practicing shifting slash-and-burn farming. With rapidly rotated fields such a population would have heavy land demands, and their intent was to change completely the vegetation of the lands they used, at least temporarily. Such practices were carried on around the Mediterranean Basin for prehistoric millennia and for some centuries before the Spanish Conquest in Central Chile. Some recognition of differing soil qualities for planting were probably recognized very early so that favored areas would be returned to, but the big difference in localization from other types of deliberate burning was that woodlands and brushlands were favored over grasslands. In fact the abandonment of a field was normally induced by the problem of grassy or herbaceous weeds. It would not be returned to for a second cycle of cultivation until brush had pretty well displaced the grass. Only in the limited irrigable lowlands would weeding be practiced and the land kept in permanent cultivation.

The cultivators around the Mediterranean Basin also had livestock, and a conflict of interest developed between the pastoralists who would regularly burn the hillsides to maintain a grass cover, and the cultivators who could see abandoned fields returning to brush as potential reserve farmland. In Libya, Morocco, Sicily, and Spain the pastoralists, for complex economic, social, and even military reasons, long tended to win the argument, effectively advancing the steppe against the sown.[20] In fact the Mesta in Spain effectively maintained regularly burned open-range grazing in what might be considered a rurally overpopulated land until the 18th century.[21]

Special Regional Considerations

Mediterranean Basin

The ecology and way of life of the hunters and gatherers who lived around the Mediterranean Sea for hundreds of thousands of years from Lower Paleolithic Abbevillean to the postglacial period just before the invention of agriculture and the beginning of the Neolithic are remarkably obscure. Postglacial rises in sea level that covered coastal sites, the heavy erosion and alluviation characteristic of Mediterranean climates and accentuated by the mountainous character of most of the Mediterranean Rim, and the activities of larger, more technically competent populations who disturbed the most favored settlement areas for thousands of subsequent years may account for the thin paleolithic record. . . .

The Neolithic must have advanced along both shores of the Mediterranean Sea to reach the Atlantic by 3000 B.C., and shortly even beyond to the Canary Islands. Our detailed knowledge of what was happening to the natural vegetation comes largely from considerably later biblical and Homeric sources. Both describe brushland wildfires in vivid terms. References to deforestation by Plato in historic times and the rising prices and more distant sources of lumber in Roman times suggest that an equilibrium had been reached in the Neolithic and Early Bronze Age that retained extensive forests on steep slopes and highlands.[22] Growing urban and maritime markets for timber as the Classical Age advanced led to forest cutting then burning that left grassy pastures or *garrigue* brushland on what would have been forest, but this is a historic rather than aboriginal product.

An intriguing problem is presented by the open oak woodlands of Southern Iberia and France, where swine are today pastured on grass and fattened on acorns in season. Swine are an ancient domesticate in this region,[23] and while today the oak woodlands are kept open by grubbing out the brush, the open woodland association might have been established and maintained by frequent burning in prehistoric times.[24] The cork oak native to the area is a striking fire adaptation. Lands south and east of the Mediterranean Sea do not maintain the association because of disinterest in pigs since the rise of Judaism and, more significantly, Islam.

Chile

Two farming systems were practiced by the Chilean Indians at the time of conquest. In the valleys of the Norte Chico from Aconcagua northward canal irrigation and sedentary farming not unlike that of the Peruvian coastal valleys had considerable antiquity. The extensive rugged interfluves were exploited for both wild animals and plants, and fires were caused by hunting and by accident. Though not especially frequent these man-caused disturbances were relatively significant because lightning-caused fires are especially rare in a region that effectively never has summer thunderstorms. Also, the area is dry, and woody vegetation does not recover easily.

The extreme denudation of the region found today, however, is the product of the colonial and modern periods when fuel for mines and towns and regular burning of communally held lands to improve pasture were especially destructive. Woodland and brushland survive only in isolated localities or where some protection was offered on large estates.[25]

From the latitude of Santiago southward, even beyond the Río Bío-Bío, commonly taken as the southern limit of the Mediterranean climate, a fairly numerous Indian population practiced slash-and-burn farming on small plots, working the soil with a weighted digging stick. On the alluvial fan–covered sides of a few valleys in the Coast Ranges the practice survives, going by the name *curbén*. Digging stick–using cultivators could not plant in a grassland so they sought wooded and brush-covered areas for their fields, often girdling the plants before burning a plot to increase inflammability. In the 50 years of Inca domination some irrigation probably was introduced by *mitimaes,* or colonists from Peru, in the Central Valley between Santiago and the Maule River, but as it was not needed for farming by the more numerous natives it probably was not of importance. Llamas and alpacas also were present but not a significant pastoral tradition.

The original vegetation of the level lands of the Central Valley has been displaced almost completely since the Spanish Conquest by crops, weeds, and ornamentals from the Old World. It is likely to have been a grassland swept by extensive and frequent fires. The most circumstantial evidence is that the mounted Spaniards could maneuver and generally defeat the Indians in combat in that region. South of the Bío-Bío, where fire-cleared fields were scattered and regrowth forest dominated, the Indians defended themselves until late in the 19th century. There regrowth forests still dominate all the uncultivated land.

Burning is still prevalent in the broken country of the Coast Ranges, but with fairly humid air and accidental topography individual fires are not extensive or particularly hot. Much woody vegetation survives a burn and probably did in aboriginal times unless it was locally extirpated for planting.

California

An appreciation of the importance and character of aboriginal burning in California may best be gained by observing documented changes in vegetation, some of them photographically documented, in the little more than a century since Indian practices ceased to dominate the landscape. Burning by ranchers to clear brush on rangelands in the central and northern parts of the state, although sometimes legally proscribed, may have modified Indian practices only slightly.[26] Burning of slash in connection with logging operations, however, introduced a new factor. The available dry fuel could often step up the intensity of a fire, causing it to kill trees left standing and even crown in untouched forests, degrading a forest to a brushland for a long time, perhaps permanently.[27]

Perhaps more significant is the modern program of fire suppression and prevention. In the southern part of the state, because of concern for watershed damage and the danger of rapidly running brush fires, and on National Forest and Park Lands for aesthetic and political reasons, burning has been fully proscribed, and considerable investment has been made in fire suppression. While such suppression has not been completely successful and great and damaging fires have occurred during this century, their frequency in a given area has been notably reduced.[28] A recent examination of fires scars on ponderosa and Jeffrey pines in the San Bernardino mountains showed a fire frequency once every 10 and 12 years, respectively, prior to 1905, when full fire suppression was instituted, and every 22 and 29 years after that date.[29]

Even more spectacular is the photographic documentation at Yosemite National Park with photos taken from the same points between 1866 and 1961.[30] Only Miwok Indians had occupied the valley until about 1854. In addition to tourists and their facilities being present, cattle were grazed on the meadows until 1924. The early photos show much more extensive meadows, open stands of conifers, and very little brush. Later ones show a continuing closing of the forest, partly with brush and deciduous trees, and a constriction of the meadows and other grassy areas. Effective fire suppression is the only reasonable explanation of the change from aboriginal pattern.

There is some evidence that the greater brush accumulation below and into the open lower edges of the coniferous forests in Southern California resulting from fire suppression is producing hotter fires that crown in the conifers, and is actually raising the lower timberline by killing the conifers at their lower margin where they have trouble reestablishing themselves.[31] In the Sierras Juárez and San Pedro Mártir in Baja California, where the burning practices of local ranchers, some of whom are Indians, may resemble the aboriginal ones and fire prevention or suppression was not attempted at all until the last few years, the coniferous forests are more open and brush free than in comparable sites in the Peninsular Range north of the border.

South Africa

The level and gently rolling surfaces in the limited Mediterranean-climate area of South Africa seem to have been covered with grass and herbs at the time of Dutch settlement in the 17th century. It is reasonable to assume that regular, end of dry season burning by the Hottentots had established and maintained this association. Such lands, except for wet meadows used to graze dairy cattle, had largely been brought under cultivation by the mid-18th century, and the precise character of their vegetation in aboriginal times is obscure.

The region is bordered on the east by a steep escarpment cut in elevated but largely horizontal sedimentary rocks with some steep subparallel ridges, the reexposed roots of ancient mountains, in between. The high level lands north and east of the escarpment have steppe climates without the Mediterranean winter rain concentration. Today a sclerophyllous brushland with many succulent *Euphorbiaceae* characterizes the lower or rain-shadowed areas of steep slopes, and a dense woodland the higher and wetter parts. In these rougher lands some nonpastoral Bushmen bands survived the Hottentot intrusion. They burned for hunting and had undoubtedly had some fire escapes. The strongly accidented terrain restricted the spread of an individual fire, so the frequency of burning of a given plot may not have differed much from the present situation.

Australia

In both Western Australia and South Australia the lands of Mediterranean climate are characterized by low relief, but it is only in the hilly areas and the dry eastern and northern margins that considerable areas of wild vegetation survive, the level and rolling lands being farmed or kept in improved pastures. Two genera, *Eucalyptus* and *Acacia*, and one family, *Proteaceae*, supply most of the woody vegetation, with some of the many individual species occurring as both trees and bushes.[32] Many also show the fire adaptations of root or crown sprouting and heat-stimulated seed germination. Lightning-caused fires are as common as in California or more so. Records of the burning practices of the relatively large aboriginal populations are scant, but the Aborigines of the North and Center were, and still are, enthusiastic burners, both to improve vegetable food supply and to attract game to the newly sprouting grasses. Similar practices in the South are probable.

There seems to have been little pure native grassland in the Mediterranean-climate parts of the continent. There is none now. Open woodland with a mixed brush and grass understory is characteristic of the hilly regions except for the very wet (1,500 millimeters of rainfall) southwest tip of Western Australia, where there is a closed eucalyptus forest. The drier parts of Western Australia carried a brushy assembly three to five meters in height with a few tall eucalyptus rising above it whereas in South Australia a savanna with scattered eucalyptus trees was characteristic. All were adapted

312

to frequent fires, which still occur or are set wherever the land is not cultivated. The tall trees are resistant.

The red tingle tree, *Eucalyptus jaksonii,* of the humid southwest tip of Western Australia merits special mention. Fire-damaged trees will experience heart rot, but the buttresses continue growing, giving hollow shells 25 feet in diameter. The habit is strikingly like that of the California redwood, a species of utterly different phylogeny but growing in a similar environment.

Conclusions

In all inhabited areas of Mediterranean climate the aboriginal populations burned the vegetation both accidentally and deliberately for many thousands of years. These areas were also subject to naturally caused fires, severe because of the long summer drought, and the resident species had probably evolved their fire-adapted characteristics over much more than human time.

Accidental man-caused burns may have roughly equaled the frequency of naturally caused fires, but their effects may have been somewhat greater as they were likely to occur under dry, windy weather conditions when their intensity could kill even somewhat fire-adapted trees. Uninhabited Madeira, with forests on now almost desertic lower slopes, may afford negative evidence. Where burning was deliberate for economic purposes it could be of annual frequency in grasslands, serving to establish and maintain them on fairly level or rolling surfaces. Brushland and forest burning was less frequent, perhaps every 7 to 10 years, but this frequency was enough to moderate the fire intensity and permit the survival of an open forest of mature trees that could only reproduce effectively when, for whatever reason, burning did not occur.

Modern practices of fire prevention and suppression, particularly in Southern California, are reducing the frequency of fires well below that of aboriginal times and possibly below their prehuman level. This permits the buildup of fuel from dead brush and new levels of conflagration intensity that are a permanent threat to marginal forest lands.

Environment and Ecology in the "Northern Tonto" Claim Area

The relict character of the clumps of trees might make one suspect that this area had been deforested recently. Whipple's records of 1853–54, however, describe the country almost exactly as it is today.

I get no impression at all that these Indians had any consciousness of pressure on the resources of a given land from other groups of Indians, so in a real sense there was none.

INDIANS REPRESENTING two entirely distinct linguistic stocks occupied a sector of the middle Verde Valley and its eastern tributaries. They were the Yavapais of the Yuman linguistic family, the other members of which are located toward the west and southwest, and the most northwesterly element of the Western Apache group.[1] The linguistic stock of the Apaches is Athapaskan and their relatives are to the east and more distantly in the far north. Both appear to be unrelated to the prehistoric Puebloan peoples whose ruins lie within the region.

In addition to the distributional pattern three lines of evidence suggest that the Yavapai were fairly long-time residents in the Verde Valley and that the Apaches were recently intrusive. The economic pattern of the Yavapais was clearly that of hunters and gatherers ranging widely to scratch a living from an extensive and relatively barren land.[2] While not dissimilar, the Western Apache groups had an economy dependent in significant measure on raiding livestock-owning settlements well to the south, a pattern that must postdate Spanish settlements in neighboring regions. The accounts of early Spanish explorers described populations that are most easily identified as Yavapais.[3]

Finally, Hoijer's careful linguistic work on the Athapaskan language, using the still experimental technique of glottochronology but based on an unexcelled knowledge of the Athapaskan languages, points clearly to a recent diversification of this widespread linguistic family that spread from Central Alaska and Northwestern Canada to Northern Mexico. Only about 1,300 years is allowed for the whole linguistic stock to diversify. All the Apachean groups, from the Lipan in South Central Texas to the Tonto in Central Arizona, have only been separated for about 400 years (see Table 3).[4] Since there are archaeologic and historic records of an Apachean people in northeasternmost New Mexico and beyond into the southern High Plains in the 1540s and earlier, and records of a *despoblado* in the present Western Apache area of Eastern Arizona at the same time,[5] the probabilities of the whole Apache entrance into Arizona and Western New Mexico being very recent are high indeed. The small linguistic diversification noted in Table 3 has taken place after their arrival.

TABLE 3. Years of Separation for the Apachean Languages

	Navajo	Chiricagua	San Carlos	Jicarilla
Chiricaguga	149			
San Carlos	279	227		
Jicarilla	279	200	335	
Lipan	335	227	419	227

Source: Harry Hoijer, "The Chronology of the Athapaskan Languages," *International Journal of American Linguistics* 22, no. 4 (1956): 226.

The distinctions among the San Carlos and the other Western Apache dialects is even smaller, though common reservation experience since 1873 makes precise assessment of linguistic separation dubious.

Although both the Yavapai and Western Apache have reputations for being unfriendly and normally at war with their neighbors, in the Northern Tonto claim area the two groups got along well. Intermarriage and common residence, or residence exchange in both directions following extratribal marriage, are reported by modern ethnographers.[6] Military cooperation against the Americans in the 1860s and '70s was so close that the American military forces commonly did not distinguish the groups, calling them all Apaches. Lt. Schuyler, evidently at Camp Verde in 1874, however, stated:

> The so called Tontos are mainly half breed Apaches and Apache Mojaves, as a rule they speak both languages and style themselves either Apaches or Apache Mojaves as the humor strikes them. On this reserve some are classed among the Apaches, but the greater number among the Apache Mojaves. They partake of the peculiarities of character and features of both tribes, and generally speak both languages, though incorrectly. They were first called Tontos by the White Mountain Apaches on account of their childish, i.e. foolish, manner of talking the language. In physical development they are better than the Apaches, but inferior to the Apache Mojaves. Almost all of the San Carlos Indians belong to this class.
>
> A band of Tontos is now on this reservation who were rated as Apaches at San Carlos a few months ago. They surrendered here as Apache Mojaves, and are under control of the Apache Mojave Chief.[7]

On reservations individuals from the two groups maintain their linguistic identity to the present day.[8] There is theoretical interest in the fact that the Yavapai and Tonto Apache did not see themselves as competitors for the same land. Their subsistence economies were almost identical and would place the same demands on the resources of a territory. This similarity, however, seems to have acted to unite them against outsiders, whether sedentary farming Indians such as the Pimas or American ranchers and miners entering their territory or residing vulnerably near their homes.

The Terrain

Verde Valley Bottoms

Between Sycamore Creek and the Cottonwood-Jerome area the valley of the Verde River is fairly broad and the river meanders in a floodplain. Downstream to the south the Verde River runs through a narrow steep canyon, through which a road for wheeled vehicles has never been constructed. West of the broad part of the Verde Valley the slopes rise steeply to the forested semiplateau on which Prescott is located. This semiplateau is composed of granitic and metamorphic rocks and has been raised to an elevation of about 6,000 feet. It is moderately dissected, but most of the drainage is toward the south, parallel to the Verde River. Consequently only short steep washes without permanent streams enter the Verde from the west. There are a number of small springs on this slope of the semiplateau, but from the hard rock their water yield is too low for them to be used for even small-scale irrigation.

The streams that enter the Verde River from the east, however, are much longer, and several have permanent flow. From north to south within the area of interest they are Oak Creek, Wet Beaver Creek, and its important tributary Dry Beaver Creek, Clear Creek, Sycamore Creek, and Fossil Creek. They head on the great sedimentary plateau of Northern Arizona and have cut steep, narrow canyons in the horizontal beds of limestone, sandstone, and shale. The previous beds yield very substantial spring flow, and as a result the tributaries and the Verde River itself have a singularly regular stream regimen for rivers rising in an area of very irregular precipitation.

The Verde bottoms are from one to three miles wide and the river has changed its course in them periodically, aggrading its meandering bed. A gallery forest of sycamores, cottonwoods, and willows interspersed with marshy meadows was the probable natural vegetation of which only remnants survive. Farms, producing principally alfalfa, have come to occupy most of the bottoms.

Except in the south, near Fossil Creek, the rise in elevation east of the Verde River is much more gradual. Terraces and benches of soft alluvial lacustrine and river valley deposits step up from the river bottom, rising irregularly about 1,000 feet in six or eight miles. The valleys of the eastern tributaries cut through these benchlands, and rise more slowly, two or three hundred feet in a similar distance. Cottonwoods, willows, and sycamores grow along these streams, but the dominant valley perennial in the side valleys is mesquite, which also forms thickets on the edges of the gallery hydrophytes of the main stream.

Where groundwater is inaccessible the lower benches, composed as they are of soft and pervious material, present a distinctly desertic aspect. The thorny plants are small and widely scattered. *Opuntia*, *Atriplex*, and *Larrea* are important genera. There is some *Yucca* but very little *Agave* (Mescal). In wet years there may be a fair cover of grass and ephemeral herbs either in late spring or following the late summer rains in fall. But most of the time the bulk of the ground surface is bare.

Climatic conditions in the Verde Valley bottom are probably best represented by the station at Montezuma Castle, a short distance up Wet Beaver Creek.[9] Its records run from 1937 to 1959 inclusive. Despite the elevation of 3,180 feet, the summers are extremely hot. July's average is 81.6°F, and the average daily maximum temperature for that month is 100.6°F. Temperatures over 100°F have been recorded in all months from May through October, and the record maximum is 115°F. The winter cold, with the Verde Valley holding a nighttime inversion layer, is fairly severe with very high diurnal ranges. An absolute minimum of 3°F has been recorded; the average daily minimum for December is 25.8°F and in the same month the average daily maximum is 60.6°F.

Average annual precipitation is 11.5 inches, this broad valley district being distinctly drier than any of the surrounding country. There is a summer maximum, with nearly five inches falling as convectional showers in July, August, and September. Nearly four inches fall in the four months from December through March; this precipitation is associated with attenuated cyclonic storms from the Pacific Coast. May and June are distinctly dry months. Rainfall is extremely variable from year to year; within the 21-year period of record, annual precipitation has ranged from 19.5 to 3.5 inches.

Where the Verde Valley narrows to enter its canyon to the south, precipitation values are notably higher despite the lower elevation. Childs, on the Verde River just north of the mouth of Fossil Creek, at 2,650 feet receives 17.3 inches,[10] distributed as and with ranges similar to those at Montezuma Castle. Its temperatures are about the same as at Montezuma Castle, except that with better air drainage the average daily minima in midwinter are higher; for December it is 32.5°F.

With high temperatures and variable and generally low rainfall agriculture is possible only with irrigation, and except for the immediate riparian areas the flora and fauna are impoverished.

Dissected Canyon Terrain

A few miles east of the Verde River north of Sycamore Creek, and immediately east of it to the south, the terrain rises irregularly from about 4,000 feet to the rim of the Northern Arizona Plateau at about 6,000 feet. The major streams flow in canyons which cut into the plateau somewhat east or north of its general front so that the regular front of the Mogollon Rim is considered to end on the west at Fossil Creek. These canyons have a longitudinal profile that is concave upward and their bottoms are often 2,000 feet below the surrounding terrain.

North of Clear Creek and Sycamore Creek this intermediate area consists of benches eroded out of the horizontally bedded sedimentary rocks, largely of Paleozoic age, that underlie the North Arizona Plateau. They are sandstones and limestones, and their differential resistance makes for a mixture of gentle slopes and nearly vertical cliffs that expose surfaces of bright red sandstone. The canyon walls along the larger streams are nearly vertical. South of Clear Creek the geologic structure is more complex, and the folded, faulted, and tilted rocks of many sorts including igneous and metamorphics form ridges that are a northern extension of the rugged Mazatzal Mountains.

Very few stations provide climatic data for this sector, and with great differences in elevation and exposure the climate surely varies greatly over short distances. Temperatures, both summer and winter, undoubtedly vary inversely with elevation at approximately the normal lapse rate (decreasing a little more than 3°F with each 1,000 feet of elevation). Precipitation effectiveness increases with increasing elevation and decreasing temperature, and above 5,000 feet the maintenance of a snow cover for weeks or months may store and release moisture only when a thaw initiates the growing season, bringing a disproportionate increase in precipitation effectiveness. In the most general terms precipitation increases with increased elevation, but local exposures are perhaps more significant. A 10-year record for the Sedona Ranger Station, in Oak Creek Canyon at 4,223 feet, but below the deeper parts of that canyon, shows an average annual precipitation of 17.1 inches, ranging from 10.6 to 23.1 inches during the period.[11] Sedona is probably one of the drier points in the dissected canyon country. Strawberry, just below the southwestern end of the Mogollon Rim at 5,875 feet is probably one of the wetter sites. A broken record for an average of 12 years gives an average annual precipitation of 24 inches.[12] At both these stations, and presumably throughout the district, there is the same pattern of a brief, late summer maximum and a longer less intense one in winter with a late spring drought.

The vegetation on the interfluves reflects the increasing moisture effectiveness with increasing elevation. Shrubs become larger and more closely spaced and the herbaceous and grassy ephemerals grow more regularly and are green for longer periods. Junipers and live oaks, together with chaparral species such as sumac and manzanita, begin to appear in shadier slopes at 4,000 feet moving to sunnier exposures at higher elevations. At about 5,000 feet piñon pine (*Pinus edulis* and *P. monophylla*) is added to the shrub assemblage, in some places forming with the junipers substantial forests on fairly level surfaces. *Pinus ponderosa* begins to appear just below the plateau rim at 5,500 to 6,000 feet. This brushy country provides good browse and sustains a substantial deer population as well as many game birds.

In the major canyon bottoms there is a moister environment, even though precipitation values are probably slightly lower than on the interfluves. Permanent, spring-fed streams and lowered evaporation because of shade in the deep canyon bottoms are factors. Cottonwoods, sycamores, and willows continue up the canyon bottoms. Similarly large live oaks stand a little away from the stream bottom. Near the head of Oak Creek Canyon aspen, Arizona ash and Arizona walnut appear, and there is often a tangle of berry (*Ribes*) and grape vines (*Vitis arizonica*). In the upper canyons there is more diversity of conifers than there is higher up on top of the plateau, and spruce, fir, and Douglas fir appear on the slopes in addition to *P. ponderosa*. This richer flora undoubtedly supported more animals than the interfluvial surfaces, but now the heavy road traffic in the accessible canyons has largely driven all but the rodents away.

The Plateau

An elevated plateau of nearly horizontally lying sedimentary rocks of Paleozoic age forms approximately the northeastern third of the state of Arizona. Almost the entire plateau is drained by the Colorado River and its tributary the Little Colorado, with the drainage divide lying very close to Mogollon Rim, which forms the plateau's southern edge. The Northern Tonto claim involves the southwestern corner of this great plateau and in this sector, almost uniquely, streams draining toward the south, from Oak Creek to Fossil Creek, have cut canyons into the edges of the plateau. The San Francisco Mountains, which lie at the northern tip of the Northern Tonto claim, are a mass of recent volcanos, which have erupted through the sedimentary strata of the plateau, building cones as much as 5,000 feet above the general 7,000-foot elevation of the surface, pouring layers of basaltic lava over their immediate vicinity, and covering a wider area with layers of lapilli and volcanic ash. These volcanic eruptions are known to have continued almost into historic times, the last having been dated by tree rings in logs in Pueblo Indian ruins to the mid-11th century.[13] The lava flows and ash falls have substantially disturbed the drainage pattern in the area south of the San Francisco Mountains, so there are a number of temporary and permanent shallow lakes and undrained basins of which Rogers Lake and Mormon Lake are the largest.

The 6,000-foot contour may be regarded as marking the western edge of the plateau, although it runs in places into the upper canyons of the streams that drain to the southwest (Figure 30). Peaks and ridges, mostly of volcanic origin and running roughly south-southeast from the San Francisco Mountains, rise as much as 1,500 feet above the median 7,000-foot plateau elevation and a few streams, notably Walnut Canyon, which drains toward the northeast, as well as those draining westward into the Verde, have begun to cut into the surface. Thus the plateau terrain is rolling and hilly and in a few places cut by canyons.

A long climatic record at Flagstaff (elevation 6,903 feet) gives an average annual precipitation of 20.3 inches. Fort Valley at 7,347 feet, seven miles northwest of Flagstaff at the foot of the main peaks of the San Francisco Mountains, records 23.6 inches.[14] At both stations the distribution involves a pronounced late summer maximum, a longer, less intense maximum in winter, and a dry late spring. In a 50-year period precipitation at Flagstaff has ranged from 9.9 to 34.5 inches. Fort Valley has been slightly less variable.

The difference in precipitation between the two stations can be correlated directly with elevation, and the stations are undoubtedly representative of the western side

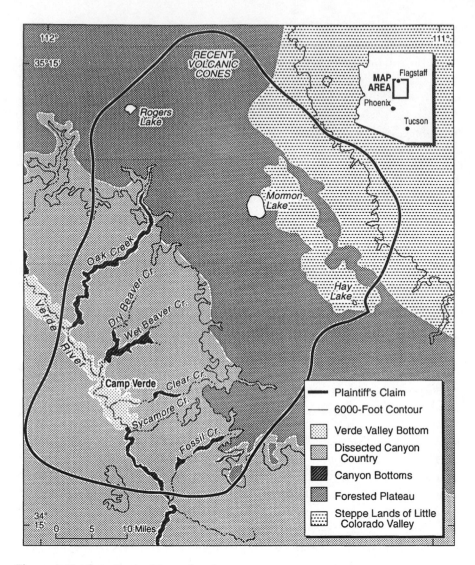

Figure 30. Northern Tonto claim and environment

of the whole plateau area. To the east, however, as the surface slopes downward slightly toward the Little Colorado Valley and is protected by the broken line of up-lands west of Mormon Lake, there is a notable falling off in precipitation, particularly that occurring in the winter months. Cosnino, 10 miles east of Flagstaff at about 6,500 feet gets only 16.8 inches.[15]

Summer temperatures on the plateau are pleasant with cool nights and warm days. The average July maximum at Flagstaff is 80.6°F and the minimum 50.5°F. Adjusting for elevation according to the normal lapse rate, these values would apply to the whole plateau area. The average temperature for January is 27.6° F, but the diurnal range is great. Most days warm up above freezing and nights and mornings can be bitterly cold. All months from October to April have had temperatures low as 4°F, and minus 30°F has been recorded.[16] If snowfall is heavy it may cover the ground for months. Although the average period between the last frost in spring and the first in fall is 116 days, summer frosts do occur. Only August has failed to record a frost and it has experienced 33°F. Fort Valley, with its slightly greater elevation, has experienced a temperature as low as 27°F in every month of the year.[17] There is archaeological evidence around the Walnut Canyon ruins of prehistoric Indian agriculture on the plateau,[18] but it must have been very risky. Modern farmers confine themselves to wild hay crops.

Pinus ponderosa is the dominant tree in the western part of the plateau. A rather small deciduous oak (*Quercus gambelii*) is subdominant in places. At elevations above 8,000 feet, Douglas fir and spruce are present as are aspens. At present the accessible parts of the extensive *P. ponderosa* forest have been logged, and the forest is in various stages of regrowth. In aboriginal times most of the forest probably was mature, though there would be some grassy glades in poorly drained spots scattered through it. Though a mature yellow pine forest is scenic and a valuable source of lumber to a modern economy, it is both florally and faunally somewhat impoverished. There is little browse for game animals and seeds of the yellow pine are not a good source of food for humans as are piñon seeds. The acorns of *Q. gambelii* were eaten by man.

An extensive area east and southwest of Mormon Lake, largely at elevations between 6,700 and 7,200 feet, now carries only scattered clumps of pines and junipers on low rocky prominences. The bulk of the areas is grassland with numerous pools and wet meadows scattered through it. The relict character of the clumps of trees might make one suspect that this area had been deforested recently. Whipple's records of 1853–54, however, describe the country almost exactly as it is today.[19] Waterfowl on the lakes and, in summer, pronghorn antelope must have been quite abundant in these grassy areas, but there are few deer. Waterfowl seem to have been taboo as food to the Yavapai and Western Apaches.[20]

The Breaks and the Little Colorado Benchlands

Just to the east of the lake-strewn prairie areas there is a two or three mile wide strip running northwest-southeast about 35 miles from Winona to Chavez Pass, in which the terrain falls off sharply toward the northeast. The drop is about 700 feet, and is erosional, a rimrock of more resistant sediments holding up the higher plateau surface. This Break terrain is wooded, with dense thickets of piñon, juniper, and live oak as well as other smaller bushy species. Only Walnut Canyon in the north cuts more than a mile or two into the plateau surface, but there are many small valleys and coves. Below the zone of steep slopes which run down to about 6,300 feet, a much broader gentler slope extends about 30 miles to the Little Colorado River, falling in that distance to about 4,900 feet. Though its slope is actually considerable this surface gives the impression of a broad, almost featureless plain, though a few streams that head on the plateau such as Canyon Diablo have cut almost vertical gorges a few hundred feet deep into the horizontal beds of sedimentary rock that underlie it.

There are no weather stations in this terrain area of the Northern Tonto claim, but Winslow, at the bottom of the slope on the Little Colorado River, permits a projection of the trends in climate from Flagstaff through Cosnino to that point. The average annual rainfall at Winslow is 8.05 inches. A late summer maximum exists, but sheltered by higher lands to the west this lowland gets only about one-half inch of rain in each of the winter months.[21] The slightly higher lands near the "Breaks" probably get from 12 to 15 inches of rain.

Summers are hot at Winslow, temperatures over 100°F occurring fairly frequently in the months of May through August, but the diurnal range is substantial. The average temperature for January is 32.8°F, but again this is produced by wide diurnal variations. Very few days in winter do not experience temperatures on both sides of the freezing point, and −19°F has been recorded at Winslow.[22] In the higher lands near the Breaks proportionately lower values are to be expected, but the hilly terrain would offer some protection against strong and cold winter winds. A few of the streams that head in the plateau to the west have a small permanent water flow, but the broad interfluvial surfaces are without water except what collects in pools during the infrequent rains.

The brushy forest of the Breaks extends out on the flats for about five miles, piñon disappears quickly, and progressively smaller and more widely spaced junipers are the dominant form. At the present time there is a major program to eradicate the juniper to improve grassy pastures right up to the Breaks, and it is possible that juniper once extended somewhat further into the plains. Below 6,000 feet, however, the vegetation is a poor short grass steppe, the grasses and herbs growing about a foot high in wet years and half that in dry ones. Replacement of native bunch grasses by Russian thistle in recent times, partly because of overgrazing, has impoverished the vegetation somewhat, but it was always light.

Subsistence Economy of the Northern Tonto

Agriculture

To some degree the Athapaskan-speaking Northern Tonto and associated Yavapais hunted, gathered wild plant foods, and cultivated crops, but there is some question as to the proportion of their sustenance which came from each of these activities. Buskirk estimates that the White Mountain and Cibecue groups of the Western Apaches got 25 percent of their subsistence from agriculture and 40 percent from gathering wild plants; hunting provided the remainder.[23] It is my opinion that these figures underestimate the significance of agriculture to these groups to some degree. The modern ethnographic evidence available, however, suggests that agriculture was less important to the Northern Tonto and Northeastern Yavapai.[24] Corbusier, who visited them shortly after 1875, for example says only "A Few of the A-Yumas [presumably Western Yavapai] formerly raised small patches of maize ti-yatch, and laid by a little for winter."[25] He also mentions raising small patches of tobacco. Gifford discusses agriculture among the Northeastern Yavapai briefly, and among the Northern Tonto,[26] but elsewhere he speaks of "their virtual neglect of agriculture."[27] He evidently found more record of agriculture than he had among the Southeastern Yavapai, where in his earliest researches in the general area he reported only the slightest trace of the activity.[28] Goodwin indicates that there was farming among only two (Fossil Creek and Oak Creek) of the four Northern Tonto bands that he identifies. Devin, writing in 1869, says that the Tontos "seldom attempt cultivation."[29] Several of the U.S. Army officers who responded to the Army Circular Letter of July 20, 1874, however, refer to agriculture by both the Apaches and Apache Mojaves (Yavapais), mentioning irrigation and maize, beans, squash, and melons as crops, though none indicate that it is extremely important.[30] These, of course, were especially troubled times.

All of the observations noted above, however, were made after the Northern Tonto had been much disturbed by U.S. military expeditions which began about 1864. Leroux seems to be the only one who penetrated the Verde Valley before that time, and only brief excerpts from his Journal are known.[31] In these he refers to prehistoric ruins and traces of cultivations, but he says nothing about farming among the Indians in the valley who attacked him. On the other hand the much earlier Spanish explorers who visited the mineralized areas around the modern mines at Jerome indicate that irrigated agriculture was being carried on. Luxán, reporting on the Espejo Expedition of 1582–83 speaks of a cienguilla flowing into a small water ditch and plantings of maize,[32] and Farfán in 1598 describes the soil in the same area as especially fertile because of the fine maize crops it yields.[33]

These early accounts, which in all probability refer to Yavapais, suggest a quite significant farming pattern, in an area where many small plots of ground could be irrigated from spring flow quite easily. It is my belief that when in 1864 American military forces, supported by irregular volunteers from the Prescott area, began a series of raids on the people of the Verde Valley, agricultural activities constituted the first casualty. Fields could easily be destroyed, and the Indians could not defend them. An example of the pressure put on farming is Krause's comment in March 1866 that the Indians he was seeking from Fort Whipple had gone east of the Verde, and his recommendation that their fields be destroyed by a scouting party in May or June.[34] At this time the more mobile food procurement enterprises of hunting, wild plant collecting, receipt of military rations, and raiding had to suffice. This was the disturbed and finally inadequate pattern of living the Northern Tonto were following in 1872 when they were settled at Camp Verde and in 1875 moved to the San Carlos reservation.

A figure of 25 percent of subsistence coming from agriculture may be approximately correct, but it should be noted that maize and beans were dried and stored for the winter.[35] This meant that these crops could and did play an absolutely critical part in sustaining life when other foods were likely to be scarce. Even in aboriginal times the Indians attended their fields only at planting and harvest time so that gathering and hunting could be pursued at a distance during the growing season.[36]

Hunting

Corbusier's evaluation of the relative importance of various animal foods among the Northeastern and Western Yavapai is singularly perceptive. "Their meat was principally rats, ma-le-ke; hare, ku-le; and cottontail rabbits, he-lo, which are numerous and can easily be captured or killed. All the men and boys frequently engage in the sport of a rat hunt. . . . Before they had firearms large game was not easily killed, but when hunters were successful there was a feast."[37]

The officers responding to the Army Circular Letter of 1874 present a similar picture. Schuyler's comment is particularly instructive: "All animals whether carnivorous or otherwise are killed and eaten. The game held in highest estimation being a large rat which makes its nest in the Spanish bayonet. A species of army worm is also considered fine."[38]

Gifford's statements that deer were the staple animal food and abundant for the Northeastern Yavapai and his accounts of successful antelope and mountain sheep hunts, Goodwin's frequent references to successful deer hunts, and Buskirk's extended discussion of the subject probably should be treated with some reservation.[39] Talking to old Apache men about the hunting experience of their youth is likely to yield the ethnographer tales of about the same validity as those he would get from asking white men of the same age to talk about hunting. On the other hand the importance of buckskin for clothing meant that there was a reasonably consistent take of deer.[40]

Again, the pattern of consumption of a wide variety of small desert animals suggests a shortage of animal foods, an animal dietary which is paralleled among the Indians in the dry areas of Southwestern North America as in the Great Basin[41] and peninsular Baja California.[42] Corbusier's statement for the Yavapai: "As to food they consume with relish many things that a more fastidious palate would reject. They devour rats, coyotes, lizards and caterpillars" is supported by Gifford's culture element list for the Apaches. The officers at Camp Verde in 1874 give the same impression about both groups.[43] There were, however, some notable animal food taboos that in the Northern Tonto area would have significantly cut into the otherwise available animal foods. The most important is the refusal to eat fish, and the Verde river and its larger tributaries could be substantial producers. The Apachean fish taboo is well established.[44] The Yavapai shared it, though from which direction this uneconomic idea diffused is not known. Among the Yavapai, and evidently the Apache as well, the taboo was extended to other aquatic life such as frogs and water birds. Gifford cites an informant: "Ducks and geese not eaten; Shampura said of them, 'I'm human; we do not eat water stuff.'"[45]

With their available hunting technology, taboos on eating animals that might have been taken easily, and the absence of herds of large easily slaughtered animals, it would seem that foods derived from animals could not have made up more than 20 to 25 percent of the native diet rather than the 25 percent suggested by Goodwin and Buskirk.[46]

Gathered Plant Foods

Field botanical identifications were made with the aid of Thomas H. Kearney and Robert H. Peebles, *Arizona Flora* (2d ed., University of California Press, Berkeley and Los Angeles, 1960), and Lyman Benson and Robert A. Darrow, *Manual of Southwestern Desert Trees and Shrubs* (University of Arizona Press, Tucson, 1944).

Acorns and *Agave* (mescal) are commonly identified as the most important plant foods of the Western Apache and the Yavapais of the highlands. Mescal had the virtue of being available at all times of the year, though considerable effort in cutting, transporting, and pit baking it was required to make it edible.[47] Gifford notes the standard patterns of the plant's use among both the Northern Tonto and the Northeastern Yavapai.[48] The slopes on either side of the Verde Valley, however, are not well supplied with *Agave*. While there is some *Agave* growing on the middle slopes around Fossil Creek, it seems likely that the Apaches got most of their mescal through visits to their Southern Tonto congeners in the Mazatzal area and the Yavapai of the Verde Valley made similar trips to the south and east as suggested by Gifford.[49] Cooked mescal could be preserved and transported rather effectively.[50]

The acorn situation was much more favorable. The slopes and upper valleys of streams on both sides of the Middle Verde carry respectively scattered and concentrated stands of live oaks, *Quercus emoryi*, and *Quercus arizonica*. On the plateau the seeds of *Quercus gambelii* were available and used as food. Gifford states that the Northern Tonto regarded acorns as their principal wild crop,[51] and elsewhere notes their importance to the Northeastern Yavapai.[52] The lack of knowledge of leaching the bitter tannin from acorns, a technique so important to the California Indians, does not seem to have been too significant to the Northern Tonto and associated Yavapai,[53] since the bulk of the acorns in their area seem to have been directly edible.

For the White Mountain Apache, the acorn season began in the latter part of July according to Goodwin, but Gifford states that the season in the Mazatzals was in the middle of August[54] and later in the Granite Mountain area.[55] Presumably *Q. Gambelii* at its higher elevation on the plateau produced its crop even later. In any event acorns were treated as a crop. [Residents of] *rancherías* visited sites where they grew, collected them in burden baskets, lived on them at the time, but also transported substantial quantities to winter quarters, where they contributed to the winter diet.

Mesquite beanpods were another important foodstuff which ripened in late summer. They were to be found at lower, hotter, localities than those which had oaks, on the better-drained banks of the Verde River and in the low parts of its tributary canyons. The Fossil Creek mouth area in the southern part of the Northern Tonto lands was an important place for them, and bands who were centered at higher elevations visited from some distance to get them. "There was travel between mountain and valley on both sides of the Verde . . . . Redrock people . . . entered Verde Valley for mesquite."[56]

Piñon seeds and juniper berries were an important and choice food. They are available in two general areas of the Northern Tonto claim area, the upper interfluvial slopes of the canyon country on either side of the Verde River at elevations from 4,500 to 6,000 feet, just below the elevation where the ponderosa forest is established. They also occur in the Breaks and just below them in the country that begins to fall off toward the Little Colorado River. These fruits ripen in October and November. While the juniper fruits consistently, a given stand of piñon will bear fruit heavily only about one year in four. Thus, unless they ranged very widely as did the Paiutes of Nevada,[57] piñon could not serve as a dependable source of food for the Indians. Piñon nuts are, however, a choice, storable, and reasonably easily collected foodstuff. The piñon stands adjacent to the Verde Valley undoubtedly were exploited fully whenever they bore, but those beyond in the direction of the Little Colorado might or might not be discovered in their short fruiting season. The Yavapai Indians observed by Leroux and Sitgreaves on the west side of the San Francisco Mountains in October, 1851 were in temporary piñon nut–collecting camps. They had brought with them food products from the lowlands, a bread of mesquite beans and cooked, dried mescal.[58] Whipple was in the area in 1853, with Leroux again serving as guide, and commented on the absence of Indians.[59] Two factors other than slight difference in locality are involved. The piñon season was over, and 1853 may have been a year of no piñon crop. Juniper berries were hardly worth coming for.

The cactus fruits, saguaro, and the tunas of the platyopuntias, which are referred to as important to both the Yavapais[60] and the Tonto Apaches,[61] are fairly abundant only in the lowest Verde and Fossil Creek areas. It is likely that the inhabitants of the Northern Tonto claim area collected these fruits farther south in territories of other groups. Gifford notes that the Northern Tonto made a cake of dried tunas of the *Opuntia*,[62] a fact that suggests planning to transport the fruit home.

The complete analysis of wild plants utilized by the Indians would be long indeed. Grapes and walnuts in the canyons, berries on the plateau, grass seeds and the seeds of herbaceous annuals at various elevations, liliaceous roots in swampy localities, *Yucca* on dry intermediate slopes, and many other plants are exploited on occasion.[63] Wild plant foods were seasonably available, roughly from July to November, except for mescal, which could be obtained at any season. Further, many sorts could be preserved and stored. It is likely that for the Northern Tonto claim well over half of the total diet was obtained from this source.

Raiding as a Source of Food

Because of their isolation from Spanish and Mexican settlements Indians in the Northern Tonto claim area had relatively little opportunity to become dependent on stolen stock for food as did the Western Apache groups living to the southeast of them. It is likely that some Northern Tonto joined raiding parties of these kindred groups and they may occasionally have brought back a few cattle or horses for a feast.[64] After 1863 raids on American mine supply trains and ranches in the Prescott area were more productive, but only for immediate consumption. Devin noted that they kept very little stock.[65] The food gained in this fashion probably barely compensated for that lost by the reduction in agricultural activities resulting from disturbed conditions.

The regular warfare against the Pimas does not seem to have yielded much economic return. Certainly that against the Walapai and the Havasupai to the west and northwest did not. Some horses and cattle might have been obtained by raids on the Pima, but the Tonto Apaches and Yavapais were not stimulated to develop even the transient herding pattern that the San Carlos and White Mountain groups had. In the considerable selection of Northeastern Yavapai war tales against all three groups recorded by Gifford there is no reference at all to any economic benefits.[66] Revenge and retaliation by both sides are the motives for all the fights described. I find no specific reference to raiding the Navajo or Hopi for sheep on the part of Indians in the Northern Tonto claim area, but Goodwin says that there were raids by both sides, that Navajo sheep were taken in some number by the White Mountain group, and that there were Navajo raids on the Northern Tonto.[67] In the late 1850s and '60s, when the Navajos were pasturing sheep far to the west,[68] the Tontos are likely to have gotten some.

Clothing, Shelter, and Implements

Both the Apache and Yavapai of this region used the low, round, dome-shaped wickiup. Constructed of poles and brush with little care, materials for such houses could be secured locally wherever there was enough water for a settlement. Similarly fuel, though needed, was not a problem at such locations for the small populations involved. Buckskin for clothing and moccasins and the skins of other animals used for such things as quivers were essential, but they were a by-product of hunting for food and seem to have been available in adequate quantity. In fact buckskins are mentioned as the principal trade item with which the Western Apache could procure grain and blankets on trading expeditions to the Navajo, Hopi, and Zuni.[69] Materials for basketry, wood for weapons, and stone for points were available in a number of places in and adjacent to the Verde Valley.

Population Density

It is Goodwin's estimate that the total population, Apache and Yavapai, of the Northern Tonto claim was about 800, of whom 450 were Apaches.[70] This figure, though perhaps overestimating the Apachean element, is in reasonable accord with Corbusier's data on the Verde reservation in 1873.[71] Some 1,000 Yavapais and 500 Tonto Apaches were placed on the reservation at that time. These numbers included Yavapais from farther west and some Southern Tontos as well, but they involved less than a complete roundup, and the military operations of the preceding year involved a considerable loss of Indian lives. Responses to the Army Circular Letter of July 20, 1874, give more specific population figures, from which Corbusier's figures evidently were taken. They show the Camp Verde Reservation population as 1,074 Apache Yumas and Apache Mojaves, and 497 Apaches.[72]

The area in the Northern Tonto claim amounts to about 2,400 square miles so that a gross population density of one person for three square miles is indicated. The area was, however, by no means of equal productivity. About 5 percent was canyon or valley bottom, 35 percent dissected canyon country, 40 percent forested plateau, and 20 percent high grassland (cf. Figure 30). The latter two categories are, in general, notably poor in terms of their capacity to support people using the Indian technology, but in the area of good to fairly productive lands the population density was less than one person per square mile. Except for plots of cultivated land, which were all privately owned, neither the Apache nor the Yavapai seemed to be concerned with the economic cost of letting others exploit the wild products of their lands. Yavapais from the Verde Valley went south and east into territories of other Yavapai bands to get mescal and saguaro fruit, to hunt, and during the 1860s to raid, and other Yavapai entered the Verde Valley for comparable purposes.[73] A similar pattern no doubt existed with Northern Tontos going into Southern Tonto or Yavapai country to get mescal and saguaro, and others visiting them for acorns and piñon nuts.[74] Tenure over territory was claimed, but right to usufruct of the wild flora and fauna seems to have always been granted if requested politely. Both the Yavapai and Apaches could be regarded as touchy and quarrelsome. Visitors who hunted and gathered in another band's territory might be attacked in revenge for some past action or even because of the insult caused by failure to request permission, but they were not attacked because of the economic damage they were doing. I get no impression at all that these Indians had any consciousness of pressure on the resources of a given land from other groups of Indians, so in a real sense there was none. A drought might cause hunger because of crop and wild plant failure, but there was no feeling of being crowded by too many Indians who used the same technology and sought the same resources.

The relatively small population that occupied the Northern Tonto area did not saturate its ecological niches. This may be accounted for in two ways. First, the Apachean element had not been there long and second, there was considerable loss of life, even if no economic gain, from the prevalent war pattern. The war tales recorded by Gifford,[75] primarily involving the Northeastern Yavapai, and a time span of not much more than two generations, acknowledge very substantial losses, in total perhaps half the population that at any time was present. Only in minor degree, if at all, does this warfare seem to be related to white men's activities in Arizona. Such a loss of life would just about cancel any natural increase in population during the period.

Historical Records in the Area of the Northern Tonto Claim

Three very early Spanish expeditions crossed the Northern Tonto area, and after that there is only dubious record that any Spaniard or Mexican visited the territory at all until after it became part of the United States. The first of these expeditions was led by Antonio de Espejo and chronicled in the journal of Diego Pérez de Luxán.[76] It reached the Little Colorado River from the Hopi villages May 1, 1582. Katharine Bartlett has reconstructed the most probable route of the expedition and plotted it on a map.[77] The Little Colorado River near Winslow was "settled by a warlike mountainous people."[78] On the basis of the distribution of the "cedars: with an edible fruit," which Alfred F. Whiting identifies as Utah Junipers (*Juniperus osteosperma*), found there, Whiting and Bartlett regard Chavez Pass as the most likely spot for climbing to the forested plateau from the steppe lands of the Little Colorado Valley.[79] Mormon Lake is identifiable, and the group descended one of the steep canyons that drain into the Middle Verde, proceeding four leagues up that river to the mines, almost certainly the Jerome-Clarkdale area. Both on the plateau and in the valley there is reference to the "Mountainous people" who inhabited the region. The latter farmed maize in the Verde Valley and apparently did not in the highlands. They seem to be a single linguistic group. There is no reference to the striking linguistic split that has characterized the area in recent time.[80] Hammond and Rey postulate that these Indians, who are reported to have used a cross as an ornament or symbol and thereby acquired the designation Cruzados (for the cross reported by American military officers) among the Spaniards, were Yavapais, as does Schroeder.[81]

The expedition of Marcos Farfán de los Godos sponsored by Juan de Oñate in November 1598 followed a very similar route from Hopi, across the Little Colorado River, up onto the forested plateau, down into the Verde Valley, and up into the mineralized area around Jerome.[82] Although snow was covering the ground on the forested plateau, several rancherías were visited there. Apparently settlement of the highlands was permanent. Again growing maize is reported from the Verde Valley. Significantly, as the explorers were guided from one ranchería to another, both on the plateau and in the Verde Valley, there is no indication of a change in language. Again Yavapais are indicated.

In October 1604 Oñate himself led an expedition through the area en route to the mouth of the Colorado River. Very possibly he followed Farfán's route.[83] The Indians of the Verde Valley were identified as Cruzados, but the Zárate-Salmeron account gives very little information on them other than the denial that they farmed,[84] almost certainly an error.

Captain Sitgreaves next visited the area in October 1851. He went down the Little Colorado River far enough so that he passed around the north and west sides of the San Francisco Mountains, so that at most he only touched the northwest corner of the Northern Tonto claim area, although his guide, Leroux, went some miles farther south, getting to a point where he could at least look in the Verde River drainage.[85] Two "encampments" of Indians were found by Leroux, who identified the Indians as Yampai or Tonto.

The first camp, visited on October 9, contained baskets, one coated with pitch to serve as a water jar, piñon and grass seeds, and cakes of mesquite pods and mescal.

The latter two food items clearly had been brought from the south, in all probability from the Verde Valley. This was a temporary piñon-gathering camp, and the Indians probably had arrived only shortly before to begin the harvest of that food. Leroux is said to have observed the headwaters of the Verde River on this trip out of Sitgreaves's Camp 15, but I believe that this report belongs with Leroux's second reconnoiter from Camp 17. Camp 15 was in the juniper country north of the San Francisco Mountains, within a single day's journey from the Little Colorado, and he was gone only a part of a day. From Camp 17, whence the second Indian camp, also containing piñon seeds, was discovered it would have been possible to reach Upper Oak Creek and return in a single day.[86]

In May 1854 Leroux traveled up the Verde River from its junction with the Salt. North of a difficult narrows in the river his party was attacked by "Tonto of the Yampais nation." This must have been in the vicinity of Clear Creek, because the Verde Valley was again broad and the road easy. The party probably left the river by way of Dry Beaver Creek. The preserved excerpts of Leroux's diary do not mention farming by the Indians living in the region, though they comment extensively on the ruins (Montezuma Castle is one of them) and evidence of irrigated farming on the part of an extinct prehistoric population.[87]

In December 1853 Lieutenant A. W. Whipple, with Leroux again acting as guide, crossed the northern part of the Northern Tonto claim. They traveled west from the Little Colorado River near modern Winslow, but had to go back to the river because they could not cross the deep trench of Canyon Diablo.[88] Then, heading southwestward from further down the river, they passed through the recent volcanic cones and layers of loose scoria and pumice in the vicinity of Sunset Crater. They then proceeded around the southern end of the San Francisco Mountains to Leroux's Spring, evidently Sitgreaves's Camp 17. They then went back in a southeasterly direction to discover some ruins in a canyon, called Cosnino Caves, which have been identified as Turkey Tank Caves in lower Walnut Canyon.[89] They then returned to Little Colorado and came back to the ruins by a direct route. From there they proceeded west roughly following the present course of Route 66 and the Santa Fe Railway.[90]

Whipple comments that when Leroux was there two years ago with Sitgreaves "the hills were covered with savages,"[91] but they discovered none. In the old and later fresh snow that covered most of the plateau they discovered not even the track of an Indian, and this involved two weeks of wandering around. Only well to the west in the vicinity of Bill Williams Mountain did they see evidence of modern Indians.[92] This was a large military party, and Indians tended to keep out of their way, but it is doubtful that any Indians residing in the area east of Flagstaff could have failed to leave some trace on the full snow cover. If any had been there in the fall they had gone to the lowlands to winter.

A number of expeditions into the Verde Valley were made out of the Prescott area from 1863 to 1873, and Camp Lincoln, later developing into Camp Verde, was established in 1864.[93] But no one deemed it worthwhile to climb the high plateau to the east, though General Crook did follow the top of Mogollon Rim from the east and descended to the Verde by way of Beaver Creek or possibly Clear Creek. Bourke indicates that while doing so he was, in 1871, in completely unexplored country.[94] Pursuit of the Tonto Apaches led southeastward into the Mazatzals and the Tonto Basin.[95]

Indian Land Use in Relation to Terrain Types

Of the five terrain types shown on Figure 30 the Indians utilized intensively the flat-bottom lands of the Verde Valley, and perhaps even more intensively the narrower bottoms of the tributary canyons. Both agriculture and intensive collecting were involved. The northwestern boundary of the Northern Tonto area in the Verde Valley will not be considered. Clearly there were close relations between the Yavapai of the Clarkdale area and Sycamore Canyon and those of Oak Creek Canyon and its mouth. Gifford indicates clearly that it was all a part of the Northeastern Yavapai area.[96] The Northern Tonto area as a district meriting separate consideration is defined by the extent of Apachean penetration among the Yavapai.[97]

The dissected canyon country on either side of the Verde similarly was used consistently, though less intensively than the valley bottoms, up to approximately 6,000 feet. That is about the elevation where piñon gives way to ponderosa pine. Mescal, cactus fruit, yucca, and various seeds came from the lower sector, juniper berries and piñon nuts from the higher. Both small and large game animals were hunted throughout. Again, the western and southwestern boundaries of the Northern Tonto claim are based on Goodwin's designation of how far Apaches were present among the Yavapai.[98] Gifford identifies a Yavapai band, Walkeyanyanyepa, as occupying the upland area just west of the Verde Valley.[99]

The forested plateau and the meadow and steppe grasslands within and beyond its eastern edge presents a different story. It seems to have been occupied permanently in the late 16th century. At least rancherías were reported there in May and November of 1582 and 1598 respectively.[100] Most students agree that these Indians were Yavapais. It should be pointed out that this was before the Navajo had entered Northeastern Arizona, and very possibly before the Western Apache had even entered the east central portion of that state.[101]

The strikingly long gap, until 1851, in the record of direct observation permits many speculations as to what transpired, but at that time Sitgreaves's guide Leroux found camps of Indians he called Tontos or Yampais (Yavapais) on the northwestern and western side of the San Francisco Mountains. These are just at and beyond the northwest corner of the Northern Tonto claim. The camps were, however, clearly occupied by piñon-nut gatherers who had recently come from some lower, warmer area, probably the Verde Valley, as is attested by their possession of food stocks of mesquite and *Agave*. The remarkable timidity of the Indians Leroux encountered is suggestive. It cannot have resulted from previous contacts with whites, since there had been none. A reasonable interpretation is that they were camping and gathering in a no-man's-land, where anyone encountered was likely to be alien and hostile.[102] In 1853, although he spent more time and saw more of the area of the Northern Tonto claim, Whipple found no evidence of Indians currently occupying the area.[103] He was there in December; snow was on the ground; and the piñon harvest season was over. Again there is a gap in the record. During the decade from 1863 to 1873, when there were many scouts through the Verde Valley and its tributary canyons, the military did not pursue the Indians in the forested plateau. That would have been relatively easy country for cavalry operations if it were an Indian center.

An annotated map from the National Archives, based on Wheeler's survey of 1871, has areas of certain Indian tribes noted;[104] some are outlines.[105] Since Cochise's band is referred to ("Cochese") one may suspect that the annotations were made by 1874, the date of Cochise's death. The area between the Mogollon Rim and the Little Colorado River from south of the San Francisco Mountains to east of Holbrook is marked "Neutral Hunting Ground." On the other hand "Yavapais" is written in large type further north and extending beyond the Little Colorado, and north of that "Cosinas Bands" along the Little Colorado in the vicinity of Leupp. Cosinas are presumably Havasupai. Finally, further north Moquis (Hopi) appears in the correct position. Tonto Apaches are shown occupying the Verde Valley up to Clear Creek and the Mogollon Rim. "Apache Mohaves sometimes called Yavapais Apaches" have the Camp Verde area and uppermost Verde Valley as well as the uplands to the west. These data are hard to interpret.

Warren E. Day, an Army Surgeon at Camp Verde, in his response to an official letter of inquiry about Indians in the Arizona Territory, dated July 20, 1874, defined, under the heading of Food, the territory of the "Apache Tontos" as follows: "The Apache Tonto country is bounded on the north by the Mongollon [Mogollon] range, on the south by the Salt River, on the east by the Mongollon Range, and West by Rio Verde. Within these boundaries are all the varieties of climate, animal, and vegetable productions."[106] His intent is clearly to indicate that the slopes but not the plateau itself was occupied and exploited by the Indians. Apache Tonto for him seems to include both Apaches and Yavapais.

An archaeological reconnaissance carried out for the plaintiffs in the Navajo Land Claim describes twelve sites in the Canyon Diablo and Clear Creek subareas, most of which on historical or typologic grounds the archaeologist, Roscoe Wilmeth, regards as Navajo, which bear on the Northern Tonto claim.[107] These sites are in three groups. The first group, Sites 68, 69, and 70, is on the east rim of Canyon Diablo in the steppe lands that lead down to the Little Colorado River. It is just beyond the edge of the Northern Tonto claim. Wilmeth, on the basis of Navajo informants' statements, typology, and amount of decay of the hogans, regards sites 68 and 69 as dating from just before to just after 1863, when most Navajos were taken to Bosque Redondo (Fort Sumner).

The second group of sites, 71, 72, 73, and 74, is located in the piñon-juniper country of the "Breaks," where the forested plateau drops off into the steppe country that goes down to the Little Colorado River. Somewhat protected from the winter winds, and with fair concentrations of wild plant foods, this is the most attractive sector of the Forested Plateau zone. There are hogan ruins, only one group of which, number 72, is specifically identified as Navajo on typologic grounds but several of which were known to modern Navajo informants. They seem to date after the mid-19th century on tree-ring evidence. These are at sites that Goodwin states were winter quarters for the Mormon Lake Band.[108]

Finally, there are a group of sites on the east and south sides of West Sunset Mountain, Sites 75, 76, 77, 78, and 79. This outlying, juniper-rimmed butte in the steppe zone is just east of the limit of the Northern Tonto claim. On the basis of Navajo informants' information and structural types Wilmeth regards the sites as Navajo and dating from the pre–Fort Sumner and Fort Sumner period.

The first 60 years of the 19th century were a period of shifting to a herding economy on the part of the Navajos. With sheep from villages in New Mexico being obtained by raids in great numbers, the pressure to seek new pastures toward the west grew steadily. These pressures were especially strong in the 1850s and 1860s, when American military pressures were strong in the Navajo homeland centering on Fort Defiance. The unoccupied grasslands south of the Little Colorado River seem to have attracted small bands and family groups, perhaps especially from 1863 to 1865, when Kit Carson was devastating the homeland area to force the tribe to surrender and go to Bosque Redondo.[109] Further, adoption of a sheep-herding economy made the Navajos land-hungry, and while sheep were vulnerable to enemy raiders, there was continuing pressure to push into any unoccupied pasturelands.

In June 1865 a party of New Mexican civilians raided a mixed group of Apaches and Navajos "at or near the San Francisco Mts.," taking from them 85 horses and about 1,000 sheep. All but the sheep consumed were later recaptured by the Indians. The report said that there were 200 Indian warriors, but it is almost certainly an exaggeration.[110]

Two modern ethnographic accounts individually give primary emphasis to only one of the two basic linguistic groups involved in the Northern Tonto claim. Gifford, considering the Northeastern Yavapai, draws their northeastern boundary at the headwaters of the streams that drain into the Verde River, leaving the bulk of the forested plateau and all the steppe to the northeast outside their area.[111] Gifford's informants make only slight reference to utilizing the plateau in the vicinity of Flagstaff for piñon nuts and hunting.[112] Goodwin, on the other hand, found at least one Apache informant who referred frequently to hunting and gathering piñon nuts and juniper berries on the plateau.[113] Furthermore he identified the Mormon Lake Band, which he says is completely Apache without Yavapai mixture, that lived entirely on the plateau and in steppe lands to the east. The account is circumstantial, mentioning a campsite east of Mormon Lake, at Mary's Lake (Lake Mary) at the foot of the San Francisco Mountains and Elden Mountain, and farther south near Hay Lake and Stoneman Lake. He further states that the Mormon Lake band wintered in the Breaks that go down to the steppe lands of the Little Colorado.[114]

Among his informants Goodwin lists: "Charlie Norman: Mormon Lake Band, Northern Tonto; old man, served as scout many times. Died, 1934."[115] He was Goodwin's only Northern Tonto informant, and unquestionably is the source of the statements about the Mormon Lake Band. Further, Goodwin's clan migration map shows *no clan* resident in the plateau part of the Northern Tonto claim,[116] though two Southern Tonto clans claim an origin just to the northeast of the area. The clan sections identified by Kaut for the period 1850–75 similarly show no class in the plateau area,[117] though he talks about clans coming from the area of Flagstaff, citing no source for his data.[118]

Two bits of data already discussed make the existence of the Mormon Lake Band dubious. One is the presence of Navajo archaeological sites on the edges of the area and also right in the heart of the winter quarters cited. The other is the failure of Whipple to find any trace of winter residents there in 1853. There is no question but that Indians of the lower canyons did climb to the plateau to hunt and collect seeds. On the other hand, at least up until the late 1860s they were under some pressure from Navajos who were using the area for the same purposes and to pasture sheep, and Havasupais (Cosinas) also ranged the area. Goodwin notes this condition and says it partly accounts for the complete absence of agriculture among the Mormon Lake Band.[119] Katharine Bartlett concluded that the whole southern drainage of the Little Colorado River, including both the forested plateau and the steppes below the Breaks, was a disputed area visited by several but not occupied by any one tribe in 1848.[120] On his rough tribal map Kaut similarly shows an extensive vacant area along the Little Colorado River ca. 1850, though he follows Goodwin on the area of the Northern Tonto.[121]

The most reasonable explanation of the Mormon Lake Band is that this band existed briefly as a refugee group during the 7 to 10 years after 1865. At the beginning of this period most of the Navajos were at Bosque Redondo, and American military action was beginning in the Verde Valley. This action built up to a maximum in 1872, when most of the Northern Tonto and associated Yavapais were settled at Camp Verde. A few might have chosen to remain free as hunters and gatherers on the plateau for a few years longer and thus constitute the basis for the Mormon Lake Band identification.

Goodwin's typescript notes from an interview with Charlie Norman support this interpretation. Charlie Norman's earliest recollections are of the period when General Crook was campaigning on the plateau, in other words about 1872, and the Mormon Lake Band of which he was a junior member is described as behaving consistently as a fugitive group scratching a meager existence from hunting and gathering.[122] Charlie Norman's statement that he was born "on this side of San Francisco Mountains"[123] cannot be given much credence because he follows it immediately with the above-noted reference to his earliest memories.

Terrain and Ecological Conditions in the Western Apache Range

The Western Apache thus settled in lands that had been a barrier to transit in prehistoric times. . . . Even today no east-west road crosses any significant portion of it.

Although the lands of the Western Apaches are by no means barren . . . [n]o one place was especially attractive and would reasonably constitute a center for development.

At elevations of 8,500 to 9,500 feet there is a park-like landscape.

THE MOGOLLON RIM is the dominant terrain feature in relation to which the settlements of the Western Apache Indian bands were located. This sharp line marks the southern edge of the high, but relatively level and only locally dissected, surface of the Colorado Plateau. Below the Rim to the south, for about 50 miles, the generally horizontally stratified sedimentary and extrusive volcanic rocks that underlie the Colorado Plateau are undergoing vigorous stream erosion and the surface is being cut into valleys and ridges of varying breadth and steepness. In general these ridges and valleys have a north-south orientation, falling off irregularly toward the valley of the Salt River, and east of the longitude of Globe, across it to the Gila River.

South of the Gila and Middle Salt River and west of Tonto Creek, one enters an entirely different physiographic province. This is an extension of the desert Basin and Range country. Distinct mountain ranges, often separated by broad structural basins but sometimes closely juxtaposed, dominate the landscape. The separate ranges are composed of varied complexes of rocks of many sorts, including almost unaltered and quite metamorphosed sediments, some notably mineralized, especially with copper, and basement granitic rocks. The ranges have a general north-northwest and south-southeast orientation and frequently offer evidence of an old, dissected, and long-inactive fault scarp on one or both sides. This Basin and Range Province extends southeastward through Arizona and forms a major part of the Mexican state of Sonora.

Although there are higher localities, especially in the San Francisco Peaks and White Mountain areas near the western and eastern ends, respectively, of the Mogollon Rim, the Rim elevation, and thus the southern edge of the Colorado Plateau, is at about 7,000 feet. The terrain then falls off irregularly to the south to below 3,000 feet in the Salt and Gila Valleys. In the Basin and Range areas to the south the drainage is to the north-northwest and the broad desert valleys rise slowly to the east and south. By the Arizona-Sonora boundary they are 4,000 to 5,000 feet above sea level. The separate mountain ranges rise anywhere from 3,000 to 7,000 feet above these broad basin-like valleys.

The Salt and Gila Rivers drain the whole region toward the west. They follow reasonably gentle gradients but run almost directly across the grain of the country. At places they must cut, as antecedent or superposed streams, right through the Basin and Range Mountains, and they have formed canyons almost impossible for a road, even one for pack animals, to follow.

As far north as the Gila River, its tributaries coming from the south-southeast provided reasonably easy access from Sonora and western Mexico. Various related Indian groups speaking Pima languages traveled and settled along these natural routes of communication. Spanish conquerors and explorers and Jesuit missionaries moved northward to convert them and establish settlements as part of Mexican polity. North of the Gila, however, the wooded, dissected lands rising toward the Mogollon Rim, and the Rim itself, constituted a barrier. From Coronado's crossing of its eastern end to reach Zuni in 1540 to the 1870s, few white men entered the region, and then only at its eastern and western ends. The fact that Coronado's big and vigorously scouting expedition encountered no Indians between the Sobaipuri Pima country at the confluence of the San Pedro and Gila Rivers and Zuni, well up on the Colorado Plateau, suggests that the Apaches had not yet reached the area.[1] Perhaps there were a few seminomadic Yavapai bands present. It is clear that the Apaches living south of the Gila, who later came to be called the Pinaleños, were recent arrivals, settling in the area after finally forcing out the sedentary agriculturalist Sobaipuri Pimas in 1762.

East-west travel was always difficult, both north and south of Gila and Salt Rivers. Until the construction of modern Highway 60 from Superior through Globe, traffic tended to swing far to the south through Tucson or remain on the Colorado Plateau, following roughly Highway 66. Since the base elevation of the desert basins is higher in the south, there are only modest rises to the passes between the several desert ranges; the Gila Purchase was obtained for the United States in order to provide a feasible railway route within national territory. Even today a few mining railroads penetrate but none cross the Western Apache area.

The Western Apache thus settled in lands that had been a barrier to transit in prehistoric times and continued to be such until far less than a century ago. Especially difficult was the broken terrain that falls off from the Mogollon Rim toward the Salt and Gila Rivers. Even today no east-west road crosses any significant portion of it. The areas to the south that were occupied by the Apaches in the 18th and early 19th centuries again involve the most difficult terrain. Broad basins were of little or no interest; the Apache settlements were in narrow valleys or canyons, or tucked into the slopes of the several mountains ranges.

A pattern of settlement seems clearly evident. The Western Apaches were infiltrating the refuge area. Since their social organization was simple only small bands had occasion to get together. The broken terrain afforded protection, and into it small groups and families could disperse when pressed by forces from larger and more highly organized societies, Pimas and Pueblos in earlier times, Spaniards and Americans more recently.

Although the lands of the Western Apaches are by no means barren they are characterized by a great diversity of widely distributed resources. No one place was especially attractive and would reasonably constitute a center for development. There were tiny plots that could be irrigated by springs on valley sides. A few level patches just below the Mogollon Rim received enough rain so that dry farming was possible though not secure. Game was widely distributed and probably in no place abundant enough to support a sedentary population. Wild plant foods were present in all zones but their seasonal availability varied from place to place. Mescal (*Agave*) and mesquite beans could be collected in hotter lowlands, respectively in spring and summer; acorns on the intermediate slopes and piñon and juniper berries higher up and also beyond the Mogollon Rim could be obtained in the fall.[2] This was a land adapted to small bands in irregular and temporary intercourse with one another. They could unite for festivals or to assemble raiding parties, but most of the year they gained subsistence in widely dispersed small groups.

It must be added that in earlier times more complex sedentary societies did exist within the area. Archaeological ruins, attributed to a sedentary agricultural people of the Mogollon culture which had disappeared long before Coronado's time, occur under the Mogollon Rim, and the Sobaipuri had settlements along the Gila and San Pedro Rivers up into the 18th century.

It seems likely that the social organization of the Western Apache was insufficiently structured and ordered to permit large numbers of people without close ties of blood and marriage to live in close contact with one another. One cannot help but be struck by the violence of fights and the resultant blood feuds described by Goodwin, certainly a sympathetic observer anxious to present his friends the Apaches in as favorable a light as possible.[3]

One other general characteristic of the Western Apache lands may be noted. These were in general wooded and mountainous districts which, while they might be hot in summer, all experienced severe cold, at least for a short period in winter. Further, winter quarters were likely to be near the piñon country at considerable elevation rather than in the lower major valleys. While the structure of their wickiups, or houses, was undistinguished, the Apaches had cultural equipment to resist the cold. They sought campsites with adequate fuel for continuous fires and made or traded for and kept blankets and protective clothing. They were mountaineers, used to living in country that got seasonally cold, and they did not seek to evade the severity of the winter by moving at that season to the warmest part of their domain. In this regard their tastes in environment seem to contrast directly with their neighbors the Pima and to some degree with the Yavapai and Havasupai as well.

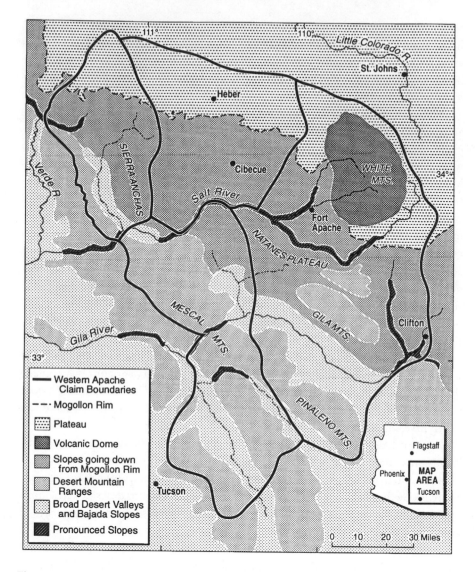

Figure 31. Western Apache claim boundaries and environment

Terrain Types

In Figure 31 the sorts of terrain characteristic of the Western Apache lands are shown in general fashion. Some comments about each type follow.

Plateau

There are two extensive relatively level surfaces at considerable elevation. The largest involves the area northward from the Mogollon Rim. Streams head almost at the lip of the Rim at elevations from 6,700 to 7,500 feet and flow down fairly gentle gradients toward the north, emptying into the Little Colorado River. In distances of from 75 to 100 miles they fall to about 5,000 feet. A few, notably Chevelon Creek, have cut canyons through the horizontal layers of sedimentary rock to depths of several hundred feet, and more resistant rock beds in places form rimrock faces, generally dropping off to the north, up to a hundred feet high. Occasionally a low, isolated butte, its edges supported by a thin rimrock, will be left standing.

In the higher, southern part of the plateau the heavy coniferous forest breaks the vistas and one gets the impression of traveling over gently rolling country with a few steeper stream-cut canyons. Farther north, with less vegetative interference, one views a vast sweep of country with flat-topped hills and broad valleys dropping gently toward the north.

Though the plateau is largely made up of successive layers of limestone and shale it is the limestone that resists erosion the best. When a stream has eroded through the top limestone layer it is likely to cut quickly through the shale to the next limestone bed. Where slopes are locally steep much bare rock is likely to be exposed. The limestone permits surface water to percolate through it so that even immediately following heavy rains or snowmelt there will be little surface water except in the larger stream courses where the bed has been cut through to the impervious shale. In a few places small closed basins, formed by solution of limestone, exist, especially in the eastern part of the plateau. The little lakes that form in them may be either transient or fairly permanent, depending on the extent of the area draining into them and whether the water they trap can percolate into another underlying limestone bed or is held by an impervious clay pan.

Despite the fact that the plateau receives fairly heavy precipitation it is an area where permanent surface water may be extremely scarce over extensive districts.

Another extensive level surface forms a double bench north of the Upper Gila River. The lower, more southerly part has an elevation of slightly over 5,000 feet and the upper area, known as Big Meadow and separated by the Natanes escarpment, is at about 6,000 feet. Both these areas present distinctly dry aspects, probably because pervious underlying volcanic materials permit rapid water loss. For the same reason the streams that drain the areas have made little progress in eroding out channels. The lower bench now supports only a short grass steppe vegetation, but Big Meadow, the higher one, gets more rain and has more taller grass and scattered trees on its surface. Again local permanent supplies of water are extremely scarce.

White Mountains Volcanic Dome

Near the eastern end of the Mogollon Rim is a district of geologically recent though long prehistoric volcanism. Volcanic activity has not occurred so recently as in the San Francisco Peaks near the west end of the Mogollon Rim, but dozens of small cinder cones still retain the characteristic shape that testifies to their origin.

Though some basalt flows were involved, the bulk of the eruptive material was lapilli or volcanic ash. Coming from a large number of separate vents spread over a circular district about 45 miles in diameter the ash and lava have been piled into a broad gently sloping dome that rises to about 10,000 feet, 3,000 feet above the plateau surface. Above the dome small cinder cone knobs rise another 500 to 1,000 feet, so that the central one, Baldy Peak, reaches 11,590 feet.

On all but the south and southwest side of this dome the land falls off irregularly but fairly gently. Many small closed basins have been filled with permanent or temporary lakes, sustained by the heavy precipitation of this elevated region. Streams are characteristically permanent. The relatively level surfaces between cinder cones have a water retentive soil derived from the volcanic ash and are covered with wet meadows. Forests are on the steeper, better-drained slopes, and they consist of mixtures of spruce and aspen. At elevations of 8,500 to 9,500 feet there is a park-like landscape.

On the south and southwest sides of the dome, tributaries of the Salt River, the White and Black Rivers, and Bonito Creek have cut deep V-shaped valleys into the relatively soft materials of the volcanic dome and the sedimentary rocks that underlie it so that the reasonably linear front of the Mogollon Rim is obliterated. The steep slopes are well wooded and even small streams carry a permanent flow of water.

Slopes Leading Down from the Mogollon Rim

In its western part, from the headwaters of Fossil Creek eastward to the head of Canyon Creek the limestone rimrock of the Mogollon Rim forms an almost continuous vertical, and in places overhanging cliff several hundred feet high. Farther east the front is less sharp and the streams draining southward have cut valleys and gorges into the southern edge of the plateau. But in all cases, below the Rim the character of the terrain changes to one of substantial local relief. Valleys and interfluves are separated by elevations of 1,000 to 2,000 feet. In general the interfluves and valleys have a north-south orientation but at various places a lower rim, such as the Natanas escarpment, cuts across the pattern in an east-west direction. Often there is a flattening of the stream courses and widening of the valleys above such sub-escarpments while steep gorges have formed in and just below them. These sub-escarpments below the plateau rim are probably the result of differential resistance of the capping rock beds. Close to the Rim the rocks in this dissected, mature topographic region are sedimentaries, still generally horizontally placed, of the several formations that underlie the Colorado Plateau. Farther south complex series of pre-Cambrian sediments and the underlying basement granitic rocks are exposed. Elsewhere, especially toward the east, layers of Tertiary lavas are interbedded with the sedimentary rocks, and at various locations mineralized thermal waters have deposited mineralized dikes through the sediments. The most notable concentration is in the Morenci-Clifton area in the southeast. Asbestos appears in minable concentrations near the southwestern corner of the Fort Apache Indian Reservation.

East of where Canyon Creek enters the Salt River this irregular falling off of the land from the Mogollon Rim extends across the Salt River nearly to the Gila. The Upper Salt River and the tributaries that form it are cut into deep isolated canyons. Most of their tributaries enter from the north and from the southern Salt River Canyon rim the drainage is again toward the south and the Gila River.

Elevations in this extensive broken region then vary from above 6,500 feet to below 2,500 feet. Sometimes high and low points will be in immediate juxtaposition and the terrain extremely broken and rugged. There are also fairly level high inter-fluval or mesa surfaces, and some of the high basins near the base of the Mogollon Rim have level areas of a few hundred acres each in which alluvial soil has been accumulated. Except for the lower San Carlos Valley the lower tributary canyons, however, tend to be steep and narrow right down to the Gila and Salt Rivers, which, because of their greater flow, have entrenched themselves more rapidly than their tributaries.

The natural vegetation in this zone of great local relief is closely correlated with elevation except where the slopes are too steep to hold any soil and bare rock is exposed. At the highest elevations, especially on the slopes of the White Mountains dome and near the western foot of the Mogollon Rim, there is a well-developed *Pinus ponderosa* forest with a few glades of deciduous oaks and other deciduous trees. From 3,000 to 6,500 feet, the area I have called *Encinal-Piñon* there is a mixed forest in which live oaks and several species of piñon pine and juniper form a low and often-open woodland. At still lower elevations desertic shrubs are dominant with the *Agave* (mescal), so important in Apache diets, often present. The lower canyons have gallery forests of mesquite thickets and cottonwoods, and the higher ones similar strips of aspens and other deciduous trees.

Desert Mountain Ranges

While the general trend of the desert ranges from the Mazatzals on the northwest to the Pinaleños in the southeast shows a consistent orientation north-northwest and south-southeast, within the individual ranges the topography may be extremely jumbled with steep or gentle slopes facing in virtually any direction. In general each range seems to have been thrown up along the sides of one or more fault lines, and the major trend of these lines is as stated. Other faults, however, seem to have run at other angles. Only a small fraction of the fault displacement has occurred in geologically recent times, so erosion of rocks of varying resistance has carved the ranges in intricate and irregular patterns. This is especially true in the tangle of mountains of intermediate height between the Salt and Gila Rivers, west of the San Carlos lowland.

Both uplift and degrees of erosion have varied for the individual ranges. Where uplift was greatest everything has been worn off to expose pre-Cambrian crystalline rocks. Elsewhere, as in the Dripping Springs, Mescal, and Galliuro Mountain alignment, Paleozoic sediments and older Tertiary volcanics have been bowed up and are now in process of dissection. Mineralization along joints and fractures in the rock systems has been widespread. Copper especially, but also lead, zinc, and silver, have been or are being mined in many places.

Crest elevations extend above 10,000 feet in the Pinaleño Range and above 9,000 feet in the Santa Catalinas. The highest peaks in most of the other ranges extend to between 7,000 and 8,000 feet. Above the basal alluvial fans, slopes are steep and intricately dissected. There is little even moderately level land either on the interfluves or in the canyons.

As is suggested by the generalized average annual precipitation map in *Arizona Climate*,[4] each of these ranges receives notably more precipitation than the surrounding lowlands. Evaporation similarly is lower and thus precipitation effectiveness greater with the lower temperatures at higher elevations. The vegetative cover has responded directly to these climatic conditions. Above 4,500 feet the encinal, with live oaks, juniper, and in places piñon develops an open woodland, kept particularly thin by steepness of slopes and prevalence of bare rock exposures. This woodland is now thinner than it was in the past in many localities because of depletion for fuel around the many past and present mining camps. Mescal (*Agave*) extends locally into the lower part of the encinal zone. In those ranges that extend above 7,000 feet a ponderosa pine forest begins a little below 6,300 feet, giving way to a fir and spruce forest above 8,000 feet.[5]

Desert Valleys and Bajada Slopes

In the region of the desert ranges it is clear that the broad intervening lowlands, generally oriented in the same northeast-southwest pattern, have their origins in structural displacements rather than having been degraded by processes of stream erosion. These were structural blocks of the earth's crust that were depressed or which failed to rise at the times when the bordering mountain ranges were uplifted. The typical valley is considerably elongated, and fills an area between two or more distinct mountain blocks. Widths vary from 10 to 25 miles. A cross section, normal to the stream course, would show a rather standard distribution of physical features between the bordering ranges.

The stream channel in the basin bottom is normally centered and has a width varying from a hundred to a few hundred yards. In some places, however, the channel may lie closer to one mountain face than the other, so the other basin features described below may be constricted or eliminated on one side while they are more broadly developed on the other. The smaller streams carry water only for brief periods, but even a permanent river such as the Gila has tremendous variability in flow. Channels are broad, sandy washes bordered by thickets of mesquite and willow and where a semipermanent bank is close to water, cottonwoods. Commonly there is a floodplain terrace a mile or so wide, located a few feet above the modern floodplain. Modern floods do not extend this high, and it may be suggested that rejuvenation of the streams in the recent past is common, probably resulting from downcutting of the harder rocks where the river, farther down its course, passes through a canyon. The same results could be produced, however, by recent fault activity.

These floodplain terraces, with loose soils, carry a meager desert shrub vegetation, but if water can be taken from the stream higher up and led along them they are subject to great agricultural development. Such irrigation works were beyond the social organizational and technical capacities of the Apaches, though the protohistoric and historic Pimas utilized them.

Beyond the floodplain terraces there are sloping surfaces that rise with gradually increasing steepness to the sharp break in slope where the desert range proper begins. In part these surfaces are alluvial fans, covered with progressively coarser cobbles and boulders as one approaches the mountain edge. Elsewhere there is only a thin alluvial cover bedrock, where mass wasting has caused the mountain slope to retreat. The term *bajada* has been applied to comprehend both types of surface, and they are quite similar in general aspect. Where the desert basin is broad these bajada slopes may be 5 or even 10 miles long and rise 1,000 to 1,500 feet above the stream channel level before reaching the mountain face.

Streams flow over the bajada surface for only a few hours after the violent desert rains, and the porous surface materials do not pond water. Reasonably permanent water sources may be found only in springs at the base of the mountains, or in the valley bottoms when the main stream carries a permanent flow from exotic sources. In the Western Apache area the desert basins are relatively high (2,500 to 3,000 feet) and do not experience the extreme drought that characterizes the lower desert basins farther west in Arizona. The lower bajada surfaces often carry a fair grass cover as well as desert shrubs. Higher up bigger desert shrubs appear and also one of the most important food sources for the Western Apaches, mescal, or *Agave*. It also grows on the rocky lower mountain slopes up into the lower encinal zone.

Pronounced Canyons

On Figure 31 a few particularly deep and narrow canyons along parts of the courses of major streams have been identified. These constitute notable barriers to crossing and are almost impossible to follow because of the steepness of their walls. It must be noted that the Apaches were troubled less by these barriers than any other group. They traveled mostly by foot (horses were for eating until reservation times) and their agility is proverbial. The presence of such deep canyons contributed to the refuge or hideout value of their homeland for Indian bands that regularly raided and consequently were on bad terms with their neighbors.

A comparison of the terrain types shown on Figure 31 and the area attributed to the four tribal groups of the Western Apache by Goodwin,[6] White Mountain, Cibecue, San Carlos, and Southern Tonto, shows clearly that each group utilized a wide variety of distinct terrain types. Even the individual bands within the groups roamed over and exploited as many kinds of land and natural vegetation as it was feasible to visit in the course of a year of seminomadic existence. The long narrow north-south strips of the several White Mountain and Cibecue band territories run across the grain of the country and include some forested plateau and mountain, broken encinal country, and some lower mountain slopes or desert basins where mescal and mesquite beans could be collected. The evidence is that the diverse terrain types were not sought for climatic advantage. Indeed most bands tended to winter in the higher parts of their range.[7] Goodwin indicates clearly that in the course of a year most if not all individual families would travel through and reside in several of the ecologic zones.[8] This mobility afforded notable protection against hostilities and punitive raids from first their Indian neighbors and in historic times from Spaniards, Mexicans, and Americans. There was virtually no center of Apache settlement that could be attacked by concentrated forces. It is clear that in the 17th and 18th centuries the Western Apaches were expanding their area of influence especially in those localities where the terrain was both diversified and rugged.

Climate

In the broken terrain of central and east-central Arizona both temperature and precipitation are strongly influenced by elevation, and within the area of the Western Apache claim elevation ranges from below 3,000 to above 11,000 feet above sea level. Thus a wide variety of climatic conditions are encountered. The general synoptic features that influence the weather are fairly uniform throughout the area, but a heat wave will be hotter at lower elevations and a cold wave colder at higher ones; winter cyclonic storms affect the whole region, but they yield far more precipitation at higher elevations; similarly summer thunderstorms can and do occur throughout the region, but their frequency, though not necessarily their intensity, is directly proportional to the elevation of the locality.

Temperature

A fairly high annual temperature range and on most days a very high diurnal temperature range are characteristic throughout the area of the Western Apache claim. These phenomena result from the considerable elevation and generally clear skies. The pattern can be illustrated by the following table (Table 4).

Comparing all five elements in the table with the normal lapse rate that indicates that the temperature should decrease 3.3°F for each 1,000 feet of increased elevation we find general agreement. Payson, Cibecue, and Whiteriver, though spread over a degree and a half of longitude at the foot of the Mogollon Rim, offer patterns almost perfectly adjusted to their respective elevations. On top of the Rim at Pinedale and McNary, notably colder conditions are prevalent. A change in the pattern appears, however, south of the Salt and Gila Rivers in the area of Desert Mountains, ba-

jadas, and canyons. Both Roosevelt and Fort Grant show average annual temperatures and average monthly minima too high for their respective elevations. Two factors seem to be involved. These are desert stations affected by high-pressure air
masses of polar continental air so infrequently that their minimum temperatures
occur only so rarely as to have little effect on monthly averages. Also both localities
because of their particular topographic situations receive unusually good air drainage
in winter, a general characteristic of canyon sites and upper bajada slopes, so their
extreme winter minima are notably raised.

TABLE 4. Temperature Data for the Western Apache Region

Station	Elevation	Average Annual Temperature	Average January Temperature	Average July Temperature	Record Maximum	Record Minimum
Pinedale	6,500'	48.9	29.4	67.3	99	−29
McNary	7,370'	46.5	30.2	64.3	93	−23
Whiteriver	5,280'	54.8	37.0	73.7	106	−13
Cibecue	5,300'	54.2	37.0	73.3	106	−14
Payson	4,848'	52.9	36.0	79.0	104	−18
Reno R.S.	2,350'	64.9	44.8	86.2	115	11
Roosevelt	2,200'	67.7	47.6	88.5	116	18
Globe	3,540'	62.4	43.9	82.5	110	10
Safford	2,900'	63.2	44.4	83.8	114	7
Oracle	4,430'	62.1	45.8	79.7	108	2
Fort Grant	4,880'	62.3	45.8	79.0	111	9

Source: Adapted from William D. Sellers, ed., *Arizona Climate* (Tucson: University of Arizona Press, 1960).
Note: All temperatures are given in degrees Fahrenheit.

In all instances one is struck by the enormous gap between the average January
temperature and the extreme minima recorded. The normal January weather involves
warm days and cold late nights and mornings, yielding a mild average, but on occasion a polar continental air mass sits over the area, bringing cold days and, with clear
skies, bitterly cold nights. Conversely, in summer a tropical continental air mass may
sit over the area, bringing spectacularly hot weather even at very considerable elevations. At the highest elevations, however, the temperature maxima are likely to be
moderated by the local development of unstable air, convectional clouds, and thunderstorms. We may state that this is an area of fairly temperate climate with considerable seasonal differentiation, punctuated by periods of a few days of notably cold
and notably hot weather. Above 6,500 feet the winter period when severely cold conditions may develop is long, and below 3,500 feet (Figure 31) unpleasantly high temperatures (over 95°F) by day affect a period of about five months with regularity.

Precipitation

To an even greater extent than temperature, precipitation is controlled by elevation in east-central Arizona. An isohyetal map could substitute quite effectively for a hypsometric one.[9] A listing of stations with their elevations and average annual rainfalls is indicative (see Table 5).

TABLE 5. Precipitation Data for the Western Apache Region

Station	Elevation	Average Annual Precipitation	Minimum Recorded	Maximum Recorded
Pinedale	6,500'	18.84"	12.3"	30.4"
McNary	7,370'	24.70"	10.7"	30.9"
Whiteriver	5,280'	17.89"	8.7"	33.2"
Cibecue	5,300'	18.61"	9.3"	30.4"
Payson	4,848'	21.48"	11.5"	29.7"
Reno R.S.	2,350'	18.73"	7.7"	33.9"
Roosevelt	2,200'	15.99"	6.9"	33.3"
Globe	3,540'	15.75"	8.0"	27.1"
Safford	2,900'	8.95"	3.0"	16.4"
Oracle	4,430'	19.35"	12.3"	34.1"
Fort Grant	4,880'	12.57"	5.1"	23.0"

Source: Adapted from William D. Sellers, ed., *Arizona Climate* (Tucson: University of Arizona Press, 1960).

Localities with major mountain barriers either to the east or to the southwest, notably Safford and Fort Grant and to some degree Pinedale north of the Mogollon Rim, seem to experience a rain shadow effect and distinctly dry conditions, but in general the higher the place the more precipitation it gets.

The annual precipitation curve is strikingly bimodal with pronounced maxima in late summer and in winter, a drought in spring, and a less pronounced one in fall. Actually the weather patterns that yield precipitation for the two seasons of maxima are entirely distinct, though there can be some temporal overlap during the fall period.

Winter storms are of the cyclonic sort, typically originating over the North Pacific and affecting areas of hundreds of thousands of square miles for one or two days. As air is forced to rise over considerable elevations more moisture is squeezed from it. On the lee (i.e., eastern) side of such an elevation, as at Safford for example, descending air will yield very little precipitation. Above 6,000 feet most of the winter storms bring snowfall; below that elevation precipitation may be either in the form of rain or snow. Throughout the Western Apache claim, snow is likely to fall every year, but only above 5,000 feet is it likely to remain on the ground for more than a day or two.

Summer storms involve local convection and affect moist, unstable air that has traveled aloft (above 10,000 feet above sea level) from the area of the Gulf of Mexico. Great heating at the surface in summer may raise a column of air into this moist mass, setting off free convection, the development of thick cumulus clouds, and violent downpours. Such cloudbursts are likely to be quite localized, affecting only a few square miles, but where they hit, several inches of rain may fall in an afternoon. Again, such storms may occur anywhere in Arizona, but they are particularly frequent at greater elevations, where the warmed air rising from the surface has only a short distance to travel before it gets into the moist, unstable Gulf air aloft.

From year to year there is great variability in both types of precipitation, and they may or may not run parallel.[10] In a given winter there may be many or few strong cyclonic storms that sweep in from the Pacific, and in a given summer, air may flow aloft westward from the Gulf of Mexico regularly or it may be notably impeded by Pacific air, which is too stable to permit significant convection. The last two columns in Table 5 indicate the wettest and driest years of record for each station and it is clear that the former is typically three or four times as great as the latter. Where the difference is less it is typically because of a short climatic record. For example, 1905 was a generally wet year, especially in winter, and several of the stations do not have records that go so far back.

Although moisture effectiveness is affected by temperature and thus elevation, and the seasonal distribution of rains in a particular year determines how well they support crops, it is probably true that 16 inches of rain are needed to get adequate crops without irrigation. If badly distributed, e.g., occurring in late summer and fall, even this may not be enough. It is clear that even the highest, wettest localities would suffer frequent crop failures due to drought if they could not be irrigated. Most localities below 5,000 feet would need irrigation in more than half of the years, and even higher up, as at Whiteriver and Cibecue, irrigation was always desirable and practiced to the extent feasible, as it still is.

Frost-free periods average about 150 days at sites just below the Mogollon Rim with elevations around 5,000 feet. Above the Rim stations average 120 to 130 frost-free days except above 7,000 feet, where the frost-free period is notably brief.[11] Whether frost danger or problems of security from other tribes kept the Apaches from planting crops in what are the wettest areas, i.e., above the Rim, is not certain, but it is clear that all their fields were and are below it.

Appreciation of Their Varied Environments by the Western Apache

Buskirk's analysis of the subsistence economy of the Western Apache suggests that while proportions varied from band to band, and no doubt from season to season, agricultural products, meat (both from wild game and captured Mexican livestock), and wild plant foods each played parts of roughly equal importance in the diets of the Western Apache bands.[12] It is my belief that the part played by plant foods was somewhat more significant, with crops rising in importance in years of good rains and high yields and collected foods rising in importance to make up deficiencies when crops failed or were poor. When there had been notably successful raids on Mexican ranches in Sonora, feasting on cattle and horses no doubt upset the normal dietary pattern. This conclusion is supported by my reading of Goodwin's account of the yearly migration cycle of certain White Mountain groups down into the Gila Valley in spring and summer and up onto the Mogollon Rim in late fall, pausing at their fields just below the Rim to harvest crops in September. Like crop plant foods, acorns, mescal, juniper berries, and piñon nuts were transported from collecting sites and stored for use in the winter season and other times of food scarcity.[13]

Buskirk makes a strong point, however, that families and clan groups were specifically attached to particular sites, and cultivable, especially irrigable lands were treated as owned property.[14] It is pertinent to note that with the possible exception of the Apache Peaks Band of the San Carlos Group, each of the bands of the Western Apaches had at least one agricultural site within its band territory.[15] The band might spend approximately half the year at such localities, and individuals incapacitated for rapid movement by age might remain at such places permanently, watching but probably not cultivating growing crops while the more vigorous members of the bands journeyed to a place where particular wild foods were regularly gathered.[16]

Similarly, in winter men would go out on extended hunting expeditions (which were probably far less productive of food than remembered), leaving women and children in camps at the agricultural sites.[17] Raiding parties might well operate under the same conditions, but, of course there are no accounts from within Apache society for the period before 1860, when raiding was so extensive, regular, and relatively unpunished as to be considered a legitimate economic activity.

Food-gathering journeys were to specific localities, used regularly by particular bands, for particular stuffs, at specific seasons.[18] With permission, however, a band or group of families might be permitted to gather food in another band's area. Goodwin mentions small groups of White Mountain Apaches going as far as the Lower San Pedro river near the Aravaipa to gather saguaro fruit (p. 156).

The several ecologic zones illustrated in Figure 31 each afford a particular food-gathering potential. The steppe lands far to the north in the Little Colorado Valley might be visited by an occasional hunting party but had little to offer in the way of plant foods. The piñon-juniper country to the south, higher up and closer to the Mogollon Rim, did yield desired plant foods in fall and Goodwin mentions visits by collecting parties to points as far north as Showlow and Pinedale (p. 157), but the impression is given that such expeditions were unusual, probably because of danger from neighboring tribes, particularly Navajos. The coniferous forest on top of and just below the Mogollon Rim and in the White Mountains was poor in plant foods, though edible roots were occasionally collected in wet meadows (pp. 12–13, 16). Most of the country was ranged on occasion by hunting parties and, of course, a large fraction of the farming sites of the Northern Tonto, Cibecue, and White Mountain groups were in this zone just below the Mogollon Rim.

The areas identified as "encinal" were, on the other hand, rich in plant food, of special value because they were storable. Acorns, juniper berries, and piñon nuts were especially important, and mescal occurs in the lower edges of this zone. Hunting of course was carried on enthusiastically in the region especially in places accessible to farming sites. It is likely that the game resources in this zone were somewhat greater than in the forest, though because of the brushy cover it may have been harder to obtain.

Except for two limited kinds of sites the desert basins that are quite extensive in the southern part of the claim area were singularly unproductive for the Apaches. The Indians did gather mesquite beans in the galeria forests along watercourses and mescal grew in the upper bajada slopes adjacent to the encinal zone. Similarly the desert basins are notably impoverished as far as game is concerned. At a few places in the lower desert zone, farming was carried on, but the Western Apache seem to have lacked both the organizational and the engineering skill to create irrigated farmland along major rivers. Their farm sites were on smaller tributary streams such as Aravaipa Creek, Lower Eagle Creek, and the San Carlos River. Only at one point where the canyon is narrow, and farmable land very limited, does Goodwin note their farming along the Gila River.

The two questionable use areas within the Western Apache claim are at its northern and southern ends. In the central part of the claim there were also lands which were scarcely used, as for example the broad desert lowlands of the Gila Valley between Bylas and Solomon. These basin lands were crossed, and hunted over with little yield, in order to get to the mescal and acorns in the Graham and Santa Teresa Mountains to the south (p. 13).

But the area north of the Mogollon Rim, though hunted in occasionally and even less frequently visited in fall to collect piñon nuts and juniper berries, is identified by Goodwin as a sort of glacis, an unoccupied area visited by both Navajo and Apache hunting parties, and after 1863 by Navajo sheep herders (pp. 15–16, 23).

The glacis on the south was more of a direct creation of the Apaches than a product of two conflicting forces. There were no Apache agricultural settlements south of Aravaipa Creek, but there are records of regular acorn gathering in the vicinity of Oracle at the north end of the Santa Catalina Mountains (p. 28). The White Mountain group similarly are reported to have gathered regularly in the Grahams and Santa Teresa Mountains (p. 13).

This southerly fringe of mountains, however, served another very important purpose. Here bands of young men would gather to go on raiding parties. From the Graham and Winchester Mountains they traveled south and southeast into Sonora. From the Santa Catalinas they could attack Tucson and the Santa Cruz Valley ranches.

And to these rugged retreats they would return with captured livestock and other booty. Because such mountains were used as raiding bases they were of course subjected to retaliatory scouts by Pimas, Papagos, and Mexicans. Women and children, the less mobile and thus more vulnerable part of Apache society, were not encouraged to travel let alone reside for any time in such exposed places. When all hostile peoples moved out of a region, as did the Sobaipuri Pimas from the Lower San Pedro Valley in the 18th century, then permanent Apache settlement might replace them, selecting topographically sheltered spots such as Middle Aravaipa Creek.

It may be legitimate to question whether exclusively male war parties assembling in a mountain district, perhaps living off game and wild plant materials for a week or two prior to a raid, and returning to these rugged lands with their livestock booty can be considered effective land occupation.

WILDLANDS

AND

WILDERNESS

Introduction

Karl W. Butzer

In settings as diverse as California, the deserts of the southwestern United States and northern Mexico, Paraguay and Chile, the Azores, and northern Australia, Homer Aschmann was a keen and impassioned observer of the impacts of modern societies on the natural world. Long before the establishment of Earth Day, he was an open environmentalist, deeply engaged in conservation issues and policy debate. This concern is plain in his presidential address to the Association of Pacific Coast Geographers in 1966, "People, Recreation, Wild Lands, and Wilderness." But before examining that wide-ranging and thought-provoking statement, it is instructive to explore his more intimate feelings on environmental issues.

A warm, outgoing, and delightfully unceremonious person in day-to-day contact, as a lecturer Aschmann was a little uncomfortable and, like most of us in the academic world, in print usually quite formal. Fortunately, his cluster of book reviews provided a less-confined outlet for his thoughts on ecological works. Collectively they represent his characteristic pithiness, his wit and his self-ironies, and his willingness to call a spade a spade. Excerpts from several of them are found throughout this volume, and one, a review of the Sierra Club's *Wildlands in Our Civilization*, is included in this section in complete form.

As a group these book reviews also show a man filled with zeal for the environment, and one who shunned affectation, polemics, and polarization. He found Dudley Lunt's *The Woods and the Sea* a little too precious for his taste. He put down Lunt's pretensions as follows: "Walks in marshes and on beaches and canoe and fishing trips . . . are described in a style deliberately, and to me a bit painfully, patterned after Thoreau's." A little less pungently with respect to Paul Brook's *Roadless Area*, he wrote: "[A couple] saw, heard, smelled and felt the environment with open receptiveness, and they liked it." In regard to "practical advice for the non-athletic visitor who seeks solitude but not discomfort" he adds, with a breath of apparent satisfaction, "The prospect pleases."[1]

Aschmann accepted the need for and ultimate scientific value of debate within the environmental movement. In reviewing Frederick J. Pohl's *Perspectives on Conservation*, he described the two fundamental oppositions cutting across the collection of papers, criticizing those on the extremes ("nonsensical or vicious viewpoints") but stressing that "It is between (the) intermediate positions that discourse is possible, and it will and should be political." Yet he remained uneasy that "conservation arguments have not reached a common consensus among those most informed and concerned."[2]

At times he was frustrated by the extremism he sensed as rampant among some segments of the environmental movement. In analyzing the Sierra Club's *Wildlands in Our Civilization*, he concluded that "the bulk of this volume is a rabid polemic by a group of fanatics" and that "an increase in scientific knowledge is not the real goal of the protagonists."[3] It was with a note of relief that he added, "not all the participants share the editor's view that absolute elimination of man's works from wild lands is the goal." Even so, he expressed some approval in regard to "growing confidence that the extreme wilderness ideal can find public support and more realistic concern about specific policy issues *rather than platitudes about educating the youth to appreciate their wilderness heritage*" (emphasis added). He subsequently changed his mind about the thrust of the Sierra Club: "once their premises are accepted, their actions and positions shift from fanaticism and anti-popularism to the most enlightened altruism. The Club can only be criticized for excessive moderation."[4]

Aschmann's concern for environmental preservation was never in doubt, but he worried about the vigor of the dialectics, fearing that extremism would undermine public support. In referring to one author, he lamented that "his declarations of omniscience and diatribes against those who doubt it will elicit little support from the uncommitted."[5]

Aschmann opposed the "authoritarian liberal" or the proponent of welfare programs as much as he did the conservative economist who claimed that accelerated depletion of natural resources was an acceptable form of management. He was a centrist. He stood for common sense, not ideology, and his position was unambiguously clear. The choice, as he expressed it in "People Are No Damn Good," was between utilized environments "modified sensibly, sensitively and diversely" and those "reshaped with bulldozers into economically efficient esthetic sterility."[6]

What, then, were his views? The concept of a *wild landscape* was first spelled out in a 1959 paper on the persistence of the wild landscape in Southern California, the last paper appearing in this volume. His premise is that there has been no significant climatic change in the region during the last 10 millennia; consequently, the character and distribution of the plant communities described across a 300-kilometer transect from the coast to the Colorado River should relate to (a) climatic and edaphic parameters, and (b) human impacts during aboriginal and historical times. Although Aschmann's preoccupation is with the environmental interdigitation of the indigenous peoples before 1769, the paper does address recent human influences on or destruction of particular biogeographic communities. Wild landscapes, although not explicitly defined, refer to surviving areas of spontaneous vegetation within various belts of potential or "natural" vegetation.

Most problematic is the *chaparral*, the tangle of evergreen shrubs and scrub oak or sumac typically found on rougher parts of the coastal piedmont at 300 to 1,200 meters elevation. That same chaparral is the key element in the continuing fire hazard for suburban communities in the Los Angeles area, and it clearly is a subclimax vegetation adapted to and maintained by periodic burning. The brushy component is resinous, and highly flammable when dry; plants such as chamise *(Adenostoma)* sprout from thickened root bases after fire destruction or recolonize from fire-resistant seeds of long viability. Fire potential is cyclical in that the vegetation builds up, particularly during wetter years, until a threshold of combustible material has accumulated; thereafter it takes a combination of a dry year, Santa Ana winds, and a spark to set off another conflagration that resets the clock. It is this potential for periodic disasters that makes the chaparral a practical, economic issue as well as a biogeographic question.

Chaparral is structurally similar to the degraded *matorral* widespread in other mediterranean-type environments.[7] All are liable to cyclic fires, but the Santa Ana phenomenon (made possible by a unique topography and climatology) creates uncharacteristically rapid, extensive, and high-temperature infernos that can kill mature oak trees. Recent work on archaeological charcoals in the western Mediterranean Basin demonstrates that matorral was first created during periods of neolithic or Bronze Age settlement, presumably as a result of the use of fire to open up deciduous oak forests for pasturage and localized cultivation.[8] Left to regenerate, Mediterranean matorral becomes almost impenetrable and of minimal pasturage value in as little as a decade, unless subject to controlled burning. Pollen studies in southwestern Spain show that between 4,000 and 500 B.C., methods of matorral control were perfected, so as to maintain permanent pastureland *(dehesa)*, with grass and widely spaced oak trees (50 percent canopy cover).[9]

Aschmann argued convincingly that the California chaparral is a fire subclimax, but the archaeological and ethnographic evidence remains slim as to its use, condition, or extent until well into the 19th century. Given the exceptional Santa Ana phenomenon and the frequency of lightning impacts, the California chaparral may possibly have evolved without deliberate burning, although reinforced by the involuntary spread of indigenous campfires. But the new evidence about Spanish dehesas adds support to the idea that California's oak parkland community, found in flat valleys or on more level parts of the piedmont with 500 millimeters or more of precipitation, evolved as a result of deliberate burning in prehistoric times, perhaps accentuated during the Hispanic ranching era. Archival research into Hispanic range-management practices remains a possibility for following up this lead.

Aschmann did not espouse a megahypothesis about the role of deliberate fire in vegetation change, insisting instead on comparative regional studies, such as between the evolution of matorral in the mediterranean climate zone of Chile and the California chaparral. He emphasized starkly different trajectories and outcomes, and his research was highly rated in the hard-core biological community, as demonstrated by his seven "invited" papers published in international symposium reports or proceedings between 1973 and 1991 (contained in the bibliography elsewhere in this volume). It is a pity that his report on forest destruction since European colonization of the Madeiras and Azores, prepared for the Office of Naval Research after his 1960 sabbatical, remains inaccessible.

Suppression of fire also had significant impacts on vegetation change, creating ecological situations in wilderness areas that did not exist in prehistoric times. Asch-mann noted the unsightly brush and undergrowth taking over in Yosemite, a prob-lem underscored by the destructive forest fires in Yellowstone National Park a few years ago. This finds strong support in more recent palynological work. American forests protected from fire today change their dominants with respect to those of 1,000 or 1,500 years ago; for example, in a section of northern Minnesota that had no indigenous disturbance, forest composition after three generations of protection from "natural" forest fire finds no structural analogs in the pollen record.[10] The quest to identify pristine climax vegetation is illusory, as Aschmann recognized, because in Southern California, fire was a key factor prior to human occupance and even more so thereafter. Holocene vegetation and human land use coevolved, as Aschmann explicated in a more pragmatic idiom.[11]

This brings us to one of the points of his presidential address: "Most wilderness advocates seem to think that the territory of the United States was a wilderness as long as only Indians occupied it and ceased to be when Europeans appeared." That misconception draws from the romantic idealists of the 19th-century environmental movement, and was polished up for the Columbian Quincentenary in the ecologi-cally correct image of the American Indian living in oneness with nature.[12] Not only is that "benevolent" stereotype incorrect, but it also is pejorative to American Indi-ans by its implicit denial of their complex and technologically sophisticated systems of land use, or of the great urban and ceremonial centers that they created. At the same time such simplistic myths reek with misinformation about ecological equi-librium and diversity that can seriously muddy the waters at a time when critical con-servationist decisions must be made for the future.

A second, noteworthy point of Aschmann's presidential address is that the frayed city dweller in search of psychological solace or aesthetic pleasure in nature need not flee to the protected wilderness of the high sierra. There are many other wildlands of natural beauty much closer, even if they are not quite pristine. He cited the Coast Ranges from Santa Barbara to Monterey as an example: "This is the California land-scape par excellence with rolling hills covered with grass and scattered oak trees . . . [where] landscape modification involves no more than 5 percent of the total surface. I would submit that such a landscape can afford the same satisfaction." The problem, he noted, was and remains the absence of qualified public access to vigorously de-fended private property, commonly used for tax havens.

Following Aschmann's train of thought, we become aware that the overempha-sis of "wilderness" can indeed distract us from seeing beauty in those humanized landscapes where nature and human works are integrated in visible harmony. From the Flemish landscape painters of the 15th century to Constable, traditional European cultural landscapes were enjoyed as redolent of beauty. The landscape painters took great satisfaction in the symbols of human husbandry and, as P. L. Wagner has ar-gued, a landscape can serve to express a society's ideal environment, symbolic of correct, harmonious behavior; that landscape can bind past and present, while also instructing and informing each generation.[13] Whereas Southern California's mechanized monocultures—the "factory in the field," as Aschmann derides them—hardly fit that bill, there are countless rural parts of California and the United States—some small, some large—that do. We cannot let ecological purists diminish our pleasure when they dismiss such vernacular landscapes as a testimonial of human despoliation, biotic extirpation, and introduction of nonindigenous plants or animals.[14] Aschmann urges us to shape our everyday environments in a manner that will provide greater aesthetic satisfaction and contact with nature.

What surprises and disappoints is that so little has changed since Aschmann's presidential address. As he predicted, the battle between the developers and the conservationists would go on unabated, a partial victory for wildlands on one of countless fronts representing little more than a line demarcated for renewed conflict. The divergent goals evident within the environmental movement that he so feared have become polarized positions, with eco-radicals castigating now-mainstream nongovernmental organizations such as the Sierra Club for their purported complacency.[15] Everyday people of goodwill are being turned off by radical positions or behavior. And the unrealistic wilderness ideal continues to distract from much broader concerns that should encompass diverse and extensive tracts of vernacular landscape as well as the remaining but undramatic wildlands.

People, Recreation, Wild Lands, and Wilderness

He who wishes merely to look at a landscape in which the works of man are not apparent will find it ever harder to satisfy his desire.

[W]e may be prepared to recognize and perhaps even restrict the range of landscape-defiling, space-consuming machinery.

[P]rojections of future pressures on wilderness can only terrify.

WE ARE DOING spectacular things to our environment without any recognizable intent or plan. Complaints arise when the aquatic life of Lake Erie is destroyed, or water foams from taps in Reno because of detergents dumped into Lake Tahoe, or the sweeping view of the Embarcadero [San Francisco] is to be cut off by a freeway. But progress usually wins over such peripheral squawks. The real problem is that we as a society have not really decided what kind of environments we want to create or preserve. Our technological engine of inordinate power has not been put in gear except tentatively, but it is becoming ever more powerful. Until we know better what we want, I can only regard it as fortunate that the moon rather than the places closer to home is receiving our attention. This paper is directed toward inducing some deliberate and relatively unrestricted thought about what the world, or the United States, or Southern California should be like a decade or a century from now.

My recent preoccupation with this question was provoked by being asked to review a recent polemic book published by the Sierra Club and entitled *Wildlands in Our Civilization* [the review follows this essay].[1] The Sierra Club's attractive propaganda does not insult the intelligence nor jar one's sensibilities, and I tend to react positively toward it. As the book title suggests, one of the Club's concern is to preserve wilderness areas for society's benefit. It is their explicitly stated belief that such wilderness areas can afford more to man in their present state than if they were developed and put to use as sources of timber or water power or made more accessible for recreational purposes. Their premise is that a human being, to enjoy the fullness of his own life, must from time to time be able to isolate himself from other members of his species and sense other aspects of the community of nature. The dedicated enthusiasm of the proponents of the wilderness area is inspiring. The ultimate hopelessness of their cause and the inadequacy of their goals, even if by fortune and effort they should all be achieved, are depressing. One is reminded of Stephen Vincent Benét's nightmare [the poem "Nightmare, with Angels"], which includes the lines

> You will not be saved by General Motors or the prefabricated house;
> You will not be saved by dialectical materialism or the Lambeth
> Conference;
> . . . In fact you will not be saved.[2]

We might add a codicil,

> We will not be saved by the wilderness idea either.

A more comprehensive approach appears in the California Public Outdoor Recreation Plan.[3] As is appropriate in an open society, the wants of essentially all interest groups—those who like to look at scenery from a car window or after a vigorous hike, those who hunt, and those who find a crowded beach attractive for reasons which are probably obvious—receive consideration. Real numbers and reasonable projections appear. Plans for implementation are not completely unreasonable, and if carried out vigorously the outdoor recreation opportunities in the State would not deteriorate too much by 1980. After that one can only hope that his own aging will dull his sensitivity to the loss. A further conclusion must come from perusal of the Plan. If the ordinary economics of politics are applied and the costs of preservation and development are divided by the number of effective users, it will go hard for the person who craves to view scenery in lonely isolation. Because of their willingness to spend money, hunters and fishermen can hope for some protection and support, and the hunter at least creates his own glacis. He who wishes merely to look at a landscape in which the works of man are not apparent will find it ever harder to satisfy his desire.

In an operational sense the problem arises subjectively, and at this point one who has lived some decades in Southern California may be especially sensitive. The theme of paradise lost is so easy to dote on that most of us seek to keep it out of our consciousness most of the time. The idea of walking to a wild or even a farmed landscape beyond the rows of houses is ridiculous for all but fringe dwellers and they know that their period of grace is limited. Reaching a wilderness in a one-day automobile trip becomes progressively more difficult and above all is likely to dump the traveler into a mob scene peopled with others with similar unsatisfied desires.

The problem of crowding in the declining number of accessible wild areas has become worse. California's growth rate is slowing but is far from stopping; work weeks are shortening; living standards are rising; projections of future pressures on wilderness can only terrify. The chance to be alone with nature on occasion that was once the American's birthright, whether or not he ever sought to exercise it, may be lost within our lifetimes, and barring nuclear catastrophe it will never be reinstated.

The question is really, does mankind as a social organism need access to wilderness? If he does, even the notations sketched above demonstrate that tremendous efforts to preserve it are required immediately. Perhaps only a small fraction of the species has this requirement, and those who have the need will be bred out rapidly as the need cannot be satisfied. The evidence is less than conclusive. Clearly, some city dwellers reach 3-score and 10 without leaving house and pavement. Rising tensions in central cities, however, cannot be overlooked, and some recent experiments with rodents, kept with abundant food and adequate sanitation but with ever higher population densities, are suggestive.[4] As you are all aware the experimental populations were afflicted with social malaise; fighting and infanticide finally reduced the population density. We, of course, are a different species, and we have culture. In fact one can develop a remarkably coherent theory that the most fundamental cultural creations had as their functional goal the preservation of privacy for the individual and for intimate groups within growing and economically interdependent societies. Properly diverse examples of such inventions are the universal incest taboos and housing. But are these inventions enough? Do we still need contact with other wild species and with land without artifact? Let me ask you to accept for consideration the proposition that we, or at least many of us, may have need for occasional access to a wild area in reasonable isolation. Crowded campgrounds, no matter how picturesque and sanitary, or scenic highway lookouts will not do.

Perhaps the most articulate advocate of the above proposition is the Sierra Club, and once their premises are accepted, their actions and positions shift from fanaticism and antipopularism to the most enlightened altruism. The Club can only be criticized for excessive moderation. By their rather strict definition of wilderness— one which allows no permanent human habitation, exploitation of resources, or facilities for mechanical vehicles—some 2.2 percent of the contiguous United States remains in that condition. The possibility of making wilderness out of currently used land is recognized as hopeless. With literacy, photographic virtuosity, and effective exploitation of media of communication they attempt to block any desecration of extant wilderness. They and their associated organizations have become a force to be reckoned with and in many cases development has been blocked. But their war has many fronts. No victory is ever achieved, only a stalemate with recurring possibilities for new fighting and final loss.

A few recent or current actions may be listed: the San Jacinto Mountain tramway, a ski lift and access road in the high heart of the San Bernardino Mountains, a better highway versus the redwood trees in Humboldt County, dams that would flood a little of the Grand Canyon National Monument, a reservoir in the California condor refuge. The list could be lengthened, but these points stand out. It is not only the greedy mining concern, grazer, or lumberman, Two-Gun Desmond, vulnerable to attack by Bernard DeVoto, or even John Collier that assaults wilderness areas. It is also the Bureau of Reclamation with a scheme to make the desert bloom, industry flourish with low-cost public power, and the whole economy of a region expand. Or it is the recreationist seeking to enlarge with access roads the number of citizens who can and do experience our natural wonders. Before the ink was dry on the Wilderness Act of 1964, Congress set in motion studies for new dams to utilize more effectively the

West's limited water resources. The greatest victory merely established a line for renewed fighting.

Perhaps you shared my shock when we read recently of the bureaucratic arrogance of the Internal Revenue Service's threat to the Sierra Club's tax exemption status because the Club is agitating against further dams on the Colorado. Distasteful as such action is, it is instructive. How can a special interest group claim tax privilege when it fights against progress and enhanced prosperity for the entire nation? The outcome of this contest is not so significant as the assurance that it will be renewed. Two-Gun Desmond is now the high-minded public servant seeking to promote national welfare and being obstructed by a few noisy wilderness fanatics.

The Sierra Club's battleground for the contest is terribly clearly defined: we will get no more wilderness. It is terribly fragile; erosion, degradation of vegetation, and extinction of animal species can come about from the most innocent access road. Once disturbed, the wilderness character of an area cannot be restored for generations, perhaps forever. Let us bend every effort to preserve, protect, and defend what we have.

In the struggle over wilderness that has raged for the past two decades, a few points have been clarified, and this presentation would be needlessly cynical if it did not note such developments as progress. Recreation planners and managers are now prepared to segregate users so that when one wants conflict, it will not be the noisiest and most obnoxious or even the biggest spenders who always take over. As a corollary we may be prepared to recognize and perhaps even restrict the range of landscape-defiling, space-consuming machinery: power boats, trail bikes, et cetera. The hunter and shooter has fought well on the solid ground of game management. He may yet have to struggle to defend the space he requires. The absolute protection of tiny plots for scientific study, especially that of an ecological and ethological nature—of wild plants and animals in a diversity of habitats—seems to be established as a legitimate social goal. Protection of a species from extinction has enough emotional appeal to justify uneconomic efforts within limits because of the absolute finality of failure to do so. Within these clarified frames of reference, the managers of publicly owned wildlands can probably improve their practices to everyone's benefit.

Certain inherent anomalies in the programs of the proponents of preserving wilderness remain. Most are individually familiar to everyone who has concerned himself with the problem. What may be new is a recognition that they make the maintenance of wilderness impossible in a free and democratic society.

Fire is a nice example. The conservationist and wilderness advocate abhors it and may have become involved in the first place because he saw the devastation resulting from man-caused fires. But not all fires are caused by man. In many climatic situations recurring fires are a part of nature and an essential element in the wild landscape. We need not confuse the bottom of Yosemite Valley with a wilderness, but there we can document the fact that the former lovely park landscape was maintained by repeated burning. Full fire protection has led to a far less appealing brush and forest tangle, though the parking lots and campgrounds are in another league for unsightliness. In less accessible areas, however, should there be access roads to make easier the limitation of naturally caused fires, or should such fires just burn themselves out?

The role of the American Indian has been deeply involved in the wilderness idea, both historically and currently. Most wilderness advocates seem to think that the territory of the United States was a wilderness as long as only Indians occupied it and ceased to be when Europeans appeared. It is true that the Indian had no bulldozers, but he did have an impact on the landscape, exploiting heavily certain plants and animals, occupying permanent or semipermanent settlements, and creating trails, some of which can still be followed. Most importantly he used fire heavily to clear land, to drive game, and probably for recreation. No zoning limited his activities except that he could not remain long in areas so cold or dry that they afforded little food. The flora and fauna that the first European explorer of any part of the United States saw had been modified by at least 10,000 years of human use. The modern plan to zone out people is creating something that never existed before, as it is now clear that Pleistocene climatic patterns prevailed until after Indian arrival in this continent.

At present a very considerable portion of the remaining wilderness and primitive area in the country is on Indian Reservations. Reasons for this situation can afford little self-satisfaction to the American society, and the most important one is certainly that the Reservations contain much low-grade land. Another is that paternalism of the Indian Service has not resulted in economic development on Reservations comparable to that enjoyed by the rest of the country. With incredible liberal self-righteousness, John Collier, the former Commissioner of Indian Affairs and now a wilderness advocate, can brag that he usually could persuade the various tribes to leave large fractions of their Reservation lands roadless.[5] It is scarcely surprising that a fair portion of Americans believe that keeping Indians permanent wards of the Indian Service and specially protecting their lands from development is an unappealing form of segregation. Museums of live people, even if the inmates are well cared for out of public funds, require someone like John Collier, who confuses himself with God, for their justification.

The greatest contradiction, of course, arises from the fragility of true and strictly defined wilderness. In the first Biennial Wilderness Conference in 1949, sponsored by the Sierra Club, the dominant question was: Can a small body of enthusiasts educate enough individuals, especially youths who would be introduced to wildlands, to create mass support for the wilderness idea? Subsequent conferences show greater confidence as success in arousing enthusiasm outstripped expectations. Wilderness areas reasonably accessible to large population centers are already experiencing so much visitation that their wilderness character is threatened in both a subjective and an objective sense. They no longer evoke the sensation of isolation, and flora and fauna are perceptibly disturbed along heavily traveled trails. Projection of three parallel trends: population growth, increased leisure, and more effective education for wilderness appreciation permit only one conclusion.

The perceptive wilderness advocates are becoming concerned, but even moderately feasible adjustments are unpalatable. Building an obstacle course so that only athletes could enter a wild area is perhaps too ridiculous, and we are not likely to tear up extant access roads, though this may yet be the best solution for Yosemite. Some form of rationing may be necessary to assure each individual access to unspoiled wilderness, should all of our society seek such access—and the announcement of scarcity by rationing is likely to stimulate demand—a week per person per year would be too much. Two weeks per decade is a reasonable but pretty thin diet.

The other prospect is de-education. Wouldn't you really rather go to Disneyland? The mosquitoes in the Sierra are terrible. My own predilection favors such an approach but the consequences are patent. Such mass support for preserving wilderness as it exists or might be generated is essentially precluded. And it must be reemphasized that the modern threat to wilderness comes not from easily maligned greedy private interests but from public servants dedicated to economic progress and increasing the nation's wealth and prosperity. Hetch Hetchy serves a million thirsty people.

The picture could be painted blacker, but this should be enough. Let me state my own credo. I am glad the Sierra Club and similarly motivated organizations exist. I hope the Dutch Boy grows enough fingers to keep plugging holes in the dike, that obstruction to the erosion of progress continues to be effective, and that we retain wilderness areas into the distant future. Wild habitats for scientific study in as wide as possible a variety of environments can and must be preserved. But to be useful they must be restricted to a limited number of scientific investigators. On moral or religious grounds many of us will regret and seek to prevent the casual extinction of any living species. Even within a few decades, however, the lonely wilderness will not be available to all those who may seek it except on a rationed, once-in-a-lifetime, basis. If man does have need for the good of his soul and the balance or humanness of his personality for direct contact with nature at reasonably frequent intervals, the wilderness area cannot provide it for him. Our world is already too crowded and it is becoming more so. Fortunately at least one other avenue toward solution of this fundamental human problem still exists.

As the human species, slowly at first but at an accelerating rate, extended its range and increased its numbers to become the biologically dominant large animal on earth, it necessarily modified ecological communities and landscapes. It had to, and there is no turning back. A wilderness world might support between 1 percent and $\frac{1}{10}$ of 1 percent of present human populations, and not especially comfortably. We can hope that the world's population growth curve will level off with no more than a doubling of the present total, though that is optimistic. These people need a modified, that is, an agriculturally productive, landscape on most of the earth's land surface, to support themselves. Their dwellings, mines, manufactories, and communication lines are similarly essential. Of such is the humanized world, and in it we must spend almost all if not all of our lives.

It need not, however, fail to provide us with rich contacts with the burgeoning diversity of nature. The cultivated agricultural landscape need be little if any less interesting than the wild one. If it be argued that the efficiencies of monocultures are greater than more diversified cultures, they also have their drawbacks and the real economic gains are remarkably small. What is essential is recognition that the space of the earth is finite and must serve all of us. If I do not work the field, I have no entitlement to a share of its yield, but I do have a right, especially if I need, to look at it and be pleased thereby. At this point two sets of claims may be pressed against our wealthy society with its enormous, unused technological competence. Let us consider the larger and more general claim first. What should the humanized world that supports us look like and be like?

In a wonderfully perceptive essay, published in 1952 and entitled "Human, All Too Human Geography," J. B. Jackson postulated that man's goal in modifying the wild landscape is ultimately to create heaven on earth.[6] The house afforded privacy to the individual and intimate family group; the fields concentrated want-satisfying fruits; and the whole interrelated community structured its works to satisfy physical and social desires. If the arrangement was less than perfect, cultural evolution would move in the direction of the desired goals. If access to the observation of other forms of life was such a goal, it might readily be supplied. The Orient, tropical Latin America, and other less developed parts of the world suggest the preservation of some of this harmony. Perhaps the private intimacy of even the tiniest Japanese garden is a peculiarly successful achievement.

It is hard to view the humanized world of the Southern California coastal plain as a whole as anyone's concept of heaven, though many of its elements clearly have that attribute. I think of the separate suburban house with its yard and lawn. Somewhere in the agricultural and industrial revolutions the economic advantages of specialization and rationalization were allowed a dominance that we have failed to control. Because they occupy more space even than suburbs, our agricultural lands reflect the unbalance more seriously. The monoculture stretching for miles, interrupted only by roads on which produce is hauled out and supplies and gear needed for production are moved in. The factory in the field affords only its produce to the man who works it. There is aesthetic gratification neither for him nor for the non-farmer.

I would suggest that the workaday environments so needed to give us physical support can also provide infinitely more aesthetic satisfaction and contact with nature than they now do. In fact they must; for there is now or shortly will be no other place to go. So great a problem can only be stated at this time, not solved. Making, by trial and error or by ingenious insight, our humanized world the closest approximation to paradise would seem a task capable of employing human creativity into the indefinite future. My only personal charge or dictum at this point would be that my paradise will certainly include diversity. I hope yours will too.

To move to a more operational level with potential immediate courses of action, almost all of you know the Coast Ranges from Santa Barbara to Monterey. This is the California landscape par excellence, with rolling hills covered with grass and scattered oak trees. Rangeland is the common designation. It is no wilderness but privately owned, agriculturally moderately productive, tax-paying estates. Houses are rare, though ubiquitous fences show the hand of man. An occasional bottom or gentle slope is cultivated, but such landscape modification involves no more than 5 percent of the total surface. I would submit that such a landscape can afford the same satisfactions to the visitor who passes through on foot as the ardently sought and protected wilderness of the High Sierra.

If you have ever tried, you know the answer. It is often impossible even to get off or slow down on the freeway. Adjacent to every country road there is an interminable barbed-wire fence. Crossing it makes one liable to arrest for trespass, and pitching a tent or unrolling a sleeping bag is a jail offense. Why? Surely it is private property and I might scare the cows. I might also start a fire, but the fire-conscious National Forests accept my bona fides of reasonable discipline in season. Well, it would be a lot of bother to the owner if people wandered over his land.

Curiously, even as a foreigner I have these hiking and camping privileges in comparable terrain across the border in Baja California. The sorts of pressures on space that afflict our public lands can only increase to intolerable levels, but here we grant sanctuary to a small number of landowners whose land could add immeasurably to public welfare. That such lands often are held in low productivity at low tax rates as a tax haven for future subdivision makes the abuse a little more overt.

It does not seem necessary to play Zapata and cry for confiscation of these estates. We have in land taxes a potential but unused social tool. They are proving to be miserably unsuccessful as a prime source of revenue. Privacy is taxable too. We pay for it rather adequately on our suburban lots, and the owner of a 160-acre farm might well choose to exclude visitors and the cost of one or five dollars per year per acre. Would the owner of 20 square miles, not a large estate in California range country, choose the same exclusiveness? The privileges requested here are only those sought in a publicly owned wilderness. Cross-country jeeping and trail bikes that abuse the terrain make excessive demands. Hunters are probably willing to pay special fees, and since they drive out anyone else with sense they must be restricted to short seasons regardless of the state of the wildlife; the random shooter, an overabundant sort of vermin, should be locked in a special and limited range.

This diatribe, Jeremiad perhaps, can only conclude with the plea that the subject is important. The quality of life for the next and all future generations is subject to serious threat. The present wilderness reserves are already inadequate. The modest palliatives here proposed probably are too, but serious consideration and creative discussion are called for.

People Are No Damn Good

Man has placed his mark on most of the world's lands; growing populations will extend and deepen the imprint.

[T]rue wilderness is a myth.

... [T]he bulk of this volume is a rabid polemic by a group of fanatics. Although "wildlands" appears in the title, the subject is "wilderness," defined strictly as extensive areas in which evidence of man's presence is not observed. To the editor and several of the contributors a small check dam that sustains stream flow and permits trout to survive in a mountain stream is excessive human interference. The chance to get away from other humans and all their works for the spiritual regeneration the experience offers, even if the opportunity can only be exploited at rare intervals and for short periods, is what is wanted. Statements are made that satisfaction may come from knowing a true wilderness exists even if it is never visited. Some effort is made to gain support for the maintenance of wilderness areas by noting the scientific value of ecologic studies in undisturbed habitats, but an increase in scientific knowledge is not the real goal of the protagonists.

The overwhelming concern to the editor and to contributors ... is a valid one. Wilderness, so strictly defined, is infinitely fragile. Very little survives in the contiguous United States, and that can be destroyed with terrible ease by a few access roads to enhance its worth for outdoor recreation. After a season's exploitation the wilderness character of an area cannot be restored for generations. The book's exhortation is to protect by legislative act and incessant outcry every bit of wilderness that remains against encroachment, regardless of how insignificant or publicly beneficial. The authors are not naive; they know that the Dutch boy will need more than his 10 fingers to protect this dike, but their enthusiasm to try is unlimited.

... [O]ne detects a growing confidence that the extreme wilderness ideal can find public support and more realistic concern about specific policy issues rather than platitudes about educating the youth to appreciate their wilderness heritage. Further, there is no concealment of the fact that not all participants share the editor's view that absolute elimination of man's works from wildlands is the goal.

The role of fire is a substantive issue that keeps cropping up and is unresolved. The ordinary wilderness lover abhors it and finds himself in a quandary as to whether or not to allow access roads for fire control. Relatively few participants seem to have comprehended Leopold's notation that modern levels of fire control are creating ecological situations in wilderness areas that did not exist in prehistoric times. The related issue that Indians have occupied the continent for at least 10,000 years and did not recognize boundaries within which the land should not be disturbed is not raised. It is true that modern wilderness areas are such because they were not attractive to settlers, and they probably received only light use by the Indians. But they have been hunted over and probably burned for millennia. Except in polar regions there probably has been no significant part of the inhabitable world untouched by man since Madeira and the Azores were burned over in the 15th century. The true wilderness is a myth.

I am glad that fanatics devote themselves to protecting a few areas from spoilation for commercial gain. Their own deliberations, however, make it clear that what they can hope to save is at most a token or symbol. It will not satisfy even the present population's need for a chance to experience nonartificial landscapes. Man has placed his mark on most of the world's lands; growing populations will extend and deepen the imprint. Our relation to our environments will be rewarding or depressing as utilized lands are modified sensibly, sensitively, and diversely or reshaped with bulldozers into economically efficient aesthetic sterility.

The Evolution of a Wild Landscape and Its Persistence in Southern California

In any event the wild landscape the European explorers found was a product of millennia of such disturbances.

[H]ow much of the region would look like home if an Indian resident of the mid-18th century could see it now? A reasonable answer is: much more than the contemporary visitor or many permanent residents might suspect.

D ESPITE EFFORTS on the part of archaeologists to work out a system of post-glacial climatic fluctuations in the Southwestern United States, the bulk of the evidence suggests that for the past 10,000 years the climates of Southern California have been as they are now. As in the historic period, wet and dry cycles lasting a decade or two probably alternated in rather irregular fashion throughout this long time. Since so much of Southern California is arid, semiarid, or subhumid even these minor climatic fluctuations would have had a substantial effect on the environment as far as its ability to support a nonagricultural human population is concerned.

Prior to 8,000 B.C., however, climatic fluctuations of far greater amplitude had occurred; specifically, the last major glacial advance (Tioga or Wisconsin) was a time of more southerly storm tracks and greater winter precipitation, although temperatures in the lowlands need not have been substantially lower. The now-dry lake basins of the interior deserts, and the oversize stream valleys of the coast provide incontestable evidence that more humid climates existed not too long ago. During such times plant communities involving forests, now largely restricted to highland areas, were more widespread; since then there has been an increase in the more xerophytic plant communities as the forest retreated to higher, wetter localities. In Southern California most of the forests are relicts of a wetter age, hanging on but often incapable of reestablishing themselves if seriously disturbed.

Throughout this 10,000-year period of modern climatic conditions, the so-called natural vegetation of many marginal districts has been moving toward a new equilibrium adjusted to drier climates. But in this shift human activities have played a continuing part, and man still affects a substantial part of even those plant environments that he does not completely dominate as lawns or cultivated fields.

What the natural vegetation of Southern California would be like if it were completely free of human influence is a question that is just as subject to an observationally supported answer as the famous "How many angels can dance on the head of a pin?" Whether or not we choose to accept the full time span for human occupation of western North America that George Carter has suggested, it is generally agreed that man has been present the whole time since the local climates attained their present character. Archaeologically demonstrated hearths show that he controlled fire. I shall attempt to show that the native population was numerous, almost ubiquitous, and that it deliberately burned, though it probably could not and did not put the fires out once they were set. The wild landscapes are products of plants and animals adjusting to reasonably stable physical environments and each other. But deliberate, extensive burning has been a continuing feature of most of the environments. Though forest and brush fires make newspaper headlines every summer and fall they may be less abundant now than they were before 1769.

The Natural Vegetation of Southern California

Although a few introduced species of plants, particularly certain grasses, have naturalized themselves in Southern California and have become extremely abundant during the last two centuries, the accounts of early explorers seem to describe accurately the same vegetative associations that we know today in the uncultivated parts of the region. The explorers' notes, however, are too fragmentary to permit an assessment of the extent of each association which might be compared with present or late 19th-century distributions.

Working from the seacoast eastward, the following are the principal plant communities found today in Southern California (Figure 32). All are mentioned in the earliest explorers' accounts in the same ecologic settings. The classification of plant communities used here is adapted from that of Munz and Keck,[1] and is an especially attractive one in that both the floristic composition and the vegetative aspects of the communities are considered.

*   *   *

1. The Coastal Strand occurs on coastal dunes and sandy beaches (Figure 33). This community occupies only a limited area, but its composition is quite specific to the peculiar ecologic situation of relatively low rainfall but high humidity and a great deal of fog. Many of the plants are succulents. The naturalized *Mesembryanthemum*, or iceplant, distinguishes the modern flora, but the general aspect of the plant community probably has not changed too much from aboriginal times. The Indian population was dense in this zone and the nitrogen-enriched soils of campsites correspond to the heavy-feeding Chenopodiaceae and similar plants in the zone. Modern picnickers, if not too abundant, may almost match the aborigines in maintaining spots attractive to weedy, heavy-feeding plants.

2. Coastal Salt Marsh still fills the numerous tidal lagoons behind barrier beaches that have not been "developed" by modern subdividers, and these lands are costly to make suitable for residences. In fact, with the saltwater infiltration associated with

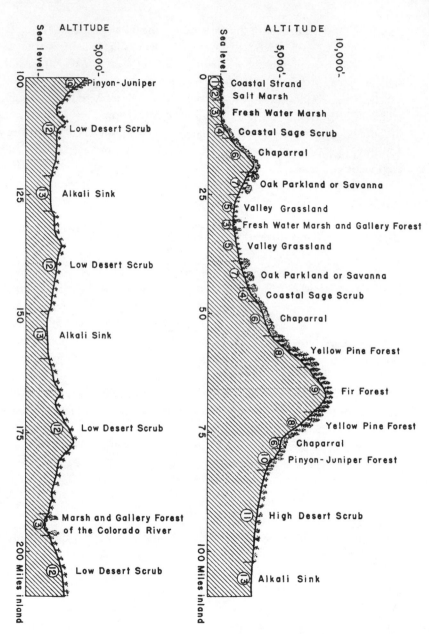

Figure 32. Plant communities of Southern California. This hypothetical transect is broken into two sections in order to show both a high desert and a low desert sector.

Figure 33. Coastal strand vegetation in the dunes west of Port Hueneme in Ventura County

falling groundwater tables, this ecologic situation, appropriate to a limited number of halophytic species, may be becoming more prevalent in the Ventura and Orange County lowlands.

3. Fresh Water Marsh, with its reeds and rushes, does not have the extensive distribution it had, and to some extent still has, in the Great Valley of California farther north. Areas regularly flooded by freshwater never were large in this dryland, and modern drainage control and water-storage projects have reduced them. This plant community is widely scattered throughout Southern California, and is dependent on local topographic and drainage conditions. A notable gallery forest often borders the reeds, rushes, and marsh grasses, frequently presenting the only native trees in an extensive lowland area. These are willows, sycamores, and poplars along the streams that drain toward the sea and willows, mesquite, and screwbean on the desert side.

4. Coastal Sage Scrub might be described as an impoverished chaparral (Figure 34). The semishrub plants are smaller and more widely spaced, but many of the same species are represented as in the chaparral. Coastal sage scrub may occur within a mile of the sea (as on Kearney Mesa north of San Diego and the seaward slopes of the Santa Monica Mountains), or as far inland as western Riverside County. It is appropriate to relatively gentle but well-drained slopes. Old, dissected marine or riverine terraces, the decomposed granite soils of the Peninsular batholith, and the areas of active alluviation at the heads of alluvial fans provide such an environment; even rather steep, south-facing slopes are likely to have a sage scrub cover. This plant association occurs primarily in situations that are climatically or edaphically dry. Sometimes the soil is notably porous, or it may have a shallow hardpan. This was a widespread but not a productive plant community for the Indians, and archaeological sites are few.

5. Valley Grassland is alluded to frequently by the earliest explorers, who noted that near various watercourses there was extensive excellent pasture.[2] While the stream courses were bordered by gallery forests, and great valley oaks were scattered

Figure 34. Coastal sage near the crest of the Santa Monica Mountains (photo courtesy of John F. Gaines)

through the lowlands, fairly solid stands of grass and herbs seem to have covered the heavy alluvial soils of the floors of the valleys in moderately humid localities. Such localities include much of the coastal lowlands and also some valleys at and above 4,000 feet in and east of the mountains. These grassland districts characteristically included the land most attractive for agricultural and urban development. Little of this wild landscape remains today. Furthermore, in the limited areas that are still uncultivated, introduced grasses and herbs such as wild oats, foxtails, and mustard have come to cover most of the surface; the native bunch grasses (*Stipa*, etc.) are notably scarce. Students of range management commonly attribute the growing dominance of the introduced grasses to the effects of grazing and overgrazing.[3]

6. Chaparral is generally considered the most characteristic plant association for all but the desert areas of Southern California (Figure 35). The evergreen shrubs and scrubby trees, forming a complete, smooth ground cover when seen from above, and developing a brownish-gray color as the dry summer progresses, still mantle extensive areas. The chaparral covers, and apparently has long covered, the steeper coastal hills, grading into coastal sage scrub on the smoother surfaces and on the edaphically drier south-facing slopes. It constitutes the typical vegetation on the lower slopes of the higher mountain ranges of the interior, especially the steeper ones. On the rainier coastal side it ranges from 1,000 to 4,000 feet above sea level, while occupying a nar-

rower belt (3,000 to 5,000 feet) on the interior sides of the various ranges. The lower margins of the chaparral on south-facing slopes are so degraded as to be more like the coastal scrub. Gentler slopes in moist localities often are covered with a park landscape of grasses and scattered live oaks. Toward the upper edge of the chaparral scattered clumps of big-cone spruce and yellow pine first take over shady, relatively moist spots and then form the dominant vegetation. The plant community possesses an extraordinary diversity of species, though chamise (*Adenostoma fasciculatum*) is almost everywhere the most abundant plant.

The chaparral is intimately related to fire, both that set by lightning and that caused by man. The brushy vegetation is resinous and highly inflammable toward the end of the long, dry summer. The component plants in the association have either or both of these characteristics: they sprout heavily and quickly from thickened root bases after even severe burning, or they produce a heavy crop of fire-resistant seeds of long viability (Figure 36). Many of these seeds seem to germinate only after being scorched or having the soil surface cleared of organic matter. Regular burning then does not eliminate the chaparral, and may cause it to spread into other plant formations.[4] One bit of evidence developed at the San Dimas Experimental Forest

Figure 35. Chaparral cover on the western slope of the Peninsular Range in Southern San Diego County. The elevation is about 1,500 feet, and the typical plant cover is 3 to 5 feet high.

Figure 36. Recovery of the chaparral association by root-sprouting and seeding following burning. This area, just west of Lake Elsinore, was burned over three years before this picture was taken.

on the south slopes of the San Gabriel Mountains suggests that this widespread plant community can only persist where there are recurring fires. The extent of the many fires in the San Gabriel Mountains during the last century has been plotted on large-scale maps. At elevations within the chaparral range only a few small spots, characteristically the outer points of spur ridges, have not been burned. These specific spots have an oak-grass plant association quite distinctive from the surrounding typical chaparral.[5]

Though a barrier to travel, and perhaps a little dangerous because of its bear population, the chaparral carried a large game population. Scrub oaks, an important element in the association, provided substantial vegetable foods to the Indians. Conversely, the chaparral offers little of direct economic worth to the modern population. It survives in substantial measure because the steep slopes it occupies are not easy to use for other purposes. In Italy they would probably be planted to olives and grapes, but labor is too expensive in California to make low-producing, dry-farmed orchards profitable. Also, chaparral resists, if it does not thrive under, repeated burning.

7. Oak Parkland or Savanna involves a park-like landscape, which for most of the year displays live oaks and other small evergreen trees, such as sumacs, scattered sparsely or densely on bright yellow fields of dry grass (Figure 37). In late winter and spring the grass is pale green, and the darker-leaved oaks stand out against it. In

the northwest corner of Southern California as here defined—northern Ventura and Santa Barbara counties—gray-needled Coulter pines are scattered among the oaks in roughly equal abundance. This plant association occurs on the leveler surfaces in the same general areas as the chaparral. It is found in the high flat valleys of the Peninsular Range, as near Warner Hot Springs, Santa Ysabel, and Vandeventer Flat, in lower valleys as at Calabasas in the Santa Monica Mountains, and formerly occupied the upper alluvial-fan surfaces along the southern edge of the San Gabriel Mountains. Most of these areas receive winter rains amounting to 20 inches or more, but all have a long, hot, dry summer. The existence of gentler slopes seems to favor the oak park as opposed to a chaparral vegetation.

Today the predominant grasses and herbs—wild oats, fescue, foxtails, and mustard—are recently naturalized introductions, and the lands are heavily and efficiently grazed. This is an economically attractive plant community and the ranchers struggle to expand it against the chaparral or at least prevent the latter's encroachment. Something like it, however, with native grasses and herbs, may go back to Indian times. The earliest accounts of localities now covered by this plant community mention good pasture, and the standing oak trees are often centuries old. Oak-park localities were economically attractive to the Indians, for, with their acorns and grass seeds, they supported large Indian villages. Annual fires running through the light grasses will scorch but not kill the oak trees (Figure 38). The grasses or their seeds sprout with the next winter's rain and, save for the problem of establishing new oak trees, an event which requires a series of years without fires, the association maintains itself.

Figure 37. Oak parkland in the Transverse Ranges near Sulphur Mountain, Ventura County (photo courtesy of Harry P. Bailey)

Figure 38. Oak parkland or savanna on the north side of the Santa Monica Mountains. The grass on the hill had been burned a few minutes before the picture was taken. Note that the mature trees were scarcely affected by the burning.

8. Yellow Pine Forest occupies the lands immediately above the chaparral on the higher mountains. On north-facing slopes Ponderosa pine and big-cone spruce may grow in favorable canyons below 4,000 feet; on south-facing ones a complete chaparral cover may extend above 5,000 feet; from 6,000 to 8,000 feet the yellow pines and incense cedar, mixed with the deciduous black oak, form a dense cover. Many chaparral shrubs, such as manzanita and *Ceanothus*, penetrate the lower, more open parts of the forest (Figure 39).

The visitors to these forests cannot escape the feeling that they are retreating upslope and being replaced by chaparral, a movement accelerated by modern logging operations. While the Indians did not cut the forests, they did start fires in the chaparral, and it is reasonable to assume that the lower edge of the forest has been rising slowly but more or less continuously during the last 10,000 years.

The yellow pine forest was not too attractive to the Indians except as hunting grounds, but the high meadows on poorly drained flats furnished vegetable foodstuffs. Here there are numerous remains of campsites, probably visited by bands from the surrounding lowlands for a few months every summer.

9. Fir Forests occur above the pines, occasionally as low as 6,500 feet but characteristically above 8,000 feet. This is a dense forest except near the windswept peaks. The relatively small areas at these great elevations were rarely visited by the Indians, and in general are little affected by the modern summer grazing and recreational visits they experience.

10. Pinyon-Juniper Woodland occurs below the yellow pine forests on the interior side of the major mountain ranges, occupying elevations roughly the same as the eastern chaparral belt, i.e., 3,500 to 5,000 feet. North and east of Southern California it is found near the crests of many desert ranges. The pinyon-juniper country has little economic attraction today. It is at best poor pasture or resort land. But it was worth much to the Indians who came up from the deserts every fall to harvest the pinyon nuts and juniper berries. This association seems to be vulnerable to fire, and after burning is likely to be replaced by chaparral. My impression is that this subarboreal vegetation now occupies the steepest and rockiest districts and lower-growing chaparral the smoother areas with deeper, more continuous soil cover (Figure 40).

A reasonable interpretation of this distribution is that the piñon-juniper only survives where, because of lack of soil, the plant cover is too thin to support extensive fires. Such a differentiation of plant associations in this zone of relatively uniform climate most likely long antedates the historic period.

11. High Desert Woodland and Scrub characterizes the western and northern Mojave Desert, particularly the long alluvial slopes north of the San Gabriel and San Bernardino Mountains (Figure 41). Typical elevations are from 4,000 down to 2,500 feet; annual rainfall is around 10 inches with high temperatures and low humidity in summer, and relatively severe cold in winter. The striking Joshua tree and juniper forest is prevalent in the higher and wetter areas, though a greater fraction of the surface is covered by more truly desert plants such as sagebrush and *Encelia*, and unusually tall creosote bushes (*Larrea*). Following rainy winters an understory of flowering herbs and grasses will provide excellent but ephemeral pasture. Only after an unusually wet series of years is the plant cover dense enough to maintain an exten-

Figure 39. A fairly open stand of pine on the western slope of the San Jacinto Mountains at about 6,000 feet (photo courtesy of U.S. Forest Service)

Figure 40. Piñon, juniper, and a few yellow pines near the rocky eastern escarpments of the Peninsular Range. The elevation is about 4,500 feet.

sive brush fire. Though not without edible seeds and small game, this region was of low attractiveness for the Indians. The scarcity of permanent water sources undoubtedly reduced their ability to exploit the region. Similarly, it is generally too dry to farm without irrigation, and is not much affected by the sporadic grazing on the ephemerals which sprout after rainy winters. Thus the association probably continues essentially unaltered from the distant past.

12. Low Desert Scrub perhaps covers more square miles than any other plant community in Southern California, occupying most of the eastern Mojave Desert and the Colorado lowlands (Figure 42). Creosote bush and burroweed (*Franseria*) are clearly the dominant plants, and over hundreds of square miles their widely spaced low bushes form almost the entire vegetation. Succulents, particularly *Opuntia* cacti, are locally abundant and in a few places there is a remarkably diverse though sparse xerophytic flora. That these deserts could support any nonfarming Indians is surprising, but they did. The desert plants in general spend more of their energy on reproduction and so provide far more concentrated nutrients suitable for animal consumption than would the same amount of vegetative matter grown in more favored regions. An Indian band occupied almost every permanent water source, at least seasonally, ranging out from these to seek game and edible seeds and roots. The edible *Agaves* may even have been seriously depleted near the waterholes by native exploitation. Today the Indians no longer seek wild plant foods, and this land is either

Figure 41. Chaparral grading into high desert scrub in Cajon Pass at about 4,000 feet. Chamise (*Adenostoma*) is the flowering plant in the foreground, with Joshua trees (*Yuccai*) in the background. (photo courtesy of Harry P. Bailey)

completely untouched or completely modified as are the extensive irrigated lands of Imperial and Coachella Valleys.

13. Alkali Sink vegetation occurs on and close to the playas of the numerous undrained basins of the Mojave Desert. This community is composed mainly of salt-tolerant species of the family Chenopodiaceae, especially of the genus *Atriplex*. Potable water is likely to be absent from these salty flats, and the historic Indians scarcely used them; nor does the modern population. During the Pleistocene, when there were permanent and overflowing lakes, however, this region was densely populated as is attested by the many archaeological sites around the extinct lakeshores. In those times, of course, a completely different plant community was present, probably freshwater marsh and gallery forest.

The Indian Occupance of Southern California

While from archaeological sources we have some knowledge of the size, distribution, and activities of the human populations that have occupied Southern California since glacial times, our data from these sources are thin. The protohistoric sites are the most abundant and productive and contain the greatest variety of cultural materials. These materials effectively supplement the contact ethnographic information provided by explorers, missionaries, and early settlers. By the time modern professional ethnographers (Alfred Kroeber and his associates) began to study the Southern California Indians, the native cultures had been seriously deformed if not destroyed, though clever reconstructions have been able to expose ancient aspects of these cultures that the earlier observers had missed completely. The Lower Colorado River and the West Coast were visited by the European explorers Ulloa, Alarcón, and Cabrillo as early as 1539–43, but, until the founding of the Franciscan mission at San Diego in 1769, exploration was sporadic and accounts of the Indians most elliptic. Such accounts as exist indicate that no major cultural changes occurred among the Indians in the 200 years preceding the beginnings of the missions, a conclusion supported by all the available archaeological evidence. In the more distant past there were substantial cultural changes, but, as Dr. Carter has suggested, Southern California has been characterized by a remarkably stable and persistent material culture. It therefore seems appropriate to examine the human geography of the region as it was at the advent of the missionaries in 1769, postulating that the ancestral patterns of life and land use since the end of the Wisconsin glaciation were generally the same.

Though its climates have remained stable, Southern California, in addition to the relatively minor coastal erosion and the construction of bay-mouth bars and the filling of coastal lagoons, has repeatedly experienced one enormous topographic change, namely the filling and drying up of the Blake Sea (Lake Cahuilla) in Imperial and

Figure 42. Desert scrub north of Mojave. The large, dark-leaved bushes are creosote bush (*Larrea*). (photo courtesy of Harry P. Bailey)

Coachella Valleys. When the Colorado River takes a northerly course across its delta, an area about five times as great as the present Salton Sea fills with freshwater, which then spills south to the Gulf of California. When the river flows directly into the Gulf, the lake dries up, its condition during the 18th and 19th centuries. The old shoreline is plainly visible along a contour just above sea level. Along the old shore there is an almost continuous line of Indian campsites, each thin midden filled with the bones of freshwater fish. Radiocarbon dates indicate that the last filling and drying of the Blake Sea occurred entirely within the last millennium. Older open sites exist, but unfortunately they are not so readily dated. The structure of the Colorado River Delta assures us that the same sequence of events has occurred many times; a filling and drying of the lake during each millennium for the last 50 would not be impossible.

That the great freshwater lake was an enormously attractive living site for the Indians is attested by the tremendous number of essentially contemporary campsites along its shore. Except where the old shoreline cuts through the recently irrigated lands of Imperial and Coachella Valleys, the sites lie in one of the most barren wastes in North America. Estimates of the Indian population when the prehistoric lake was full range from 20,000 to 100,000. Until modern irrigation systems were established, the lakeless region of historic times did not support 1,000 people. As the Blake Sea ceased to overflow, its salinity increased, and at some point all the freshwater fish died. This enormous human population had to leave or starve. The last great emigration probably occurred between 1000 A.D and 1500 A.D. The Indians could move to the Colorado River, where flood farming was practiced, or to the more humid lands west of the mountains, where acorns were the major food source.

The social and economic organization of the Indian tribes in the areas of immigration must have been strongly affected. Curiously, in the Colorado River Valley, where farming could have been extended to support a vastly greater population, an extraordinarily vicious war pattern which effectively killed, kept out, or drove out newcomers was in existence from 1540 to 1850. In the coastal areas the Indians who depended on gathering, hunting, and fishing were notably peaceful at the time of European contact, and each village recognized sharply delimited territories. Some sort of peaceful accommodation of an enormous influx of former desert dwellers must have been accomplished.

Three diverse linguistic families, Yuman, Shoshonean, and Chumash, were represented in Southern California, with the fourth, the Penutian Yokuts, extending south from the San Joaquin Valley almost to its watershed (Figure 43). It is probable that the lakeshore dwellers around the former Blake Sea involved both Shoshonean- and Yuman-speaking villages, and that when the sea dried up emigration was to the coastal lands of their respective linguistic congeners.

The Yuman and Chumash families are both included in the Hokan stock, but their linguistic relationship is more distant than that between English and Russian. Otherwise the several families bear no detectable relationships with one another. Within

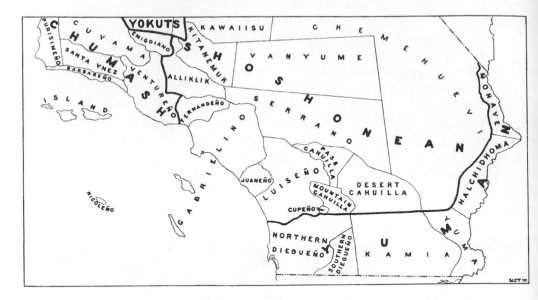

Figure 43. Aboriginal linguistic families of Southern California (after A. L. Kroeber, "Handbook of the Indians in California" [Washington, D.C.: Bureau of American Ethnology Bulletin 78, 1925], 880–91)

the linguistic families there is, further, a tremendous separation of languages and dialects. In the extreme instance, Cupeño was spoken by the inhabitants of two villages occupying no more than 50 square miles around Warner Hot Springs. Cupeño differs from its Shoshonean neighbors by about as much as does English from German and is completely alien to Diegueño spoken to the south. In the desert areas, where something approaching nomadism was economically necessary, single languages such as Chemehuevi had far more extensive distribution, though they were spoken only by small populations.

This linguistic situation affords strong support to certain inferences concerning Indian history and the economic and political organization of the region in aboriginal times. First, the distribution and diversification of languages suggests that the three major linguistic families had been in the region for many millennia and that they had diversified into dialects and languages in situ. Some thousands of years ago the Shoshoneans may have intruded to the coast between the two Hokan groups, moving from their desert homeland in the northeast; since then they have not expanded but have lived in socially semi-isolated and localized communities which permitted distinctive dialects to develop. Secondly, the basic political unit, except along the Colorado River, was the village or at most a few nearby villages. Social contacts were closely restricted to one's home and neighbor villages, and linguistic variations between villages could establish and intensify until they created separate languages. Generally speaking, the more productive the territory, the smaller would be the extent of a village's lands, and the smaller would be the area over which a single language or dialect was spoken. Finally, there was essentially no tribal loyalty. Each village attempted to maintain its right on the lands that supported it, but there was no pattern of a group of villages uniting to dispossess their neighbors of land.

Except for the Yuman tribes along the Colorado River, the only people within the state for whom farming was an important economic activity, the foregoing statements apply to all of California and to much of the Pacific Northwest. Along the Colorado real tribes such as the Mojave and Yuman existed, and people moved freely from one small village to another, keeping their main loyalty to the whole linguistic unit and its lands. These tribes could and did unite for ceremonies and to wage aggressive or defensive warfare. Their failure to expand their small territories is related to the complete unattractiveness of the adjacent lands.

Aboriginal Population Density

Before considering the Indian economy and ecology, it is vital to learn how many Indians occupied Southern California in pre-contact times. A consideration of the total size of the Indian population must begin with Kroeber's estimates.[6] He gives an aboriginal population figure of 133,000 for the state of California as a whole, of which between 35,000 and 40,000 lived in Southern California, about 31,000 if we exclude the farming tribes of the Colorado River. The prestige of Kroeber's scholarship is not easily shaken, but a number of more recent studies in adjacent areas make it likely that this figure should be at least doubled, and almost the entire increase will be placed in the coastal areas. The crux of the problem is that Kroeber explicitly refuses to accept the mission population figures and estimates;[7] by the time censuses he could accept were available in the heavily missionized coastal regions, the native populations had almost vanished.

Meigs's excellently documented study in the northwest corner of Baja California, and Cook's and my work farther south on the peninsula demonstrate that mission data can be used to make thoroughly consistent population estimates, and these are twice as large as Kroeber's for similar areas.[8] Meighan and Eberhart surveyed archaeologically the tiny, unattractive offshore island of San Nicolas; their estimate of the population of this 32-square-mile section of Southern California is 600 to 1,200.[9] Extrapolated to the other islands and coasts of Southern California occupied by fishing people with similar economies, figures of a higher order of magnitude than Kroeber's are obtained. In 1943 Cook more or less accepted Kroeber's figures for California's aboriginal population, raising them about 7 percent,[10] but more recently he has begun reworking the aboriginal population data from various parts of the state other than Southern California,[11] and his population estimates at least double Kroeber's for those areas. A total population for Southern California of 75,000 in 1769, then, seems like a conservative estimate. This figure, or even Kroeber's lower one, is striking when we note that the total pre-Columbian Indian population of the United

States is generally estimated at less than one million.[12] In precontact times Southern California held at least 7.5 percent of the total population of the territory that became the United States; its burgeoning population today [c. 1960] amounts to only 4.5 percent of the country's total! [And about 6 percent in 1995.]

The Indian Economy

The bulk of the California Indians did not farm, but they did exploit the wild flora and fauna with remarkable thoroughness. The critical factor in California's Indian population being so large in comparison with that of the remainder of the United States seems to be the remarkable peacefulness of the people and the stability of their communities. Each group stayed in its home territory and learned of its resources and how to exploit them for food until the group substantially filled the area. Furthermore, since all sorts of land and sea resources were exploited, the sorts of famine that might result from the failure of a single staple crop did not occur.

The technology of the California Indians never has received a very good press from the white man, perhaps because the implements that might be used for war were undeveloped. For hunting they used a relatively weak bow, a good throwing stick, and various sorts of snares. In the game-rich districts, however, they did get a small steady take of rabbits and other rodents, deer, and birds, apparently without depleting the breeding stock even in densely populated localities. Land animals provided a small but fairly dependable fraction of the food of all groups. On ceremonial occasions when much food was needed, productive local fire drives for small game would be undertaken.

Much more attention was given to the gathering and preparation of wild vegetable foods, and the gathering of insects, grubs, and small reptiles is best included in this kind of economic activity. Excellent basketry was made for this purpose (an art in which these Indians excelled by any standards), as well as crude seed beaters and digging sticks. More critical was their comprehensive knowledge of the potential utility of each plant in the local floras and how to render it edible or otherwise useful.[13] The acorn was the most important single plant food, and its abundance was the principal factor in determining population densities in the valleys and hill lands back of the coasts. It was pounded to a flour in deep stone mortars, leached of its bitter tannin with warm water, and boiled into gruel or pinole or baked as a flat bread. The seeds of many grasses and such herbs as amaranths and chia (sage) were ground on flat stone slabs or metates and similarly cooked. On the desert the seeds or whole pods of mesquite, screwbean, and palo verde (*Cercidium*) served as substitutes for acorns. Various wild roots were dug and boiled or roasted as were *Agave* and *Yucca* hearts and buds. Berries, nuts, especially piñon and other pine seeds, and fruits were collected as available, and the pits of the wild prune, manzanita, and other fruits were pounded to flour and leached to make them edible. On the desert the fruits of all sorts of cactus were eaten. Except for the coastal and island villages, gathering vegetable foods and making such utensils as baskets from wild plants to use in gathering were the principal economic activities, both in terms of investment of time and energy and in terms of rewards obtained.

Such intense exploitation of the wild flora could not help but alter it. In process of collection, seeds were continually being dispersed into new habitats. The refuse of campsites brought into existence concentrations of nitrogen-rich soil, on which, on occasion at least, tobacco was grown. Above all, the Indians would burn the landscape to promote the growth of desired grasses and herbs in the following season. Modern authorities are still uncertain of the long-range effects of repeated burning in specific situations. Did it cause the degradation of a complex chaparral to the less useful chamise or coastal sage association or did it expand the oak-grassland parks? Most likely, shifts in both directions occurred in different climatic and ecologic situations. In any event the wild landscape the European explorers found was a product of millennia of such disturbances.

The coastal and island peoples fished, collected shellfish, and hunted sea mammals with technologically more refined equipment. A result of the concentration of available food resources along the shore is the large Chumash villages with populations of more than 600 persons reported by Fages[14] and Crespi[15] on the coasts of Ventura and Santa Barbara counties. These Chumash, with their sewn-plank, asphalt-caulked canoes, their fine shell fishhooks, and three-pronged harpoons seem to have been most successful in developing a technology for the efficient exploitation of marine resources. All the coastal peoples also hunted and gathered and prepared plant foods, but the marine resources provided additional food security and permitted far greater population concentrations.

Style and Focus of the Indian Cultures

The Indian cultures of Southern California, as those of all human groups, merit some consideration in their own terms. What were these people seeking in life, and how successful were they in finding it?

An assessment of the "style" of their civilization may come closest to answering this basic question, and clearly there were two distinctive styles represented in Southern California, that of the Colorado River farmers, and that of all the other groups.

Our widely held, "folklore-like" opinions about the differences in attitudes toward the world held by farmers and by nomads are almost perfectly contradicted by the California data. The Mojave and Yuman farmers are reported, both by early explorers and modern ethnographers, to have been outgoing, inquisitive, individualistic people. Shyness and reserve did not mark their personalities. Their loyalties were to a small nuclear family and to the tribe as a whole, not to villages or local communities. Dream experiences were important to them as individuals, not as part of a community ritual process. On the river and with the tribes to the east they waged a kind of national warfare which involved close fighting with potato masher clubs, resulting in numerous fatalities. Individual Mohaves would journey as far west as the Santa Barbara coast, apparently impelled only by curiosity, and the foreign villages through which a man passed deferred to him as a 13th-century Russian village would defer to a representative of the Mongol Khan. In their sexual behavior, both sexes displayed a freedom and virtuosity that can be compared only with that of classical Greece. Finally, they worked at planting and harvesting crops, at hunting, fishing, and fighting in intense spurts of great activity interspersed with long periods of nearly complete indolence.

The gathering societies to the west present an almost opposite impression. These were quiet, reserved, and intellectually conservative peoples. Their loyalties were to their village and its immediate territory. People seldom traveled beyond the villages of their adjacent neighbors. The procurement of sustenance involved long and patient, but not violent, effort in gathering tiny grass seeds and grinding them or in pounding and leaching the acorn meal, and such activity went on incessantly. Making a coiled basket, perhaps their most developed craft, involves easy work but scores of hours for each product. The peaceful life was sought, and codes of laws were really concerned with manners that would permit all members of a village or band to get along in close and continuing association with a minimum of friction. The intoxicating Jimson weed and tobacco were taken in order to bring the recipient into touch with the supernatural, but he did so in connection with ceremonies and rituals concerned with the welfare of the group rather than with the exaltation of the individual. There must be substantial correlation between these attitudes and the population concentrations in California, which were so much greater than those in other parts of the United States.

Some regional variation in "style of civilization" did exist within this area. In the first place the poorer lands of the interior uplands and deserts had both smaller bands and more extensive territories to be exploited by each. More time and energy were devoted to gleaning subsistence and less to noneconomic group activities than in the larger, richer, and more closely spaced villages on the coast and in the rich alluvial valleys. In the latter areas large, permanently resident populations encouraged the elaborations of rituals. Further, there were distinctions between representatives of the three major linguistic groups, even where all lived in closely comparable environments. Some of the apparent diversity, however, may stem from the uneven quality of the records: the Chumash had no Father Boscana.[16] The Chumash were better craftsmen and, because of their boats and effective fishing gear, were richer. They could afford more effort on fine ornaments as well as fine tools. The Shoshoneans seem to have put their creative energy into spiritual questions and observances and literary invention. The clear evidence of borrowing by the Diegueño of Gabrielino songs, stories, and religious ceremonies marks the Gabrielino land as a center of invention and diffusion for this aspect of culture.[17] The Diegueño may have held a poorer land, wherein getting a living occupied more of their effort, but their enthusiastic borrowing of stories, religious concepts, and ceremonies from the Shoshonean peoples is striking. Relatively speaking, these were culturally backward pioneers who gratefully received from the sophisticates north of them.

The differential reaction to missionization of the three linguistic groups gets us into a later time period, but may further our understanding of basic cultural attributes. The relatively rich, technologically progressive Chumash seemingly welcomed the missionaries; they became extinct culturally within a century. The more backward Diegueño resisted Christianity and survive in substantial numbers, even though they were the first group to be contacted. The various Shoshonean groups had an intermediate history of cultural survival; their survival was aided by the fact that some groups such as the Cahuilla, Luiseño, and Serrano lived east of the area of strong mission influence.[18]

In terms of the subsequent development of the Southern California landscape it is pertinent to note that the Indians were concentrated on the same lands that the Europeans found attractive. Though the resources sought by the two civilizations differed, their distributions were similar. Where the Indians had been most numerous and prosperous they could not resist the flood of Europeans and were substantially obliterated by the time California became part of the United States. Indian communities survive as such only in the interior uplands and on the edge of the desert, areas that until very recently were not sought by Europeans for intensive agricultural, residential, or industrial development.

Distinctive Zones of Indian Settlements

From the Indian standpoint Southern California might be divided into seven ecological zones (Figure 44), with some villages or communities exploiting two of them at different seasons, others only one. It may be noted that there is almost no correspondence between a zone and any of the major linguistic families present in Southern California. From the coast to the interior these zones are:

*   *   *

1. The Coastal Zone involved a strip extending about one-half day's journey inland, and included all the offshore islands. People this close to the shore depended heavily on marine and estuarine animals. These were an almost inexhaustible food source, although a week of very stormy weather might make them inaccessible and produce severe hunger. The size and proximity of villages reported both by explorers and by modern archaeologists make it clear that this was the zone of densest settlement. Almost every permanent water source close to the beach has a large campsite nearby; some have several, though probably not all were occupied concurrently. They

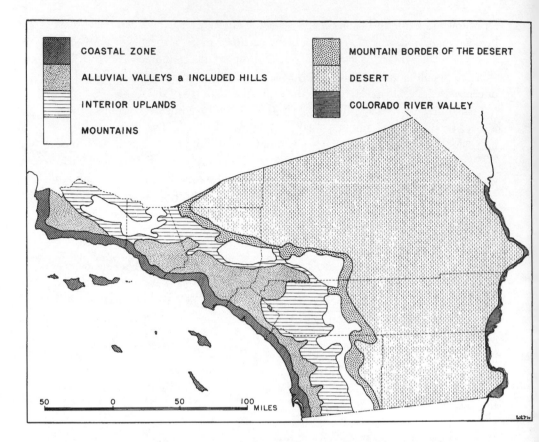

Figure 44. Ecologic zones in Southern California. The zones are outlined in terms of the distinctive opportunities which each afforded to the Indian economies.

represent large and apparently permanent settlements, and many, such as at Point Sal, Malaga Cove, and the La Jolla Beach and Tennis Club, show substantial stratigraphic depth.

2. The Alluvial Valleys and Included Hills included both the extensive coastal lowlands, such as those around Los Angeles or Oxnard, and narrower valleys, such as that of the Santa Margarita River. The immediately adjacent and included hills were exploited by Indians from this zone. These districts also were notably populous; the gallery forests and the marshy vegetation along the streams, with its edible roots, stems, and seeds, and the animal life attracted to such localities, were important resources. The combination of valley grasses and herbs and valley oaks growing on alluvial soils was similarly sustaining. Where the stream valleys cut through more broken country the chaparral and sage scrub on the hills could be exploited from more or less permanent village sites near the streamcourses. Villages were fairly large,

ranging in population from 150 to 400 persons, and rather closely spaced wherever permanent water was present.

3. The Interior Uplands. Within the coastal drainage of the Peninsular and Transverse Ranges there are extensive denudational surfaces, some rough and some quite smooth. The lower and drier of the smooth surfaces are likely to be covered with sage scrub, the higher ones by an oak-grass parkland. The rougher surfaces have a chaparral cover, often with a substantial scrub oak component. This country was more sparsely populated and, if accessible, likely to be exploited from villages in the alluvial lowlands. But there were also a substantial number of bands or villages who had their home bases at the scattered springs within the interior upland zone. For example, the entire Cupeño and Mountain Cahuilla tribes were based here. Gathering the wild plant foods was their main occupation, but they hunted as they could. Firing the grass or brush was a standard practice, and in late summer a fire might burn long and far. Villages were relatively small and ordinarily occupied only seasonally. Typical bands involved 100 individuals, and in summer and fall they would regularly camp in the higher mountains collecting pine seeds of all sorts as well as the acorns of the deciduous black oak (*Quercus kelloggii*). Clearly, population densities in this ecologic zone were lower than in the ones previously considered, but as this area was less affected by the missions far more of the natives survive, especially the Eastern Diegueño, Luiseño, and Cahuilla groups (Figure 45). Modern ethnographers know of more native villages in the interior uplands than in the richer lowlands.[19]

4. The Mountains. There do not seem to have been any permanent Indian residents in the high forested mountains. These areas are singularly cold in winter, and both ethnographic accounts and archaeological evidence indicate that the ill-clothed Indians invariably moved down to lower, warmer camps after the fall pine-nut harvest.

The enormous bear population (primarily the great California grizzlies) seems to be another reason why the Indians did not establish permanent settlements high in the mountains. Men and bears place remarkably similar demands for sustenance on their environment. The intense competition which this provokes is likely to result in somewhat mutually exclusive distributions for the two species. The Indian technology was such that a lone hunter was nearly helpless against a grizzly, but groups of men or villages, protected by fire, were fairly secure. Effectively, the bear was the better-adapted creature in the mountains, the Indians in the lowlands. During the short "fat" season of the piñon harvest an Indian band could camp together and protect itself, but the individual hunting or gleaning that would have been necessary to maintain a human population in the mountains permanently was too dangerous. The relative positions of the two species changed abruptly when Mexican and American ranchers with horses and firearms entered the area, and the bears were deliberately and promptly exterminated.[20]

5. The Mountain Border of the Desert. The bulk of the desert-dwelling Indians ac-

Figure 45. An Indian *ranchería* near Santa Catalina in Baja California. Within the United States the Indian settlements in the interior uplands have recently acquired somewhat better housing, but dwellings like these were typical less than a generation ago. The scrub vegetation close to the settlement has been stripped for firewood; the general vicinity is a piñon-juniper woodland.

tually lived in a narrow zone close to the foot of the mountains that form the western border of the Southern California deserts. Villages of from 100 to 200 persons were located at the permanent springs at the bases of various canyons, as at Palm Springs, Borrego, the middle course of the Mojave River, and many others. The Cahuilla even dug wells for drinking water to depths as great as 16 feet in the washes at the base of the San Jacinto and San Bernardino Mountains.[21] From their canyon headquarters bands would range out over the adjacent open desert and lower mountain slopes in winter and spring, hunting and gathering vegetable foods. In summer and fall the Indians would move up into the mountains to get the harvest of piñon nuts and other seeds and perhaps to enjoy the cooler weather. Despite the apparent barrenness of the landscape, this narrow zone in the edge of the desert supported an aboriginal population density comparable to that on the seaward sides of the mountains.

Barrows long ago pointed out the notable concentration of nutrients in the buds and seeds of desert plants which makes a relatively small vegetative mass capable of supplying a large amount of food for men.[22] The pods of the mesquite, screwbean, and palo verde, which grow in great thickets where the water table is high, were the most important foodstuff, but palm seeds, cactus fruits, agave buds, and the flowers and seeds of many other desert shrubs also were utilized.

390

6. The Desert. On the open desert away from the mountain canyons, springs are few and widely scattered, and in this vast region a single spring would rarely have sufficient food resources in its neighborhood to support a permanently resident band. Such groups, typically involving 50 persons or less, ranged from water source to water source, though generally within a defined territory, exploiting the thin supply of plant foods. The animal foods available to the desert dwellers were scarce and, in this area especially, reptiles, insects, and the smallest rodents supplied the bulk of the animal foods. It may be added that the extradesert residents in Southern California also did not hesitate to consume these forms of animal life. The sparseness of the human population on the open deserts of California is less surprising than that anyone could learn enough about the potential food resources to eke out a living in this barren and inhospitable land.

7. The Colorado River Valley. Here a completely un-Californian economy prevailed. The Yuma and Mojave planted maize, tepary beans, and squash on the mud banks left when the river receded after its late spring floods. Heavy yields were obtained from the fertile soil, and these were supplemented by mesquite pods from great forests along the river, and by roots and seeds of rushes, marsh grasses, and other riverine plants. There is no question but that the agricultural potential of this areally small region was not fully exploited; here, almost uniquely in California, the Indian population was kept down by a pervasive and lethal war pattern comparable to the war patterns in the Eastern Woodlands of the United States. Fish taken from the river and game from the bordering thickets probably gave a smaller proportion of animal food to the diet than elsewhere in Southern California, but substantial surpluses of vegetable foods were normal. On the other hand, when spring floods failed to occur or late ones washed out the plantings, acute famine is known to have afflicted these flood farmers. Such famines did not affect the gatherers of other parts of Southern California, who might be chronically hungry, but who drew on such diverse resources that not all could fail at once.

The Persistence of the Wild Landscape

An interesting, though not definitively answerable, question concerns how much of the wild landscape remains in Southern California. That is, how much of the region would look like home if an Indian resident of the mid-18th century could see it now? A reasonable answer is: much more than the contemporary visitor or many permanent residents might suspect. Our road-bound travels show us freeways with their great road-cuts and clearings along the rights of way, ribbon commercial and residential developments along roads, and the great spread of suburbs and intensive agriculture in the lowlands, which the roads tend to follow. But away from these roads much broken country is utilized only as extensive grazing land or it is protected as a National Forest and closed to visitors from May to December. The intent of this closing is to protect the natural vegetative cover from fire in the belief that its destruction by burning would increase runoff damage during severe winter floods.

In considering this survival of the wild landscape we may take as a basis the ecologic zones recognized by the Indians, with some reference to the plant communities which characterize each zone.

In the coastal zone proper, the large native population died early, but it was not immediately replaced by comparably intensive white exploitation of the land. On the islands it has not yet been replaced. Ports did develop at the few good natural harbors, and within this century beach resorts have gradually been filling up all the strands that are proximate to the larger cities. Two decades from now there may be no wild beach in Southern California, but at present there is a nearly wild strand vegetation along two-thirds of the sandy beach, though commonly the Coast Highway cuts it off from its normal plant neighbors. The Mission Bay salt marsh north of San Diego has just been converted into a completely artificial playground with lawns, bare sand, and dredged channels, but perhaps half of the 30 or so coastal lagoons are still essentially wild salt marshes. The rising popularity of small boat harbors may result in their disappearance within a decade. With the exception of the scrub-covered Otay and Kearney Mesas south and north of San Diego, which possess unusually unattractive soils, most of the marine terraces have been farmed or subdivided or both. Even if not readily irrigable, their relatively level surfaces invite mechanized farming. Where steeper slopes approach the sea, however, the coastal sage scrub or, in the case of the Santa Monica Mountains, chaparral prevails until the pretentious house is built and an approach road cut.

The alluvial valleys, of course, have been affected most completely. They were attractive to agriculture and commonly readily irrigable. So little of the wild landscape of this zone survives that only historical documents can suggest the appearance of these valleys in 1769. Interestingly enough, the tremendous residential-tract construction of the last 15 years has concentrated on just the areas that had been cultivated during the preceding century, so that it is principally the cultivated landscape that is being further altered, not the wild one.

The steeper hills, however, still retain much of their chaparral cover, even near the heart of the Los Angeles metropolitan area. Unfarmable, these hills are also expensive to develop as residential sites. Homes tend to be pretentious and estates large. A growing number of their owners are coming to prefer a wild vegetative cover on the land upon which they do not build rather than the formal gardens they might afford, and the storied palaces of the Hollywood Hills and along the Arroyo Seco in Pasadena are often half-hidden in chamise, sumac, and scrub oaks. The survival of the mammalian fauna is particularly suggestive. Rabbits play nightly on the small lawns and gardens, and householders within a mile of Hollywood and Sunset Boulevards complain that the deer eat their rosebushes. The chaparral-covered steep slopes of the more distant hills and mountains are even more untouched, the accidental fires caused by passing motorists perhaps having almost exactly the same effect as the deliberately set fires of the Indians in maintaining the chaparral association.

Within the general coastal lowlands there is another sort of terrain which has not supported a chaparral cover in recent times. This involves hills composed of soft shales and characterized by gently rounded contours. Baldwin Hills, the Puente Hills, Signal Hill, and Palos Verdes are typical examples, all of course rapidly being subdivided for residences. Before that, they were grass covered on the south slopes, with a park landscape on the north ones, as the name Palos Verdes might suggest. Their original cover may have been similar, but coastal sage scrub or even chaparral may have been present before the hills were occasionally dry-farmed in small grains, thus permanently altering the normal vegetative cover.

The park landscapes of the interior uplands have in some places been plowed for wheat and barley; elsewhere they are grazed and often accidentally or deliberately burned. The oaks are the same, but introduced grasses have generally replaced the native ones. In the grazed areas the vegetational aspect, if not the floristic composition, probably varies little from what it was in aboriginal times.

There is little doubt but that there has been a decrease in the area covered by coniferous forest in the mountains. In the Peninsular Range, the San Bernardino Mountains, and the Transverse Ranges to the northwest, logging and the burning of slash have exerted a more intense pressure on the lower margins of the yellow pine forest than did fires set by the Indians. On south-facing slopes there are scores of localities where chaparral covers charred pine stumps, but no seedling pines are present. This recent upward retreat of the forest is only an acceleration of a process that began with the end of the Pleistocene. Unfortunately the process seems to be tragically irreversible. At higher elevations and in less sunny sites, however, seedling pines quickly recover logged areas. The roads and clusters of resort cabins in the mountains affect only a tiny, though growing, fraction of the forested areas. Furthermore, the owners of resort cabins have preferred to set them into the forested landscape rather than cutting the trees indiscriminately and creating completely artificial landscapes.

As far as surface area is concerned, the bulk of all desert landscape and vegetation types scarcely have been altered during the last two centuries. The exceptions are striking. Irrigated agriculture has completely re-created the flora and aspect of Imperial and Coachella Valleys and smaller oases along the Colorado and Mojave Rivers. Resort and residential communities at the eastern and northern bases of the mountains that mark the desert's western edge, such as Palm Springs and Lancaster-Palmdale, show that the urban sprawl which has taken over the coastal lowlands threatens to leap the mountains. Ragged "jackrabbit homesteads" intrude over more extensive areas, though almost all the surface of the typical two and one-half acre plot remains in desert scrub. In Antelope Valley substantial areas of high desert are being dry-farmed, or sprinkler-irrigated if adequate groundwater supplies can be tapped.

Other than this, only along transportation routes is the desert affected. The gas-station oasis and the elongated beer-can midden along the highway is with us, but the desert highway also creates a strip of slightly more verdant land along its edge. The runoff after a light summer shower from the highway's impervious surface will often promote the growth of summer annuals when there is insufficient moisture in an open spot to produce germination. Though they undoubtedly would support some domestic grazing animals, the deserts of Southern California are rarely if ever used as pastures. With the high costs for labor, it has never been profitable to herd the small flocks of sheep or goats which this poor and sparse pasture can normally maintain. Mining activities have been widespread in the past, but on the California deserts the pattern of uprooting every nearby bush for fuel, so characteristic in Mexico, was never established; only the actual locus of mining has been seriously disturbed by man. Thus, nearly half of Southern California is still covered by wild desert landscape.

A very substantial part of this almost uninhabited land has been, and continues to be, used for military airfields, air gunnery and missile ranges, and desert training centers. Aside from truck tracks, airstrips in valley bottoms, and the occasional un-exploded shell or discarded casing, this activity has affected the great expanse of the desert's surface only moderately, but the effects on the fauna have been more severe. It is reported that wells and springs were filled during World War II in order to pre-vent soldiers in training from supplementing their sharply restricted water rations. Local extinction of the less mobile of those species that have to drink is almost cer-tain. The military personnel have guns and are bored. The last pronghorn antelope herd in the Mojave Desert has not been reported since World War II, and mountain sheep have disappeared from many desert ranges they once occupied. Even humbler forms of animal life, such as chuckwallas, have suffered.[23] We do not yet understand just what part each element of the fauna plays in maintaining the ecologic balance in the desert, but we are progressively more certain that the balance is a delicate one. Subtle but cumulative changes in the flora of these extensive military reservations may even now be occurring.

Table 6 is an attempt to represent numerically the extent of the various sorts of landscape discussed in this paper. It is offered only as a series of estimates in the be-lief that it presents at least the right orders of magnitude, and in the hope that through the internal cancellation of errors of ignorance and misinformation it may project a substantially correct picture. Southern California is interpreted as having an extent of some 45,000 square miles with its northern boundary being that of Santa Barbara County, the Tehachapi Mountains, and the Garlock Fault across Kern and San Bernardino counties. I assume a total Indian population of 75,000 persons for that area in 1769.

TABLE 6. Extent and Utilization of the Ecological Zones in Southern California

Zone	Percent of the Total Surface in Aboriginal Wild Landscape	Indian Population Density per Square Mile	Percent of Total surface Present in Wild State	Modern Population Density per Square Mile[1]
Coastal zone[2]	4	13	1	1,000
Alluvial valleys and adjacent included hills	18	3	3	600
Interior uplands	14	2	8[3]	40
Mountains[4]	6	—[5]	5	9
Desert slopes of major mountain ranges[6]	5	2	4.5	18
Desert[7]	51	.1	48	3.5
Colorado River valley[8]	2	5.5	.5	30
Wild landscape = 100		Surviving to the present = 70		

1. No distinction has been made between urban and rural residents; obviously the high density figures refer primarily to urban settlement.

2. Includes all islands and the strip of land immediately accessible to the coastal Indian villages.

3. Almost all of this area is grazed at present, but unless it has been plowed or deliberately seeded for pasture it is assumed that the character of the wild landscape has not been altered significantly.

4. The generally forested areas in the major mountain ranges. This involves areas more than 4,000 feet above sea level, but does not include the barren desert ranges which exceed this elevation.

5. This area had practically no permanent Indian villages, but it was exploited in summer and fall by Indians from the interior uplands and the desert slopes of the mountain ranges. Its productivity accounts in part for the relatively high aboriginal population density in the two adjacent zones.

6. The eastern and northern slopes of the Peninsular Range, San Bernardinos, San Gabriels, and Tehachapis. From piñon country down to and including the canyon bottom oases.

7. Includes the scrubland in both the low and the high deserts.

8. An area of roughly equal extent and equal aboriginal and modern population density lies across the river in Arizona.

Conclusion

One final consideration is suggested by a comparison of population densities in Southern California and in the United States as a whole during pre-Columbian times with those of the present. Then and now the coastal areas of Southern California stand among the most densely peopled districts in the continent north of Central Mexico, and in the earlier period this region did not enjoy the economic advantages of farming, present in much of the rest of North America. Two sorts of conditions can be used to explain this phenomenon: the extraordinarily favorable environment, and peculiar local developments in human history. It is my belief that both explanations are appropriate, but I should like to examine briefly the environmental one.

It is well known that the mass of vegetative cover bears only a slight relation to the mass of the animal life that can be supported in a given area. The faunal poverty of tropical rain forests contrasted with the faunal abundance of certain grasslands is an obvious illustration. Two somewhat interrelated factors seem to be involved: the proportion of the vegetative matter that is in the form of nutritious starches, sugars, and proteins as opposed to less edible cellulose, and mineral content of the soil and its parent rock which will support a nutritious or a less nutritious plant cover. The pronounced seasonality of vegetative growth, which in Southern California depends largely on winter and early spring rains, seems to favor those plants which invest much of their vital energy in storing concentrated food in their reproductive parts and these plants maintain a rich fauna, with many species included therein. Furthermore, the complex rock types in Southern California's crystalline mountains, and the depositional shales and alluvial soils derived therefrom, provide to the plants an adequate supply of the scarce and rare minerals needed to keep this life pattern circulating at a high level.

An examination of the wild landscape then suggests that Southern California is truly a favored land for the higher types of animal life of which man and the once abundant but now extinct bears are good examples. Two hundred years ago it supported lots of both. Today many more men live here. At the same time we should recognize that in paving the best part of this surface with roads and ranch-type houses we are making inaccessible one of the choicest spots on earth.

Notes

An Iconoclast on the Loose, by Martin J. Pasqualetti

1. Review of *The Woods and the Sea*, by Dudley Camett Lunt, *Landscape* 15, no. 1 (1965): 42.

2. Maynard Weston Dow. Homer Aschmann interviewed by John Fraser Hart. *Geographers on Film*, April 19, 1981, Los Angeles. GOF Transcription 1993: 1–4.

3. Ibid.

4. "Geography in the Liberal Arts College," *Annals of the Association of American Geographers* 52, no. 3 (1962): 284.

5. Richard A. Crooker, personal communication, 1986.

6. William L. Thomas, personal communication, 1995.

7. From a review of *The Cave Paintings of Baja California* by Harry W. Crosby, *Southern California Quarterly* 69, no. 3 (1985): 324.

8. "Geography in the Liberal Arts College," 287.

9. "Geography and the Liberal Arts College," *Bulletin of the California Council of Geography Teachers* 4, no. 3 (1957): 7–10. Reprinted in the *Professional Geographer*, 10, no. 2 (1958): 5.

10. "Geography in the Liberal Arts College," 288.

11. "Indian Societies and Communities in Latin America: A Historical Perspective," in *Geographic Research on Latin America: Benchmark 1970* (Muncie, Ind.: Ball State University, 1971), 132.

12. Ibid., 131.

13. "Geography in the Liberal Arts College," 284.

14. The quotation is taken from the frontispiece in *Land and Life: A Selection from the Writings of Carl Ortwin Sauer*, ed. John Leighly (Berkeley: University of California Press, 1965). In a 1981 interview, Aschmann told John Fraser Hart that he considered himself a "full-dressed disciple of Sauer." *Geographers on Film*, April 19, 1981, Los Angeles. GOF Transcription 1993: 1–4.

15. "Geography in the Liberal Arts College," 290.

16. Aschmann attributed this thought to Carl Sauer in the paper "Carl Sauer: A Self-Directed Career," in *Carl O. Sauer, A Tribute*, ed. Martin S. Kenzer (Corvallis: Oregon State University Press, 1987), 139.

17. Leland Pederson, personal communication, 1995.

18. "Carl Sauer," 139.

19. "People, Recreation, Wild Lands, and Wilderness" (presidential address), *Yearbook, Association of Pacific Coast Geographers*, 28, 1966 (1967), 1.

20. "The Evolution of a Wild Landscape and Its Persistence in Southern California," in *Man, Time, and Space in Southern California*, ed. William L. Thomas, Jr., and published as a supplement to the *Annals of the Association of American Geographers* 49, no. 3, pt. 2 (1959): 34–56.

21. From a review of *Wildlands in Our Civilization*, ed. David Brower, in *Landscape* 16, no. 1 (1966): 25.

22. Ibid.

23. "People, Recreation, Wild Lands, and Wilderness," 14.

24. Review of *Wildlands in Our Civilization*, 25.

25. See list in this volume of Aschmann's publications.

26. Some readers may be interested in the structure and content of Aschmann's graduate examination. HIs language exam consisted of two parts. Part 1 required translating without a

dictionary any passage Aschmann chose from 100 pages of a foreign language book he had approved in advance. Part 2 required translating any passage he chose from a book the student had probably not seen, this time with a dictionary. Some students never finished their graduate work because they did not pass this test. Aschmann once failed a native speaker of Spanish for not translating the Spanish passages well enough. Written comprehensive examinations had five parts: physical geography, cultural geography, history and methodology of geography, a regional specialty, and a topical specialty. These tests usually were completed over a two-day period. Virtually no one passed all five parts on the first attempt.

27. "Geography in the Liberal Arts College," 290.

28. Ibid., 288.

29. Ibid., 286.

30. "Can Cultural Geography Be Taught?" in *Introductory Geography: View Points and Themes* (Washington, D.C.: Commission on College Geography, Association of American Geographers, 1967), Publication 5, 66.

31. Ibid., 69.

32. "Carl Sauer," 139.

33. "Can Cultural Geography Be Taught?" 65.

34. Ibid., 73.

35. "Carl Sauer," 139.

36. "Geography in the Liberal Arts College," 289.

37. "The Subsistence Pattern in Mesoamerican History," *Middle American Anthropology*, vol. 2, Social Science Monographs 10 (Pan American Union, 1960), 1.

38. Ibid., 65.

39. Ibid., 74.

40. "Can Cultural Geography Be Taught?" 67.

41. "Geography in the Liberal Arts College," 287.

42. Ibid., 292.

43. William Thomas, personal communication, 1995.

Southern California: Introduction, by Daniel D. Arreola

1. David E. Sopher, "The Landscape of Home: Myth, Experience, Social Meaning," in *The Interpretation of Ordinary Landscapes: Geographical Essays,* ed. D. W. Meinig (New York: Oxford University Press, 1979), 145.

2. Carey McWilliams, *Southern California: An Island on the Land* (1946; Santa Barbara: Peregrine Smith, 1973).

3. Louise Fortmann and John W. Bruce, eds., *Whose Trees? Proprietary Dimensions of Forestry* (Boulder, Colo.: Westview Press, 1988), 63–67.

4. Henry Bruman, "Sovereign California, The State's Most Plausible Alternative Scenario," in *Early California: Perception and Reality* (Los Angeles: William Andrews Clark Memorial Library, 1981), 1–41.

Purpose in the Southern California Landscape

Originally published in *Journal of Geography* 66, no. 6 (1967): 311–17. Reprinted with permission.

1. Edward T. Price, "The Future of California's Southland," in "Man, Time, and Space in Southern California," *Annals of the Association of American Geographers* 49, no. 3 (1959): 101–16.

2. J. B. Jackson, "Human, All Too Human Geography," *Landscape* 2 (1952): 2–7.

A Late Recounting of the Vizcaíno Expedition and Plans for the Settlement of California

Originally published in the *Journal of California Anthropology* 1, no. 2 (1974): 174–85. Reprinted with permission.

1. Michael W. Mathes, ed., *Californiana I and II: Documentos para la Historia de la Demarcación Comercial de California: 1585–1632*, 2 vols. in 4 parts (Madrid: José Porrúa Turanzas, 1965 and 1970); Álvaro del Portillo y Diez de Sollano, *Descubrimientos y Exploraciones en las Costas de California* (Madrid: Escuela de Estudios Hispano-Americanos de Sevilla, 1947).

2. Michael W. Mathes, ed., *Vizcaíno and Spanish Expansion in the Pacific Ocean* (San Francisco: California Historical Society, 1968); Juan de Torquemada, *De los Veinte i un libros rituales i Monarchía Indiana . . . Madrid* (1723). A facsimile edition was published in 1943 in Mexico: Editorial Salvador Chávez Hayhoe.

3. Mathes, *Vizcaíno and Spanish Expansion*, 108–14.

4. Torquemada, *De los Veinte i un libros.*

5. Mathes, *Documentos para la Historia*, Documento 80.

6. Portillo, *Descubrimientos y Exploraciones*, 419–35; Mathes, *Documentos para la Historia*, Documento 177.

7. Ibid.

8. Mathes, *Documentos para la Historia*, Documento 188.

9. Ibid., Documento 185.

10. Ibid., Documento 38.

11. Torquemada, *De los Veinte i un libros.*

12. The measured latitudes given in this document are remarkably accurate. This is in sharp contrast with those of the Jesuits in Baja California a century and a half later, which were commonly a degree or more too high. Observations from shipboard with an oceanic horizon retained their advantage over land observations. Homer Aschmann, "The Central Desert of Baja California: Demography and Ecology" (Berkeley: University of California Press, *Ibero-Americana* 42, 1959), 36–38; Miguel Leon-Portilla, *Historia Natural y Crónica de la Antigua California de Miguel del Barco* (México: Universidad Nacional Autónoma de México, Instituto de Investigaciones Históricas, 1973), 341, 352.

13. The volcanic peaks that shelter the modern port of Mazatlán are now tied to the mainland by sandbars. This is an aggrading coast and quite different from the one shown in Enrico Martínez's sketch map of 1603 (Portillo, *Descubrimientos y Exploraciones*, 358; Mathes, *Vizcaíno and Spanish Expansion*, 63).

14. [Neither] the name San Bernabé [nor San Bernavé is any] . . . longer in use; both the bay and the town on it being called San Lucas. The bay continued to be known as San Bernabé, at least to the end of the Jesuit period. Homer Aschmann, *The Natural and Human History of Baja California: from Manuscripts by Jesuit Missionaries* (Los Angeles: Dawson's Book Shop, Plate 5).

15. This locality is as sterile and desolate as any in North America. Fr. Ascención's desire to depict Baja California as an attractive place seems to be aided by a failing memory.

16. In this instance, the observation of mineralization may have been valid. In the late 19th century, there was a small gold rush in the Sierra de Santa Clara or Sierra Pintada (Gustaf Eisen, "Explorations in the Central Part of Baja California," *Bulletin of the American Geographical Society* 22 (1900): 397–429.

17. Now known as Turtle Bay, or Bahía Tortugas, but called San Bartholomé as late as 1900 (Eisen, "Explorations in the Central Part of Baja California," 397–429).

18. Here and at San Diego, Fr. Ascención reports the existence of ambergris. In his 1620

memorial, Ascención gave a further description of the material, noting its softness, but none was brought back to New Spain. Since, however, he uses the terms *ambar* and *ambar gris* interchangeably, it is likely that he believed he was dealing with amber, useful for jewelry, rather than the whale secretion used in perfumery (Portillo, *Descubrimientos y Exploraciones*, 425).

19. The embayment south of Bahía San Quintín.

20. Mathes, *Vizcaíno and Spanish Expansion*, 161, translates this phrase as golden flowers, but the word in this manuscript and in the 1620 memorial is clearly *margajita* (Portillo, *Descubrimientos y Exploraciones*, 425).

21. A clear identification of this river is not possible from this or any of the other Vizcaíno documents. The Rogue River, Coos Bay, and the Umpqua River seem most likely.

22. Although the fairly important silver-mining district of Santa Ana was later discovered in the Cape Region, Fr. Ascención never came close to it. This, and most of his references to mineral wealth, come from an optimistic imagination.

23. This topographic detail is highly accurate, but because of lack of water San Lucas remained a minor settlement and no fort was built. A few years ago a luxury tourist hotel was constructed at the site Ascención recommends.

24. The copied text says *crapularan* (referring to becoming drunk), but in a side note the scribe says that the word might be read as *tripulacion* (crew). The latter makes sense.

25. The 20 percent tax levied by the king on the gross production of precious metals and stones, including pearls.

26. The legends of the early 16th century, despite the failure of Coronado's expedition, are still vital, at least for Fr. Ascención.

27. The Colorado River.

28. The reference is unclear, perhaps the sixth *foja* [page] of the original manuscript.

29. Although Baja California seems far out of the way for a trip from Peru to Acapulco, northward-bound ships bore far to the west to avoid headwinds and headed east well north of their destinations, knowing that they would have following winds as they traveled southeastward.

30. "And there shall be one flock and one shepherd." John 10:16.

31. Fr. Ascención's aggressive and contradictory tone is not the normal form for addressing the *oidores* of the Audiencia. He had some protection from the cloth but more from his age. He was justifying his views rather than seeking preferment.

32. Puebla.

33. [Two unrelated paragraphs from notaries that Aschmann translated have been deleted in the present version.—Ed.]

Proprietary Rights to Fruit on Trees Growing on Residential Property

Originally published in *Man*, 63, no. 84 (1963): 74–76. Reprinted in *Whose Trees? Proprietary Dimensions of Forestry*, ed. Louise Fortmann and John W. Bruce (Boulder: Westview Press, 1988), 63–67. Reprinted with permission.

1. The failure of the loquat to establish itself as a commercial fruit in the distant but climatically similar Balearic Islands is noted in a recently published short sketch (Frances Weismiller, "The Misbro Tree," *Atlantic*, March 1955, 57–60). On Majorca the rights to disposal of the fruit would seem to be similarly uncertain, with the first comer getting away with what he takes. This particular story turns on the right of a pregnant woman to any food she craves, and, of course, her right to loquats was established as soon as she made it known.

2. Jean Anthelme Brillat-Savarin, *The Physiology of Taste*, trans. M. F. K. Fisher (New York: The Heritage Press, 1949), 457.

3. The word *loquat* actually comes from the Cantonese Dialect of Chinese, and it is likely that Brillat-Savarin, like Shakespeare (*As You Like It*, 3.2), meant *Mespilus germanica* when he said medlar. The medlar looks quite a bit like the loquat and both belong to the Rosaceae, but the former is much hardier and is better known in all but the southernmost parts of Europe.

Baja California: Introduction, by Conrad J. Bahre

1. Maynard Weston Dow. Homer Aschmann interviewed by John Fraser Hart. *Geographers on Film*, April 19, 1981, Los Angeles. GOF Transcription 1993: 1–4.

2. Homer Aschmann, "The Central Desert of Baja California: Demography and Ecology" (Berkeley: University of California Press, *Ibero-Americana* 42, 1959). Reprinted as *The Central Desert of Baja California: Demography and Ecology* (Riverside, Calif.: Hugh Manessier, 1967).

3. Ibid, xvi.

4. Edward Price, personal communication, 1995.

5. *The Natural and Human History of Baja California*, trans. and ed., with introduction (Los Angeles: Dawson's Book Shop, 1966).

Desert Genocide

Originally published in *El Museo*, n.s., 1, no. 4 (1953): 3–15. Reprinted with permission. This is the title, as published, although Aschmann was furious that the editor tacked on the title without consulting him. (See the Introduction to Baja California for more details about this incident.)

The following publications do not contain all the materials used in the article, but are offered as an introduction to the literature on the Indians and early history of Baja California. Several of the papers mentioned carry bibliographies: Johann Jakob Baegert, *Observations in Lower California*, trans. M. M. Brandenburg and Carl L. Baumann (1772; Berkeley: University of California Press, 1952); J. Ross Browne, *Resources of the Pacific Slope* (New York, 1887); Francisco Javier Clavigero, *The History of (Lower) California*, trans. Sara E. Lake and A. A. Gray (1789; Stanford: Stanford University Press, 1937); S. F. Cook, "The Extent and Significance of Disease among the Indians of Baja California, 1697–1773" (Berkeley: University of California Press, *Ibero-Americana* 12, 1937); Zephyrin Englehardt, *The Missions and Missionaries of California*, vol. 1, "Lower California," 2d ed. (Santa Barbara, Calif., 1929); Paul Kirchhoff, "Las Tribus de la Baja California y el Libro del P. Raegert" (Mexico: Antigua Librería Robredo de José Porrúa e Hijos, 1942) (published as the introduction to the Spanish translation of the book by Raegert cited above; this relatively inaccessible work is mentioned because of its great merit as a summary of ethnographic knowledge of Baja California); Peveril Meigs, "The Dominican Mission Frontier of Lower California," *University of California Publications in Geography* 7 (1935); Edward W. Nelson, "Lower California and its Natural Resources" (Washington, D.C.: Memoirs of the National Academy of Sciences, vol. 16, 1921); Miguel Venegas, *Noticia de la California*, 3 vols. (Mexico: Editorial Layac, 1943). (The original edition was published in Madrid in 1757. It involved the rewriting of a manuscript by Venegas written in 1739 by Andrés Burriel, who might better be known as the author of this work.)

Historical Sources for a Contact Ethnography of Baja California

Originally published in *California Historical Society Quarterly* 44, no. 2 (1965): 99–121. Reprinted with permission.

1. Edward F. Castetter and Willis H. Bell, *Yuman Indian Agriculture* (Albuquerque, 1951); A. L. Kroeber, "Handbook of the Indians of California" (Washington: *Bureau of American Ethnology Bulletin* 78, 1925), 726–803.

2. William C. Massey, "Tribes and Languages of Baja California," *Southwestern Journal of Anthropology* 5 (1949): 272–307.

3. William C. Massey, "Brief Report on Archaeological Investigations in Baja California," *Southwestern Journal of Anthropology* 3 (Winter 1947): 344–59; Brigham A. Arnold, "Late Pleistocene and Recent Changes in Land Forms, Climate and Archaeology in Central Baja California," *University of California Publications in Geography* 10 (1957): 201–317.

4. Herman Frederik Carel Ten Kate, *Reizen en Onderzoekingen in Noord Amerika* (Leiden, 1885); Massey, "Brief Report on Archaeological Investigations," 348–49, 355–56, and personal communication.

5. Homer Aschmann, "The Central Desert of Baja California: Demography and Ecology" (Berkeley: University of California Press, *Ibero-Americana* 42, 1959).

6. W. J. McGee, *The Seri Indians* (Washington, D.C.: Bureau of American Ethnology, Annual Report, 1895–96, no. 17, 1898).

7. Alfred L. Kroeber, "The Seri" (Los Angeles: Southwest Museum Papers 6, 1931).

8. Johann Jakob Baegert, *Observations in Lower California*, trans. M. M. Brandenburg and Carl L. Baumann (1772; Berkeley: University of California Press, 1952), 72, 86.

9. Fernando Ocaranza, *Crónicas y relaciones del Occidente de México*, vol. 5 of Biblioteca Histórica Mexicana de Obras Inéditas (Mexico City, 1937), I, 113–23.

10. Miguel Venegas and Andrés Marcos Burriel, *Noticia de la California y de su Conquista temporal y espiritual hasta el tiempo presente. Sacada de la historia manuscrita, formada en Mexico año 1739 por el Padre Miguel Venegas de la Compañía de Jesús; y otro Noticias y Relaciones antiguas y modernas* (Madrid, 1757).

11. The full title of the early English edition is informative as to the work's content. Miguel Venegas, *A Natural and Civil History of California: containing an accurate description of that country, its soil, mountains, harbours, lakes, rivers and seas; its animals, vegetables, minerals and famous fishery for pearls. The customs of the inhabitants, their religion, government and manner of living before their conversion to the Christian religion by the missionary Jesuits. Together with accounts of the several voyages and attempts made for settling California and taking actual surveys of that country, its gulf, and coast of the South-sea. Illustrated with copper plates and an accurate map of the country and the adjacent seas. Tr. from the original Spanish of Miguel Venegas, a Mexican Jesuit*, 2 vols. (Madrid, 1758 and London, 1759).

12. 3 vols. (Mexico City: Editorial Layac, 1943). Spellings are modernized, but the text has been changed only minimally.

13. Francisco Javier Clavigero, *The History of Lower California*.

14. The Translators' Introduction to the recent American edition contains a biographical sketch of the author Johann Jakob Baegert, *Observations in Lower California*, xi–xx.

15. Ibid.

16. Johann Jakob Baegert, *Noticias de la Península Americana de California*, trans. Pedro R. Hendrichs (Mexico City: Porrúa, 1942).

17. Luis Sales, O. P., *Observations on California 1772–1790*, trans. and ed. Charles N. Rudkin (Los Angeles: Glen Dawson, 1956). The foreword, pp. ix–xiii, provides a brief biography of Father Sales, and offers a tentative explanation for the extreme rarity of the original edition.

18. Pedro Alonso O'Crouley, "Idea Compendiosa del Reyno de Nueva España . . . año de 1774." The original manuscript is in the Archivo General de Indias, Sevilla; a photostatic copy is in the Carl Sauer collection, Department of Geography, University of California, Berkeley.

19. Manuscript no. 1295, Huntington Library, Pasadena, California.

20. *Documentos para la Historia de México*, 21 vols. in 4 ser. (Mexico City, 1853–57), ser. 4, vol. 5.

21. Lesley Byrd Simpson, *California in 1792: The Expedition of José Longinos Martínez* (San Marino, 1938).

22. Richard Hakluyt, *The Principal Navigations, Voyages, Traffiques and Discoveries of the English Nation*, 3 vols. (London, 1598–1600). An excellent recent edition was published by J. MacLehose (Glasgow, 1903–5). Preciado's account appears in vol. 3 of the original and vol. 9 of the more accessible Glasgow one. H. R. Wagner, *California Voyages, 1539–41* (San Francisco, 1925), 25–60.

23. *Empressas Apostolicas de los PP. Missioneros de la Compañía de Jesús . . . en la Conquista de Californias . . .*, paragraphs 1170–1218 of the Bancroft Library copy.

24. Henry R. Wagner, *The Cartography of the Northwest Coast of America to 1800*, 2 vols. (Berkeley, 1937), 1:41–52.

25. Juan de Torquemada, *Libros rituales i Monarchia*, 3 vols. (Madrid, 1724; Mexico: Editorial Salvador Chávez Hayhoe, 1943), 1:700–21.

26. William C. Massey, "The Survival of the Dart-Thrower on the Peninsula of Baja California," *Southwestern Journal of Anthropology* 17 (Spring 1961): 81–93.

27. Ibid., 86–91; Massey, "Brief Report on Archaeological Investigations," 349.

28. Joseph Stöcklein, *Das Neuen Welt-Botts. Allerhand so Lehr als Geistreiche Brief. Schrifften und Reis-Beschreihung, Welche von denen Misionariis der Gesellschaft Jesu aus Beyden Indien und Anderen über Meer Gelegenen Ländern . . .* (Augsburg and Graz, 1726–63). Published in 38 fascicles, which were grouped and bound in volumes at the discretion of their purchasers.

29. Manuscript no. 4097, Huntington Library, Pasadena, California.

30. Misiones, XXII, 182–89, Manuscript at the Archivo General de la Nación, Mexico City.

31. Miguel Venegas, *Noticia de la California*, vol. 3.

32. José de Ortega (and Juan Antonio Balthasar), *Apostólicos Afanes de la Compañía de Jesús: Escritos por un padre de la misma sagrada religión de su provincia de México* (Barcelona, 1754). A recent edition exists: Mexico City, Álvarez y Álvarez (1944). A less than satisfactory English translation of both the 1746 and 1751 diaries is in M. D. Krmpotic, *Life and Works of the Reverend Ferdinand Konscak, S.J.* (Boston, 1923).

33. Peter Gerhard (of Tepoztlán, Mexico), personal communication.

34. California Transcripts, November 4, 1773 (Berkeley: University of California, Manuscripts, Bancroft Library). (This collection of copies of manuscripts from various sources is filed by the earliest date on each manuscript.)

35. California Transcripts, December 9, 1796 (Berkeley: University of California, Manuscripts, Bancroft Library).

36. Aschmann, "The Central Desert of Baja California," 54–55, 276; Homer Aschmann, "The Ecology, Demography, and Fate of the Indians of the Central Desert of Baja California" (Ph.D. diss., University of California, Berkeley, 1954), Appendix B.

37. Albert S. Gatschet, "Der Yuma Sprachstamm," *Zeitschrift für Ethnologie* 9 (1887): 365–418.

38. Massey, "Tribes and Languages of Baja California," 302–5.

39. Baegert, *Observations in Lower California*, 94–104.

40. Massey, "Tribes and Languages of Baja California," 275–87.

41. Aschmann, "The Central Desert of Baja California," 58–132, 177–80.

42. Peveril Meigs, "The Dominican Mission Frontier of Lower California," *University of California Publications in Geography* 7 (1935).

43. Julian Steward, "Basin-Plateau Aboriginal Socio-political Groups" (Washington, D.C.: *Bureau of American Ethnology Bulletin* 120, 1938).

44. Baegert, *Observations in Lower California*, 88. Stöcklein, *Das Neuen Welt-Botts*, letter 763. Consag, Manuscript no. 1095, 3, Huntington Library, Pasadena, California.

45. Peveril Meigs, "The Kiliwa Indians of Lower California" (Berkeley: University of California Press, *Ibero-Americana* 15, 1939), 50, 59.

46. Krmpotic, *Life of Konscak*, 69.

47. Ortega, *Apostólicas Afanes de la compañía*, 404.

48. Arnold, "Late Pleistocene and Recent Changes," 253–61, 271–72; A. E. Treganza and C. G. Malamud, *The Topanga Culture: First Season's Excavation at the Tank Site, 1947* (*University of California Publications*, Anthropological Records 12, 1950).

49. O'Crouley, "Idea compendiosa del Reyno de Nueva España . . . año de 1774," 69–70. The custom is also reported by many other authorities who wrote after 1740 when missionary activity was occurring north of San Ignacio: Baegert, Clavigero, Sales, Consag, Longinos Martínez, but not by Venegas.

The Baja California Highway

Originally published in *Brand Book V* (San Diego Corral of the Westerners: San Diego, 1978), 170–76. Reprinted with permission.

1. It is possible that a wagon road ran over the 25 miles between the silver mines of Santa Ana and La Paz. The mines were opened in 1748 and worked sporadically for several decades. No mention of such a road has been discovered, however (Zephyrin Englehardt, *Missions and Missionaries in California*, vol. 1 [Santa Barbara: 1929]). The drawings of Fr. Ignacio Tirsch, presumably describing Baja California in 1767, the time of the Jesuit expulsion, offer two views of San José del Cabo, Plates VIII and IX, and several other scenes in the Cape area. San José is shown as a busy port, but the paths in and out of town are only for riding animals, and many of them, but no wheeled vehicles appear in his several scenes. *The Drawings of Ignacio Tirsch: a Jesuit Missionary in Baja California*, narr. Doyce B. Nunis, Jr., trans. Elsbeth Schulz-Bischof (Los Angeles: Dawson's Book Shop, 1972).

2. Homer Aschmann, "Recuperación de la vegetación desértica," *Calafia* 3, no. 3 (1976): 52–57.

3. Jorge Engerrand and Trinidad Paredes, "Informe relativo a la parte occidental de la región Norte de la Baja California," in *Memoria de la Comisión del Instituto Geológico de México que exploró la región Norte de la Baja California*, vol. 4 (Parergones del Instituto Geológico de México, 1913), 277–306.

4. Carl H. Beal, *Reconnaissance of the Geology and Oil Possibilities of Baja California* (Mexico: Geological Society of America, Memoir 31, 1948).

5. A copy of the typescript, evidently the original, is in my possession. [This manuscript has not been located.—Ed.]

6. *Report of the trip made by C. B. Salisbury and J. E. McLean of the Automobile Club of Southern California from Los Angeles into Lower California for the Purpose of Ascertaining Road Conditions As Well As Outing and Hunting Possibilities, and to Take the Necessary Notes and Data with Which to Compile a General Map, Particularly of the West Coast Portion* (Los Angeles: Automobile Club of Southern California, 1926).

7. G. P. Parmalee (Automobile Club of Southern California, retired), personal communication.

8. Manuscript of lecture given by G. P. Parmalee, June 1967, entitled "History of Road Signing in California."

9. Personal communication from Paul Jacot from San Diego, California, who trucked onyx for his father's mine in the 1940s.

10. Reference must be made to the *Baja California Guidebook*, Peter Gerhard and Howard E. Gulick (Glendale, Calif.: Arthur H. Clark Co.), four editions beginning in 1956. With its accurate discussions of road conditions and mileages to the tenth of a mile, becoming lost— even on side roads in uninhabited areas—was no longer an unavoidable risk.

11. Ulises Irigoyen, *Carretera Transpeninsular de la Baja California*, 2 vols. (Mexico: Editorial America, 1943).

12. The dates are from my own observations and personal communications from Howard E. Gulick of Glendale, California.

Latin America: Introduction, by James J. Parsons

1. "You sat on this clutch a long time," Sauer wrote Aschmann about the Baja work in April 1954, "but I think you may be a proud mother hen as to what you have hatched out." Personal communication. It was to be nearly five more years before the pruned-down dissertation was to appear in the *Ibero-Americana* series of the University of California Press.

2. *Estudio general de alta y bajo Guajira* (Bogotá: Instituto Geográfico Agustín Codazzi, 1978).

3. Personal communication from Aschmann to Sauer while the former was conducting fieldwork in Colombia.

4. Deborah Pacini Hernández, *Resource development and indigenous people: the El Cerrajón coal project in Guajira, Colombia* (Cambridge, Mass.: Cultural Survival Inc., 1984). Ernesto Guhl et al., *Indios y blancos en la Guajira* (Bogotá: Tercer Mundo, 1963). Ardila C. Gerardo, ed., *La Guajira: de la memoria al porvenir, una visión antropológica* (Bogotá: Universidad Nacional de Colombia, 1990).

5. Leland Pederson, *The Mining Industry of the Norte Chico, Chile* (Evanston, Ill: Northwestern University Studies in Geography, no. 11, 1966).

6. See "Paraguay: A Bilingual Country," in "Person, Place and Thing," *Geoscience and Man* 31 (1992).

The Cultural Vitality of the Guajira Indians of Colombia and Venezuela

Originally published in Akten des 34. Internationalen Amerikanistenkongresses, Vienna, 18–25 July 1960 (published 1962), 592–96.

The Subsistence Problem in Mesoamerican History

Originally published in *Middle American Anthropology* 2, Special Symposium of the American Anthropological Association, Social Sciences Monographs, X. Copyright 1960, Organization of American States. Reproduced with permission of the General Secretariat of the Organization of American States.

1. Angel Palerm, "La distribución del regadío en el area central de Mesoamerica," *Ciencias Sociales* 5 (1954): 2–15, 64–74; Eric R. Wolf and Angel Palerm, "Irrigation in the Old Acolhua Domain, Mexico," *Southwestern Journal of Anthropology* 11, no. 3 (1955).

2. O. F. Cook, *Vegetation Affected by Agriculture* (U. S. Department of Agriculture, Bureau of Plant Industry, Bulletin 145, Washington, D.C., 1909); Wolf and Palerm, "Irrigation in the Old Acolhua Domain, Mexico."

3. O. F. Cook, "Milpa Agriculture, A Primitive Tropical System" (Annual Report of the Smithsonian Institution, Washington, D.C., 1919).

4. Sherburne F. Cook, "The Historical Demography and Ecology of the Teotlalpán" (Berkeley: University of California Press, *Ibero-Americana* 33, 1949).

5. Oscar Schmieder, "The Settlements of the Tzapotec and Mije Indians," *University of California Publications in Geography* 4 (1930).

6. Isabel Kelly, "Excavations at Chametla, Sinaloa" (Berkeley: University of California Press, *Ibero-Americana* 14, 1938); C. O. Sauer, "Aztatlán" (Berkeley: University of California Press, *Ibero-Americana* 1, 1932).

Hillside Farms, Valley Ranches: Land-Clearing Costs and Settlement Patterns in South America

Originally published in *Landscape* (Winter 1955–56): 17–24. Reprinted with permission.

1. O. F. Cook, "Milpa Agriculture, a primitive Tropical System" (Annual Report of the Smithsonian Institute, Washington, D.C., 1919).
2. N. I. Vavilov, "The Origin, Variation, Immunity and Breeding of Cultivated Plants," *Chronica Botanica* (1949–50); Carl Sauer, "American Agricultural Origins: A Consideration of Nature and Culture," in *Essays in Anthropology Presented to A. L. Kroeber* (Berkeley: University of California Press, 1936); Robert J. Braidwood, "From Cave to Village," *Scientific American* (October 1952); V. Gordon Childe, "Old World Prehistory: Neolithic," in *Anthropology Today*, ed. A. L. Kroeber (Chicago: University of Chicago Press, 1953).
3. Carl Sauer, "The Personality of Mexico," *Geographical Review* 31 (1941): 353–64.
4. Emil W. Haury and Julio César Cubillos, "Investigaciones arqueológicas en la Sabana de Bogotá, Colombia (Cultura Chibcha)," *University of Arizona Social Science Bulletin* no. 22 (1953).
5. Gerardo Reichel-Dolmatoff, *Datos histórico-culturales sobre las Tribus de la Antigua Gobernación de Santa Marta* (Bogotá: Banco de la República, 1951).
6. Carl Sauer, "Colima of New Spain in the Sixteenth Century" (Berkeley: University of California Press, *Ibero-Americana* 29, 1948); Leslie Simpson, "Exploitation of Land in Central Mexico in the Sixteenth Century" (Berkeley: University of California Press, *Ibero-Americana* 36, 1952).
7. Simpson, "Exploitation of Land."

The Natural History of a Mine

Originally published in *Economic Geography* 46, no. 2 (1970): 172–89. Reprinted with permission.

This paper was written during a sabbatical year spent in Chile, a country whose economy has been and remains closely dependent on the exploitation of its mineral resources. For this reason most of the mines referred to by way of examples are Chilean. It is my belief that equally cogent examples would be found in Canada, Australia, Peru, or any other country in which mineral production for export constitutes a significant part of the economy. I wish to acknowledge indebtedness for both intellectual stimulus and informational support to Pederson's notable monograph, "The Mining Industry of the Norte Chico, Chile" (Evanston, Ill.: Northwestern University Studies in Geography, no. 11, 1966), 5.

1. Ibid.
2. Compare E. Zimmermann, *World Resources and Industries* (New York: Harper, 1951), 698–705; and D. W. Fryer, *World Economic Development* (New York: McGraw-Hill, 1965), 400–402.
3. Data concerning the profitability of famous mines during their early history are singularly hard to procure. An unusual record from the isolated Caracoles district east of Antofagasta was assembled by Pedro Lucía Cuadra in 1875. In the preceding three years 17 mines had been opened, but 8 of them showed no production or were operating at a loss. Total expenses for the 9 better mines amounted to 4.114 million pesos. Their silver production was valued at about 16 million pesos. See B. Vicuña Mackenna, *El Libro de Plata Santiago* (1882), 399–401.
4. Pederson, *The Mining History of the Norte Chico, Chile*, 118–21, 228.
5. K. Segerstrom, "Regional Geology of the Chañarcillo Silver Mining District and Adja-

cent Areas, Chile," *Economic Geology* 57 (1962): 1247–61. Segerstrom comes to the conclusion that almost all the production of the Chañarcillo district came from 15 or 20 mines.

6. See maps in Pederson, *The Mining History of the Norte Chico, Chile*, 83, 181.

7. H. C. and L. H. Hoover, trans., *G. Georgius: De Re Metallica* (London, 1912).

8. J. R. Bourgeois, *Derecho de minería: Apuntes de clases* (Santiago, 1932).

9. An example of how uninhibited these late demands can be is provided by the discussions and resolutions of the recent seminar on problems of the coal industry in Chile: "El Carbón, sus problemas y posibles soluciones" (Seminario electuado los días 26, 27, y 28 de Abril, Santiago, 1967).

10. It is believed that the models presented in this paper are basically valid for all but one set of the minerals, metallic or nonmetallic, of commerce. The exception is petroleum and natural gas. In their case the higher costs of exploration and initial development are distinctly disproportionate to the minimal cost of exploitation, though the end of the field's life is similarly determinate. A distinctive and simpler model for the history of an oil field that suggests a specific public policy in relation to its exploitation is called for, but that subject is not treated here.

Indian Societies and Communities in Latin America: A Historical Perspective

Originally published as the *Proceedings of the National Conference of Latin Americanist Geographers*, Muncie, Indiana, April–May 1970, 173–91. Reprinted in *Geographic Research on Latin America: Benchmark 1970* (Muncie, Ind.: Ball State University, 1971), 124–37. Reprinted here with permission of the National Conference of Latin Americanist Geographers.

1. C. O. Sauer, *The Early Spanish Main* (Berkeley: University of California Press, 1966). Bartolomé de Las Casas, *Apologética historia de las Indias* (Madrid: Biblioteca de Autores Españoles, 1909).

2. John P. Augelli, "The Rimland-Mainland Concept of Culture Areas in Middle America," *Annals of the Association of American Geographers* 52 (1962): 119–29. Robert C. West and John P. Augelli, *Middle America: Its Lands and Peoples* (Englewood Cliffs, N.J.: Prentice-Hall Inc., 1966).

3. Carl O. Sauer, "The Road to Cíbola" (Berkeley: University of California Press, *Ibero-Americana* 3, 1932); Sauer, "The Personality of Mexico," 353–64.

4. Homer Aschmann, "The Central Desert of Baja California: Demography and Ecology" (Berkeley: University of California Press, *Ibero-Americana* 42, 1959).

5. George Kubler, "The Quechua in the Colonial World," in "Handbook of South American Indians," vol. 2 (Washington, D.C.: *Bureau of American Ethnology Bulletin* 143, 1946), 331–410.

6. S. F. Cook and W. Borah, "The Indian Population of Central Mexico: 1531–1610" (Berkeley: University of California Press, *Ibero-Americana* 44, 1960); S. F. Cook, and L. B. Simpson, "The Population of Central Mexico in the Sixteenth Century" (Berkeley: University of California Press, *Ibero-Americana* 31, 1948).

7. Kubler, "The Quechua in the Colonial World," 331–410.

8. Sol Tax, *Penny Capitalism: A Guatemalan Indian Economy* (Washington, D.C.: Smithsonian Institution, Institute of Social Anthropology, no. 16, 1953).

9. Karl M. Helbig, "Das Stromgebiet des oberen Río Grijalva: Eine Landschaftsstudie aus Chiapas, Süd-Mexiko," *Mitteilungen der Geographischen Gesellschaft in Hamburg* 54 (1961): 5–274. Also translated as *La cuenca superior del Río Grijalva: Un estudio regional de Chiapas, Sureste de Mexico*, trans. Felix Heyne (Tuxtla Gutiérrez: Instituto de Ciencias y Artes de Chiapas, 1964); David A. Hill, *The Changing Landscape of a Mexican Municipio; Villa las Rosas, Chiapas, NAS-NRC Foreign Field Research Program Report* no. 26 (Chicago: University of Chicago, Department of Geography Research Paper no. 91, 1964).

10. Gerardo Reichel-Dolmatoff, *Los Kogi—Una Tribu de la Sierra, en Colombia*, vol. 1, in *Revista del Instituto del Etnológico Nacional* 4 (Bogotá: Editorial Iqueima, 1951).

11. Allan R. Holmberg, *Nomads of the Long Bow: The Siriono of Eastern Bolivia* (Washington, D.C.: Smithsonian Institution, Institute of Social Anthropology, no. 10, 1950); Julian Steward, ed., *Handbook of South American Indians*, vol. 1, *The Marginal Tribes*, vol. 3, *The Tropical Forest Tribes* (Washington, D.C.: *Bureau of American Ethnology Bulletin* 143, 1946, 1948); G. Tessmann, *Die Indianer Nordperus* (Hamburg: Friedrichsen, de Gruyter, 1930).

12. Edward H. Spicer, *Cycles of Conquest* (Tucson: University of Arizona Press, 1962).

13. Louis C. Faron, *Mapuche Social Structure: Institutional Reintegration in a Patrilineal Society of Central Chile* (Urbana: University of Illinois Press, University of Illinois Studies in Anthropology, no. 1., 1961).

14. H. Aschmann, "Indian Pastoralists of the Guajira Peninsula," *Annals of the Association of American Geographers* 50 (1960): 408–18; H. Aschmann, "The Cultural Vitality of the Guajira Indians of Colombia and Venezuela" (Akten des 34. Internationalen Amerikanistenkongresses, Vienna, 18–25 July 1960; published 1962), 592–96.

15. Charles F. Bennett, "The Bayano Cuna Indians, Panama: An Ecological Study of Livelihood and Diet," *Annals of the Association of American Geographers* 52 (1962): 32–50; Peveril Meigs, "The Kiliwa Indians of Lower California" (Berkeley: University of California Press, *Ibero-Americana* 15, 1939); Campbell D. Pennington, "The Tarahumar of Mexico: Their Environment and Material Culture" (Salt Lake City: University of Utah Press, 1963); B. L. Gordon, "Human Geography and Ecology of the Sinú Country of Columbia" (Berkeley: University of California Press, *Ibero-Americana* 39, 1957).

16. Donald D. Brand, "Quiroga: A Mexican Municipio" (Washington, D.C.: Smithsonian Institution, Institute of Social Anthropology, no. 11, 1951); Webster F. McBryde, "Cultural and Historical Geography of Southwest Guatemala" (Washington, D.C.: Smithsonian Institution, Institute of Social Anthropology, no. 4, 1947); Karl Sapper, "Beitrage zur Ethnographie des südlichen Mittelamerika," *Petermanns Geographische Mitteilungen* 47 (1901): 25–40; Karl Sapper, "Die Zukunft der mittelamerikanischen Indianerstämme," *Archiv für Rassen-und Gesellschaftsbiologie* 2 (1905): 383–413; Oscar Schmieder, "Settlements of the Tzapotec and Mije Indians Oaxaca, Mexico," *University of California Publications in Geography* 4 (1930); Franz Termer, "Die Ethnischen Grundlagen der Politischen Geographie von Mittelamerika," *Zeitschrift der Gesellschaft für Erdkunde zu Berlin* (1943): 148–71; Robert C. West, "Cultural Geography of the Modern Tarascan Area" (Washington, D.C.: Smithsonian Institution, Institute of Social Anthropology, no. 7, 1948).

17. George M. Foster, *Empire's Children: The People of Tzintzuntzan* (Washington, D.C.: Smithsonian Institution, Institute of Social Anthropology no. 6, 1948); Manning Nash, *Machine Age Maya: The Industrialization of a Guatemalan Community* (Glencoe, Ill.: The Free Press, 1958); Tax, "Penny Capitalism."

18. Ximena Aranda, "San Pedro de Atacama. Elementos diagnósticos para un plan de desarrollo local," *Informaciones Geográficas* 11–14 (1964): 19–61; Dan Stanislawski, "The Anatomy of Eleven Towns of Michoacán" (Austin: University of Texas, Latin American Studies no. 10, 1950); Philip L. Wagner, "Nicoya, a Cultural Geography," *University of California Publications in Geography* 12, no. 3 (1958): 195–250.

19. Rafael Baraona, "Informe sobre los maices de Socaire" (Publicación del Centro de Estudios Antropológicos de la Universidad de Chile, no. 5, 1958), 36–41; Charles F. Bennett, "Human Influences on the Zoogeography of Panama" (Berkeley: University of California Press, *Ibero-Americana* 51, 1968); Daniel W. Gade, "Plant Use and Folk Agriculture in the Vilcanota Valley of Peru: A Cultural-Historical Geography of Plant Resources" (Ph.D. Diss., University of Wisconsin, 1967); Carl Johannessen, "Man's Role in the Distribution of the Corozo Palm (*Orbigyna* spp.)" (*Yearbook, Association of Pacific Coast Geographers* 19, 1957), 29–33; Carlos

Keller, "Introducción" to José Toribio Medina's *Los Aborígines de Chile* (1882; Santiago: Fondo Histórico y Bibliográfico José Toribio Medina, 1952): vii–lxxii; C. O. Sauer, "Cultivated Plants of South and Central America," in *Handbook of South American Indians*, vol. 6 (Washington, D.C.: *Bureau of American Ethnology Bulletin* 143, 1950), 487–543; C. O. Sauer, *Agricultural Origins and Dispersals* (New York: American Geographical Society, 1952); Jonathan Sauer, "The Grain Amaranths: A Survey of Their History and Classification," *Annals of the Missouri Botanical Garden* 37 (1950): 561–632.

20. Clinton R. Edwards, "Aboriginal Watercraft on the Pacific Coast of South America" (Berkeley: University of California Press, *Ibero-Americana* 42, 1965).

21. George M. McBride, *Agrarian Indian Communities of Highland Bolivia* (New York: American Geographical Society, 1921); George M. McBride, *The Land Systems of Mexico* (New York: American Geographical Society, 1923).

22. Rafael Baraona, Ximena Aranda, and Rómulo Santana, *Valle de Putaendo: Estudio de estructura agrícola* (Santiago: Instituto de Geografía, Universidad de Chile, 1961); Jean Borde, and Mario Góngora, *Evolución de la propriedad rural en el Valle del Puangue*, 2 vols. (Santiago: Instituto de Sociología, Universidad de Chile, 1956).

23. John F. Goins, "An Ethnographic Study of a Quechua Indian Community in the Highlands of Southern Ecuador." Preliminary Report on Research Accomplished on Sabbatical Leave from University of California, Riverside, 1961. (Dittoed.)

24. Harold C. Brookfield, "Questions on the Human Frontiers of Geography," *Economic Geography* 40 (1964): 283–303.

The Immortality of Latin American States

Originally published in *Geographic Research on Latin America: Benchmark, 1980*. Proceedings of the Conference of Latin Americanist Geographers, ed. Tom L. Martinson and Gary S. Elbow, vol. 8, 1981, 323–29. Reprinted with permission.

Flora and Fauna

Man's Impact on the Southern California Flora

Originally published in *Plant Communities of Southern California*, special publication no. 2, California Native Plant Society, 1976, 40–48. Reprinted with permission.

Human Impact on the Biota of Mediterranean-Climate Regions of Chile and California

Originally published in *Biogeography of Mediterranean Invasions*, ed. R. H. Groves and F. DiCastri (New York: Cambridge University Press, 1991), 33–41. © Cambridge University Press 1991. Reprinted with permission of Cambridge University Press.

1. C. J. Bahre, "Destruction of the Natural Vegetation of North-Central Chile," *University of California Publications in Geography* 23 (1979): 1–116.

2. G. M. McBride, *Chile: Land and Society* (New York: American Geographical Society, 1936).

3. H. Aschmann and C. Bahre, "Man's Impact on the Wild Landscape," in *Convergent Evolution in Chile and California: Mediterranean Climate Ecosystems*, ed. H. A. Mooney (Stroudsberg, Pa.: Dowden, Hutchinson & Ross, 1977), 73–84.

4. McBride, *Chile*.

5. R. H. Dana, *Two Years before the Mast* (New York: Harper and Bros, 1840); M. W. Donley, S. Allan, P. Caro, and C. P. Patton, *Atlas of California* (Culver City, Calif.: Pacific Book Center, 1979).

6. H. Aschmann, "The Evolution of a Wild Landscape and Its Persistence in Southern California," in *Man, Time, and Space in Southern California,* ed. W. L. Thomas, Jr., and published as a supplement to the *Annals of the Association of American Geographers* 49, no. 3, pt. 2 (1959): 34–56.

7. Aschmann and Bahre, "Man's Impact on the Wild Landscape," 73–84.

8. McBride, *Chile.*

9. W. W. Robbins, M. K. Bellue, and W. S. Ball, *Weeds of California* (Sacramento: State of California Printing Division, 1951).

10. McBride, *Chile.*

11. Donley, *Atlas of California.*

12. McBride, *Chile*; Aschmann and Bahre, "Man's Impact on the Wild Landscape," 73–84.

13. W. W. Winnie Jr., "Communal Land Tenure in Chile," *Annals of the Association of American Geographers* 55 (1965): 67–86.

14. E. R. Fuentes and E. R. Hajek, "Patterns of Landscape Modification in Relation to Agricultural Practice in Central Chile," *Ecological Conservation* 6 (1979): 265–71.

15. Donley, *Atlas of California.*

16. *Geografía económica de Chile* (Santiago: Corporación de Fomento de la Producción de Chile, 1967); Donley, *Atlas of California.*

17. R. A. Minnich, "Fire Behavior in Southern California Chaparral before Fire Control: The Mount Wilson Burns of the Turn of the Century," *Annals of the Association of American Geographers* 77 (1987): 599–618.

18. L. F. Howard and R. A. Minnich, "The Introduction and Naturalization of *Schinus molle* (Pepper Tree) in Riverside, California," *Landscape and Urban Planning* 18 (1989): 77–89.

19. Aschmann and Bahre, "Man's Impact on the Wild Landscape," 73–84.

The Introduction of Date Palms into Baja California

Reprinted by permission from *Economic Botany* 11, no. 3, 174–77. Copyright 1957, the New York Botanical Garden.

1. Peter J. M. Dunne, *Black Robes in Lower California* (Berkeley: University of California Press, 1952), 227–28.

2. Arthur W. North, *Mother of California* (San Francisco: P. Elder and Co., 1908).

3. Johann Jakob Baegert, *Observations in Lower California* (Mannheim, 1772; 1952), 130.

4. Provincias Internas 166 (Mexico City, Manuscript collection of the Archivo General de la Nación).

5. Mission Statistics (Berkeley: University of California, Manuscript collection, Transcripts, vol. 61, Bancroft Library). The arroba is a measure of weight roughly equaling 25 pounds.

6. Ulises Urbano Lassepas, *De la colonizacion de la Baja California y decreto de 10 Marzo de 1857* (1859).

7. Roy W. Nixon, "North America's Oldest Date Garden," *Pacific Discovery* 6: 18–24.

8. Woodrow Borah, "Early Colonial Trade and Navigation between Mexico and Peru" (Berkeley: University of California Press, *Ibero-Americana* 38, 1954).

9. Ralph H. Gray, "Status of Horticulture in Peru, 1930" (Riverside, Calif.: Manuscript, Citrus Experimental Station Library).

10. There is a curious and completely isolated reference to the introduction of date palms from Spain into Central Mexico in a book written between 1536 and 1541. Francis Borgia Steck,

Motolinia's History of the Indians of New Spain (Washington, D.C.: Academy of American Franciscan History, 1951), 276. Motolinia states that the date-bearing palm trees began to yield fruit a short time after they were planted; he does not say where. No part of Mexico then under Spanish control was at all suitable for date production; it is likely that this attempted plant introduction failed long before Baja California was colonized successfully. Since Motolinia speaks of the fruits of several species of native American palms as dates, he may even have been referring to an introduction of *Phoenix canariensis*, a palm with wider climatic tolerance than *P. dactylifera*.

11. Roy W. Nixon, *Imported varieties of dates in the United States* (Washington, D.C.: U.S. Department of Agriculture, Circular 834, 1950).

Recovery of Desert Vegetation

Originally published in *International Geography: 1972*, vol. 1, ed. W. Peter Adams and Frederick M. Helleiner, 27th International Geographic Congress, Montreal, 631–33. Reprinted with permission.

1. J. Engerrand, and T. Paredes, "Memoria de la Comisión del Instituto Geológico de México que exploró la región Norte de la Baja California," *Parergones del Instituto Geológico de Mexico* 4 (1913): 277–306; M. McCarthy, "Promising Copper, Silver, and Gold Mines in Baja California," *Pacific Coast Miner* 7, no. 21 (1903); E. Wisser, "Geology and Ore Deposits of Baja California, Mexico," *Economic Geology* 49 (1954): 44–76.

2. J. R. Hastings, "Climatological Data for Baja California" (University of Arizona: Inst Atmospheric Phys., Technical Reports on Meteorology and Climatology of Arid Regions, no. 14, 1964).

Man's Impact on the Several Regions with Mediterranean Climates

Originally published in *Ecological Studies 7: Mediterranean Type Ecosystems* (New York: Springer-Verlag, 1973), 363–71. Reprinted with permission.

1. C. O. Sauer, *Agricultural Origins and Dispersals* (Cambridge: M.I.T. Press, 1969), 19–39.

2. E. Hahn, *Die Haustiere and ihre Beziehungen zur Wirtschaft des Menschen* (Leipzig: Drucker and Humblot, 1896). F. Isaac, "On the Domestication of Cattle," *Science* 137 (1962): 195–204.

3. J. Kolars, "Locational Aspects of Cultural Ecology: The Case of the Goat in Non-Western Agriculture," *Geographical Review* 56 (1966): 577–84.

4. M. W. Mikesell, "Northern Morocco: A Cultural Geography," *University of California Publications in Geography* 14 (1961): 19–31, 95–116.

5. J. Montane, "Paleo-Indian Remains from Laguna de Tagua, Central Chile," *Science* 161 (1968): 1137–38.

6. C. Keller, *Introducción a los aborigines de Chile por José Toribio Medina* (Santiago: Fondo Histórico y Bibliográfico José Toribio Medina, 1952), 47–59.

7. S. Sepulveda, *El Trigo Chileno en el Mercado Mundial* (Santiago: Editorial Universitaria, 1959), 11–50.

8. A. L. Kroeber, "Cultural and Natural Areas of Native North America," *University of California Publications in American Archaeology and Ethnology* 38 (1939); H. Aschmann, "The Evolution of a Wild Landscape and Its Persistence in Southern California," published as a supplement to *Annals of the Association of American Geographers* 49, no. 3, pt. 2 (1959): 34–56.

9. S. F. Cook, "The Conflict between the California Indian and the White Civilization: I.

The Indian versus the Spanish Mission" (Berkeley: University of California Press, *Ibero-Americana* 21, 1943).

10. H. L. Shantz, *The Use of Fire as a Tool in the Management of the Brush Ranges of California* (Sacramento: California Division of Forestry, 1947).

11. D. W. Meinig, *On the Margins of the Good Earth* (Chicago: Rand McNally, 1962).

12. F. H. Bauer, "A Pinch of Salt: Trace-Element Agriculture in Australia," *California Geographer* 4 (1963): 55–62.

13. E. Anderson, *Plants, Man and Life* (Berkeley: University of California, 1967), 3–15.

Linguistics: Introduction, by William G. Loy

1. "Kriegie Talk," *American Speech* (October–December 1948): 217–22.

2. He later published a paper on just such a linguistic characteristic, something he found of great interest apart from its reflection in the landscape. See "Paraguay: A Bilingual Country," in "Person, Place and Thing," *Geoscience and Man* 31 (1992).

Athapaskan Expansion in the Southwest

Originally published in the *Yearbook, Association of Pacific Coast Geographers* 32 (1970): 79–97. Reprinted with permission.

This paper was presented as part of the symposium "Cultural Geography" honoring Carl Sauer at the meeting of the American Association for the Advancement of Science in Berkeley, California, December 27, 1965.

1. Sifton Praed and Co. (London, 1936).

2. D. H. Hymes, "Lexicostatistics So Far," *Current Anthropology* 1, no. 1 (1960): 3–44, with the included commentary, provides a balanced treatment of the methodological problems and the utility of the results of a large body of lexicostatistical work. Special attention is given to glottochronology.

3. John Wesley Powell, "Indian Linguistic Families of America North of Mexico" (Washington, D.C.: Seventh Annual Report of the Bureau of Ethnology [BAE] 1885–86, 1891), 1–142; Edward Sapir, "Indian, North America," *Encyclopedia Brittanica*, 14th ed. (1929); Edward Sapir, "Time Perspective in Aboriginal American Culture: A Study of Method" (Ottawa: Geological Survey of Canada, Memoir 90, Anthropological Series no. 13, 1916); Alfred L. Kroeber, *Cultural and Natural Areas of Native North America* (Berkeley: University of California Press, 1939).

4. Melville Jacobs, "Historic Perspectives in Indian Languages of Oregon and Washington," *Pacific Northwest Quarterly* 28, no. 1 (1937): 55–74.

5. Harry Hoijer, "The Chronology of the Athapaskan Languages" *International Journal of American Linguistics* 22, no. 4 (1956): 219–32.

6. Ibid.

7. Tribal and linguistic distributions, culture content, and ecological relations of North American Indians are treated by Alfred L. Kroeber, *Cultural and Natural Areas of Native North America* (Berkeley: University of California Press, 1939), and Harold E. Driver and William C. Massey, "Comparative Studies of North American Indians," *Transactions of the American Philosophical Society*, n.s., 47, pt. 2 (1957). Their maps form the source for Figure 18 and refer to a time of intense, well-documented contact. In the southern Great Plains and Southwest this seems to be mid-19th century rather than the time of Coronado (Kroeber, *Cultural and Natural Areas of Native North America*, 8–9; Driver and Massey, "Comparative Studies of North American Indians," 165).

412

8. John L. Champe, "White Cat Village," *American Antiquity* 14, no. 4 (1949): 285–92; [Author unknown] "White Cat Village" (Research Report, University of Nebraska, vol. 3, no. 1, Spring, 1950).

9. George P. Hammond and Agapito Rey, *Narratives of the Coronado Expedition 1540–1542* (Albuquerque: University of New Mexico Press, 1940).

10. Hoijer, "The Chronology of the Athapaskan Languages," 219–32.

11. George P. Hammond and Agapito Rey, ed., trans., *The Espejo Expedition into New Mexico Made by Antonio de Espejo, 1582–1583, as Revealed in the Journal of Diego Pérez de Luxán* (Los Angeles: Quivira Society, 1929), 85–86.

12. George P. Hammond and Agapito Rey, *Juan de Oñate: Colonizer of New Mexico 1595–1628* (Albuquerque: University of New Mexico Press, 1953), 345–485.

13. Albert H. Schroeder, "Documentary Evidence Pertaining to the Early Historic Period of Southern Arizona," *New Mexico Historical Review* 27, no. 2 (1952): 137–67.

14. Hammond and Rey, *The Espejo Expedition*, 104–8.

15. William F. Corbusier, "The Apache-Yumas and the Apache-Mojaves," *The American Antiquarian and Oriental Journal* 8 (1886): 276–84 and 325–39.

16. Hammond and Rey, *Juan de Oñate: Colonizer of New Mexico 1595–1628*, 408–14.

17. Katherine Bartlett, "Oñate's Route across West Central Arizona," *Plateau* 15, no. 3 (1943): 33–39.

18. Edward H. Spicer, *Cycles of Conquest* (Tucson: University of Arizona Press, 1962): 265–66.

19. Rufus K. Wyllys, ed., "Padre Luis Verlardés Relación of Pimería Alta, 1716," *New Mexico Historical Review* 6, no. 2 (1931): 111–57.

20. *Rudo Ensayo by an Unknown Jesuit Padre, 1763*, trans. Eusebio Guiteras (Tucson: Arizona Silhouettes, 1951), 79.

21. Spicer's *Cycles of Conquest* presents a comprehensive and balanced though by no means complete discussion of the subject.

22. Albert H. Schroeder, "Navajo and Apache Relationships West of the Rio Grande," *Plateau* (Fall 1963): 5–23.

23. Spicer, *Cycles of Conquest*, 214–28.

24. Grenville Goodwin, *The Social Organization of the Western Apache* (Chicago: University of Chicago Press, 1942), 43–62.

25. There is not unanimous agreement among students of the Southwest that the entry of Athapaskan speakers into the region is as recent as here indicated. All, however, recognize the Athapaskans as the last linguistic group to enter a long-settled area. Stephen C. Jett, "Pueblo Indian Migrations: An Evaluation of the Possible Physical and Cultural Elements," *American Antiquity* 29, no. 3 (1964): 281–300, and Betty H. and Harold A. Huscher, "Athapaskan Migration via the Intermontane Region," *American Antiquity* 8, no. 1 (1942): 80–88, postulate an Athapaskan entry to the Southwest perhaps as early as the 12th century by way of the Great Basin and Colorado Plateau. The principal basis for this conclusion is their need of an invading group to cause the abandonment of most of the Western Pueblos during the 13th and 14th centuries. Roscoe Wilment, "The Present Status of Athapaskan Archaeology," mimeo (February 2, 1967), finds only a single dubious identification of an Athapaskan site anywhere in the Plains or the Southwest that is earlier than the 17th century.

Miracle Mile

Originally published in *American Speech* (May 1957): 156–58. Copyright The University of Alabama Press. Reprinted with permission.

1. *American Speech* 31 (1956): 230–31.

2. Pt. 5, p. 3, col. 2, *Los Angeles Times*, 2 April 1939.

3. Midwinter Number, p. 14., *Los Angeles Times*, 3 Jan 1938.

4. According to information from the association; see also pt. 5, p. 1, col. 5, *Los Angeles Times*, 2 March 1952.

Calendar Dates as Street Names in Asunción, Paraguay

Originally published in *Names* 34, no. 2 (1986): 146–53. Reprinted with permission.

1. *Plano Turístico de la Ciudad de Asunción,* Dirección (General de Turismo, 1978). Asunción, scale 1:40,000; the four-sheet topographic map of Gran Asunción, scale 1:25,000 of the Instituto Geográfico Militar.

2. "Datos de la Ciudad-Nomenclatura Barrios, Avenidas, Calles, Pasajes" (Municipalidad de Asunción, Sección Catastro Libro III, June 1970).

3. Osvaldo Kallsen, *Asunción y sus calles: Antecedentes históricos* (Asunción: Imprenta Comuneros, 1974).

4. *Historia Edilicia de la Ciudad de Asunción* (Municipalidad de Asunción: Departamento de Cultura y Arte, 1966), 55–57, end paper insert map. On the map, 16 major buildings are identified, but not the passages connecting them.

5. Ibid., 93–99.

6. Ibid., 101.

7. *Rutas de Venezuela* (Caracas: Langoven, 1980).

8. *Chile: Guía Turística y Plano de Santiago* (ESSO, 1963).

9. *Mapa Carretero de la República Argentina* (ESSO, 1968).

10. *República Argentina, Carta Turística* (Hojal: Automóvil Club Argentino, 1970).

11. *Mapa de la República Oriental del Uruguay* (ESSO, n.d.).

Coromuel and Pichilingue

Originally published in *Names* 40, no. 1 (1992): 33–38. Reprinted with permission.

1. *Bob Ferris News*, KNX, Los Angeles (transcript), 5 December 1955.

2. Fernando Jordán, *El otro México: Biografía de Baja California* (Mexico City: Biografías Gandesa, 1951).

3. Peter Gerhard, *Pirates on the West Coast of New Spain, 1575–1742* (Glendale, Calif.: Arthur H. Clark, 1960).

4. Ibid., 117–18, 205.

5. *Los Angeles Star*, 6 June 1857, 3. David Schulman's article on Spanish words in American English led me to this reference. David Schulman, "Spanish Words in American English," *American Speech* 30, no. 3 (1955): 227–31.

6. Gustave Aimard, *The Freebooters*, trans. of *Les francs-tireurs* (London: Ward and Lock, 1861).

7. Francisco J. Santamaría, *Diccionario general de americanismos*, 3 vols. (Mexico City: Pedro Robredo, 1942), 3:399.

8. Engel Sluiter, "The Word Pechelingue: Its Derivation and Meaning," *Hispanic American Historical Review* 24, no. 4 (1944): 683–98.

9. Manuscript by Miguel Venegas, entitled "Empressas Apostólicas de los P.P. Missioneros de la Compañía de Jesús de la Provincia de Nueva España Obradas en la Conquista de Californias," 1739. A fine copy is in the Bancroft Library, University of California, Berkeley. Gerhard, *Pirates on the West Coast of New Spain 1575–1742*, 194.

10. Sluiter, "The Word Pechelingue: Its Derivation and Meaning," 695.

Deserts

Desertification—A World Problem

Originally published in Section E in *Drought: Our Heritage*, 9th Annual Land-Use Symposium, Albuquerque, 1978.

Divergent Trends in Agricultural Productivity on Two Dry Islands: Lanzarote and Fuerteventura

Delivered, in slightly different form, at the International Geographical Union Meeting in London, 1964.

Evaluations of Dryland Environments by Societies at Various Levels of Technical Competence

Originally published in *Civilizations in Desert Lands*, University of Utah Anthropological Papers, no. 62, December 1962. Reprinted with permission.

1. F. J. Clavigero, *The History of (Lower) California*, trans. S. E. Lake and A. A. Gray (1789; Stanford: Stanford University Press, 1937), 33–35.
2. N. I. Vavilov, "The Origin, Variation, Immunity and Breeding of the Cultivated Plants," trans. K. S. Chester, *Chronica Botanica* 13 (1949–50).
3. Personal communication.
4. E. F. Castetter and Willis H. Bell, *Yuman Indian Agriculture* (Albuquerque: University of New Mexico Press, 1951).
5. A. L. Kroeber, "Handbook of the Indians of California" (Bur. Am. Ethnology, Bull. 78, 1925), 729–31; 751–53.
6. N. Ahmad, "Soil Salinity in West Pakistan and Means to Deal with It," in *Salinity Problems in the Arid Zones, Proc. Teheran Symp., Arid Zone Research* 14 (1961): 117–25.
7. E. Hahn, *Die Haustiere und ihre Beziehungen zum Menschen* (Leipzig: Duncker and Humblot, 1896).
8. K. A. Wittfogel, *Oriental Despotism: A Comparative Study of Total Power* (New Haven: Yale University Press, 1957).
9. H. Bobek, "Die Hauptstufen der Gesellschafts-und Wirtschafts-entfaltung in geographischer Sicht," *Die Erde II* (1959): 259–89.

The Turno in Northern Chile: An Institution for Defense against Drought

Originally published in *Geoscience and Man* 5 (1974): 97–110. Reprinted with permission.

1. R. Baraona, X. Aranda, and R. Santana, *Valle de Putaendo, Estudio de estructura agraria* (Santiago: Instituto de Geografía de la Universidad de Chile, 1961).
2. D. L. Stewart, "Aspects of Chilean Water Law in Action" (Ph.D. diss., University of Wisconsin, 1967), 1–55.
3. *Las Siete Partidas*, ed., trans., S. P. Scott (Chicago: American Bar Association, Comparative Law Bureau, 1931), 856–60.
4. Scott, *Las Siete Partidas*, 822.
5. I. Bowman, *Desert Trails in the Atacama* (New York: American Geographic Society, 1924).

6. X. Aranda, "Evolución de la agricultura y el riego en el Norte Chico, Valle del Huasco," *Informaciones Geográficas* 16 (Santiago, 1969), 9–41.

7. R. Baraona, X. Aranda, and R. Santana, *Valle de Putaendo*, 103.

8. Stewart, "Aspects of Chilean Water Law," 99, 106; R. Baraona, X. Aranda, and R. Santana, *Valle de Putaendo*, 86.

9. Stewart, "Aspects of Chilean Water Law," 48.

10. L. R. Pederson, "The Mining Industry of the Norte Chico, Chile" (Evanston, Ill: Northwestern University Studies in Geography, no. 11, 1966).

11. Corporación de Fomento de la Producción (CORFO), *Geografía Económica de Chile* (Santiago: Corporación de Fomento de la Producción [CORFO], 1965).

12. X. Aranda, "Evolución de la agricultura y el riego," 20–22.

13. W. W. Winnie, "Communal Land Tenure in Chile," *Association American Geographers*, Ann., vol. 55 (1965): 67–86.

14. R. Baraona, X. Aranda, and R. Santana, *Valle de Putaendo*, 76–77; Stewart, "Aspects of Chilean Water Law," 63, 165–67.

15. W. W. Winnie, "Communal Land Tenure in Chile," 75.

16. R. Baraona, X. Aranda, and R. Santana, *Valle de Putaendo*, 72.

17. Ibid., 80–87.

18. Ibid., 87–88.

19. X. Aranda, "Evolución de la agricultura y el riego," 26–34.

20. Stewart, "Aspects of Chilean Water Law," 76–83.

21. Ibid., 148–54.

22. Ibid., 189–306.

23. Ibid., 306–28.

24. P. Denis, *La république Argentine* (Paris: Armand Colin, 1920).

25. Corporación de Fomento de la Producción (CORFO), *Geografía Económica de Chile* (Santiago: Corporación de Fomento de la Producción [CORFO], 1966), Primer Apendice.

26. G. M. McBride, *Chile, Land and Society* (New York: American Geographic Society, 1936).

The Head of the Colorado Delta

Originally published in *Geography as Human Ecology*, ed. S. R. Eyre and G. R. J. Jones (1996): 231–63. Copyright Edward Arnold (Publishers) Ltd 1966. Reprinted with permission.

1. Fred B. Kniffen, "Lower California Studies IV; the Natural Landscape of the California Delta," *University of California Publications in Geography* 5, no. 4 (1931). Chester R. Longwell, "History of the Lower Colorado River and the Imperial Depression," in California Division of Mines, *Bulletin 170*; "The Geology of Southern California," chap. 5, Geomorphology (Sacramento, 1954). Godfrey Sykes, *The Colorado Delta*, American Geographical Society, Special Publications, no. 19, 1937, New York, and Carnegie Institution of Washington, Publication no. 460.

2. Sykes, *The Colorado Delta*.

3. United States Department of Agriculture, *Soil Survey of the Yuma-Wellton Area* (Washington, D.C.: Arizona–California Series, no. 20, 1929).

4. Ibid.

5. Aldo Leopold, *A Sand County Almanac and Sketches from Here and There* (New York: Oxford University Press, 1949).

6. Albert H. Schroeder, *A Brief Survey of the Lower Colorado River from Davis Dam to the International Border* (Washington, D.C.: U.S. Bureau of Reclamation, Reproduction, 1953).

7. Daryll C. Forde, "Ethnography of the Yuma Indians" (Berkeley: *University of California Publications in American Archaeology and Ethnology*, vol. 28, no. 4, 1931). Alfred L. Kroeber,

"Handbook of the Indians of California"; Charles L. McNichols, *Crazy Weather* (New York: Macmillan, 1943), a novel presenting a remarkably penetrating insight into the ethos of a riverine Yuma Indian tribe, in this case the Mohave.

8. Leslie Spier, *Yuman Tribes of the Gila River* (Chicago: University of Chicago Press, 1933).

9. Edward H. Spicer, *Cycles of Conquest* (Tucson: University of Arizona Press, 1962).

10. Forde, "Ethnography of the Yuma Indians"; Kniffen, "Lower California Studies III; The Primitive Cultural Landscape of the Colorado Delta," *University of California Publications in Geography* 5, no. 2 (1931); Spier, *Yuman Tribes of the Gila River.*

11. S. P. Heintzelman, *Indian Affairs on the Pacific*, U.S. House of Representatives, Thirty-fourth Congress, Third Sess., 1857, Executive no. 76, 34–58.

12. Eugene Herbert Bolton, *Anza's California Expeditions*, 5 vols. (Berkeley: University of California Press, 1930), 3:321.

13. Kniffen, "Lower California Studies III."

14. George F. Carter, "Plant Geography and Culture History in the American Southwest" (New York: Viking Fund Publications in Anthropology, no. 5, 1945); Edward F. Castetter and Willis H. Bell, *Yuman Indian Agriculture.*

15. Sykes, *The Colorado Delta.*

16. The data on agriculture in the allotted areas were obtained from the files of the Bureau of Indian Affairs and by interview at the Indian Agency Office at Fort Yuma.

Native Environments: Introduction, by George F. Carter

1. From a review by Aschmann of *Warriors of the Colorado: The Yumans of the Quechan Nation and Their Neighbors,* by Jack D. Forbes. *Southern California Quarterly* 48, no. 1 (1966): 88.

Aboriginal Use of Fire in Areas of Mediterranean Climate

Originally titled "Aboriginal Use of Fire." Proceedings of the Symposium on the Environmental Consequences of Fire and Fuel Management in Mediterranean Ecosystems, USDA Forest Service, General Technical Report WO-3, November 1977, 132–41.

1. Omer C. Stewart, "Fire as the First Great Force Employed by Man," in *Man's Role in Changing the Face of the Earth,* ed. William L. Thomas, Jr. (Chicago: University of Chicago Press, 1956), 115–33.

2. Omer C. Stewart, *Notes on Pomo Ethnography* (*University of California Publications on American Archaeology and Ethnography* 40, 1943), 29–62.

3. Omer C. Stewart, "Burning and Natural Vegetation in the United States," *Geographical Review* 41, no. 2 (1951): 317–20. Omer C. Stewart, "The Forgotten Side of Ethnography," in *Methods and Perspectives in Anthropology: Essays in Honor of William D. Wallis,* ed. R. Spencer (Minneapolis: University of Minnesota Press, 1954), 221–48.

4. Personal communication and unpublished manuscripts by Omer C. Stewart, Department of Anthropology, University of Colorado.

5. John H. Rowe, "Inca Culture at the Time of the Spanish Conquest," in *Handbook of South American Indians,* vol. 2, *The Andean Civilizations,* ed. Julian H. Steward (Washington, D.C.: Smithsonian Institution, Bureau of American Ethnology, Bulletin 143, 1946), 183–330, illus.

6. Carlos Keller, "Introducción" to José Toribio Medina's *Los Aborigines de Chile,* vii–lxxii (1882; Santiago de Chile: Fondo Histórico y Bibliográfico José Toribio Medina, 1952).

7. Isaac Schapera, *The Khoisan Peoples of South Africa* (London: Routledge and Kegan Paul, 1930).

8. William J. Talbot, "Land Utilization in the Arid Regions of Southern Africa Part I: South Africa," in *A History of Land Use in Arid Regions*, ed. L. Dudley Stamp (Paris: UNESCO, 1961), 299–331.

9. John Lowell Bean and Harry W. Lawton, "Some Explanations for the Rise of Cultural Complexity in Native California with Comments on Proto-agriculture and Agriculture," in *Patterns of Indian Burning in California: Ecology and Ethnohistory*, ed. Henry T. Lewis (Ramona, Calif.: Ballena Press, 1973), v–xlvii; Henry T. Lewis, *Patterns of Indian Burning*, 101. Homer L. Shantz, *Fire as a Tool in Management of the Brush Ranges of California* (California Division of Forestry, 1947), 156.

10. Karl W. Butzer, "Climatic Change in Arid Regions Since the Pliocene," in *A History of Land Use in Arid Regions*, ed. L. Dudley Stamp (Paris: UNESCO, 1961), 31–56.

11. Schapera, *The Khoisan Peoples of South Africa*.

12. Julio Montané, "Paleo-Indian Remains from Laguna de Tagua-Tagua, Central Chile," *Science* 161 (1968): 1137–38.

13. Manuel Juvenal Pita Ferreira, *O Arquipélago da Madeira: Terra do Senhor Infante* (Portugal: Funchal Madeira, Junta Geral do Distrito Autónomo do Funchal, 1959), 413.

14. Lewis, *Patterns of Indian Burning*.

15. Mona Stuart Webster, *John McDouall Stuart* (Melbourne, Australia: Melbourne University Press, 1958), 319.

16. Lewis, *Patterns of Indian Burning*.

17. Bean and Lawton, "Some Explanations for the Rise of Cultural Complexity."

18. Schapera, *The Khoisan Peoples of South Africa*; Talbot, "Land Utilization in the Arid Regions."

19. Julius Klein, *The Mesta: A Study in Spanish Economic History, 1273–1836* (Cambridge: Harvard University Press, 1920), 444.

20. J. Despois, "Development of Land Use in Northern Africa, with References to Spain," in *A History of Land Use in Arid Regions*, ed. L. Dudley Stamp (Paris: UNESCO, 1961), 219–37; George P. Marsh, *Man and Nature; Or Physical Geography as Modified by Human Action* (New York: Charles Scribner, 1869), 577; R. O. Whyte, "Evolution of Land Use in South-Western Asia," in *A History of Land Use in Arid Regions*, ed. L. Dudley Stamp (Paris: UNESCO, 1961), 57–118.

21. Klein, *The Mesta*.

22. Ellen Churchill Semple, *The Geography of the Mediterranean Region: Its Relation to Ancient History* (New York: Holt, 1931), 737.

23. C. O. Sauer, *Agricultural Origins and Dispersals* (Cambridge: M.I.T. Press, 1969), 110.

24. James J. Parsons, "The Acorn-Hog Economy of the Oak Woodlands of Southwestern Spain," *Geographical Review* 51, no. 2 (1962): 211–35, illus.

25. Conrad J. Bahre, "Relationships between Man and the Wild Vegetation in the Province of Coquimbo, Chile" (Ph.D. diss., University of California, Riverside, 1974), 311.

26. Shantz, *Fire as a Tool*.

27. Emanuel Fritz, "The Role of Fire in the Redwood Region" (Berkeley: University of California, Agr. Exp. Stn., Circular 323, 1932), 23, illus.

28. John Marvin Dodge, "Vegetational Changes Associated with Land Use and Fire History in San Diego County" (Ph.D. diss., University of California, Riverside, 1975), 216.

29. Joe R. McBride and Richard D. Laven, "Scars as an Indicator of Fire Frequency in the San Bernardino Mountains," *California Journal of Forestry* 74, no. 7 (1976): 439–42, illus.

30. Robert P. Gibbons and Harold F. Heady, "The Influence of Modern Man on the Vegetation of Yosemite Valley" (University of California: Division of Agricultural Science, Manual 36, 1964), 44, illus.

31. John Marvin Dodge, "Forest Fuel Accumulation—A Growing Problem," *Science* 177 (1972): 139–42; Dodge, "Vegetational Changes Associated with Land Use."

32. Norman Hall, R. D. Johnston, and G. M. Chippendale, *Forest Trees of Australia* (Canberra: Australian Government Publishing Service, 1970), 334, illus.

Environment and Ecology in the "Northern Tonto" Claim Area

Originally published in *Apache Indian V*, American Indian Ethnohistory Series (New York: Garland Publishing, 1974), 167–232. Reprinted with permission.

1. Harold E. Driver and William C. Massey, "Comparative Studies of North American Indians," *Transactions of the American Philosophical Society*, n.s., 47, pt. 2 (1957): 169–71.

2. Edward W. Gifford, *Northeastern and Western Yavapai* (*University of California Publications in American Archaeology and Ethnology*, vol. 34, no. 4, 1936), 254–88.

3. George Peter Hammond and Agapito Rey, trans., ed., *Expedition into New Mexico by Antonio de Espejo 1582–1583: As Revealed in the Journal of Diego Pérez de Luxán, a Member of the Party* (Los Angeles: Quivira Society, 1929), 104–8; George Peter Hammond and Agapito Rey, *Juan de Oñate Colonizer of New Mexico 1595–1628* (Albuquerque: University of New Mexico Press, 1953), 408–14; Albert H. Schroeder, "A Brief History of the Yavapai of the Middle Verde Valley," *Plateau* 24, no. 3 (1952): 112–13.

4. Harry Hoijer, "The Chronology of the Athapaskan Languages," *International Journal of American Linguistics* 22, no. 4 (1956): 226, 231–32.

5. John L. Champe, "White Cat Village," *American Antiquity* 14, no. 4 (1949): 289–92; Donald J. Lehmer, "The Sedentary Horizon of the Northern Plains," *Southwestern Journal of Anthropology* 10, no. 2 (1954): 144, 150; Waldo R. Wedel, "Culture Chronology in the Central Great Plains," *American Antiquity* 12, no. 3 (1947): 151–52; George P. Hammond and Agapito Rey, *Narratives of the Coronado Expedition 1540–1542* (Albuquerque: University of New Mexico Press, 1940), 74–75, 79–80, 166, 186–87, 207–8, 235–38, 258–59, 261–62, 271–72.

6. Gifford, *Northeastern and Western Yavapai*, 252–53; Grenville Goodwin, *The Social Organization of the Western Apache* (Chicago: University of Chicago Press, 1942), 44–47.

7. Walter Schuyler, Apparent response to Army letter of July 20, 1874, Prescott, Arizona (Berkeley: University of California, Bancroft Library, Manuscript letter), 40–41.

8. Goodwin, *The Social Organization of the Western Apache*, 47; Eric Mierau, "Concerning Yavapai-Apache Bilingualism," *International Journal of American Linguistics* 29, no. 1 (1963), 1–3.

9. William D. Sellers, ed., *Arizona Climate* (Tucson: University of Arizona Press, 1960), figs. 3, 14, 15, Montezuma Castle.

10. Ibid., Childs.

11. United States Weather Bureau, *Climatic Summary of the United States—Supplement for 1931 through 1952* (Washington, D.C.: GPO, no. 2, "Arizona"), 31, 56.

12. United States Weather Bureau, *Climatic Summary of the United States* (Washington, D.C.: GPO, Bulletin W, sec. 25, "Northern Arizona," 1930), 14.

13. Harold S. Colton, *Black Sand: Prehistory in Northern Arizona* (Albuquerque: University of New Mexico Press, 1960), 24–29.

14. Sellers, *Arizona Climate*, Flagstaff, Fort Valley.

15. United States Weather Bureau, *Climatic Summary of the United States*, 7.

16. Sellers, *Arizona Climate*, Flagstaff.

17. Sellers, *Arizona Climate*, Fort Valley.

18. U.S. National Park Service, *Walnut Canyon National Monument* (Washington, D.C., reprint 1961).

19. A. W. Whipple, *Report of Explorations for a Railway Route, Near the Thirty-fifth Parallel of North Latitude, from the Mississippi River to the Pacific Ocean, 1853–4*, U.S. House of Rep-

resentatives, 33d Congress, 2d sess., Pacific Railway Surveys, vol. 3, 1856, Executive no. 91, 80 and facing plate.

20. Gifford, *Northeastern and Western Yavapai*, 10, 90; Gifford, *Culture Element Distributions: XII. Apache-Pueblo* (University of California Press, Anthropological Records, vol. 4, no. 1, 1940), 255–56, 267–68; Winfred Buskirk, "Western Apache Subsistence Economy" (Ph.D. diss., University of New Mexico, Albuquerque, 1949), 281.

21. Sellers, *Arizona Climate*, Winslow.

22. Ibid.

23. Buskirk, "Western Apache Subsistence Economy," 189.

24. Grenville Goodwin, "The Social Divisions and Economic Life of the Western Apache," *American Anthropologist* 37, no. 1 (1935): 62.

25. William F. Corbusier, "The Apache-Yumas and Apache-Mojaves," the *American Antiquarian and Oriental Journal* 8 (1886): 326.

26. Ibid., 328. Gifford, *Northeastern and Western Yavapai*, 262–63. Gifford, *Culture Element Distributions*, 16–20, 100–105.

27. Gifford, *Northeastern and Western Yavapai*, 254.

28. Edward W. Gifford, *The Southeastern Yavapai* (*University of California Publications in American Archaeology and Ethnology*, vol. 29, no. 3, 1932), 214.

29. Goodwin, *The Social Organization of the Western Apache*, 43–46. Lt. Col. Thomas C. Devin to Lt. Col. Roger Jones, April 1869 (National Archives, Manuscript letter, Record Group 98), 7.

30. J. W. Mason, Response to Army letter of July 20, 1874, Prescott, Arizona; signed by A. H. Nickerson, Camp Verde, August 2, 1874 (Berkeley: University of California, Bancroft Library, Manuscript letter), 50; George F. Price, Apparent response to Army letter of July 20, 1874, Prescott, Arizona; Newburg, N.Y., August 14, 1874 (Berkeley: University of California, Bancroft Library, Manuscript letter), 31; Schuyler, Apparent response to Army letter of July 20, 1874, 16.

31. A. W. Whipple, Thomas Ewbank, and Wm. M. Turner, *Report upon the Indian Tribes*, 1855, U.S. House of Representatives, Pacific Railway Surveys, vol. 3, 1856, Executive no. 91, 14–15.

32. Hammond and Rey, *Expedition into New Mexico by Antonio de Espejo*, 106–7.

33. Ibid., 413.

34. David Krause to John Green, March 29, 1866 (National Archives, Manuscript letter, Record Group 98), 183.

35. Gifford, *Northeastern and Western Yavapai*, 263; Gifford, *Culture Element and Distributions*, 18–19.

36. Buskirk, "Western Apache Subsistence Economy," 191.

37. Corbusier, "The Apache-Yumas and Apache-Mojaves," 328.

38. Schuyler, Apparent response to Army letter of July 20, 1874, 13.

39. Gifford, *Northeastern and Western Yavapai*, 264–65. Goodwin, *The Social Organization of the Western Apache*, 157–58, 475–76; Goodwin, "The Social Divisions and Economic Life of the Western Apache," 61–63. Buskirk, "Western Apache Subsistence Economy," 200–203, 220–34, 279–80.

40. Gifford, *Culture Element Distributions*, 37–40.

41. Julian H. Steward, "Basin-Plateau Aboriginal Socio-Political Groups" (Washington, D.C.: *Bureau of American Ethnology Bulletin 120*, 1938), 33–34, 40, 231–32.

42. Alfred L. Kroeber, *Cultural and Natural Areas of Native North America* (*University of California Publications in American Archaeology and Ethnology*, vol. 38, 1939), 41–42; Homer Aschmann, *The Central Desert of Baja California: Demography and Ecology* (Berkeley: University of California Press, *Ibero-Americana* 42, 1959), 95–96.

43. Corbusier, "The Apache-Yumas and the Apache-Mojaves," 326. Gifford, *Culture Element Distributions*, 6, 9–10. Warren E. Day, Response to Army letter of July 20, 1874, Prescott, Ari-

zona; signed by A. H. Nickerson, Camp Verde, September 10, 1874 (Berkeley: University of California, Bancroft Library, Manuscript letter), 4; Price, Apparent response to Army letter of July 20, 1874, 28; Schuyler, Apparent response to Army letter of July 20, 1874, 13.

44. Gifford, *Culture Element Distributions*, 90; Erhard Rostlund, "Freshwater Fish and Fishing in Native North America," *University of California Publications in Geography* 9 (1952): 208, 304; Corbusier, "The Apache-Yumas and Apache-Mojaves," 326; Leslie Spier, "Havasupai Ethnography" (New York: American Museum of Natural History, Anthropological Papers, vol. 29, pt. 3, 1928), 123; Albert B. Regan, "Notes on the Indians of the Fort Apache Region" (New York: American Museum of Natural History, Anthropological Papers, vol. 31, pt. 5, 1930), 295; Buskirk, "Western Apache Subsistence Economy," 281; Price, Apparent response to Army letter of July 20, 1874, 76; Schuyler, Apparent response to Army letter of July 20, 1874, 30–31.

45. Gifford, *The Southeastern Yavapai*, 255–65, 268.

46. Goodwin, "The Social Divisions and Economic Life of the Western Apache," 61–62; Buskirk, "Western Apache Subsistence Economy," 279.

47. Day, Response to Army letter of July 20, 1874, 12; Price, Apparent response to Army letter of July 20, 1874, 74–75; Schuyler, Apparent response to Army letter of July 20, 1874, 29–30.

48. Gifford, *Culture Element Distributions*, 12, 93. Gifford, *Northeastern and Western Yavapai*, 259–60.

49. Goodwin, *The Social Organization of the Western Apache*, 44; Gifford, *Northeastern and Western Yavapai*, 259–60.

50. Gifford, *Northeastern and Western Yavapai*, 259–60; Corbusier, "The Apache-Yumas and Apache-Mojaves," 327; L. Sitgreaves, *Report of an Expedition down the Zuni and Colorado Rivers*, U.S. Senate, 32d Congress, 2d Sess., 1953, Executive no. 59, 10.

51. Gifford, *Culture Element Distributions*, 92.

52. Gifford, *Northeastern and Western Yavapai*, 255–57.

53. Ibid., 254.

54. Goodwin, *The Social Organization of the Western Apache*, 156–57. Gifford, *The Southeastern Yavapai*, 208.

55. Gifford, *Northeastern and Western Yavapai*, 255.

56. Ibid.

57. Steward, "Basin-Plateau Aboriginal," 27–28.

58. Sitgreaves, *Report of an Expedition*, 9–11.

59. Whipple, *Report of Explorations*, 80.

60. Gifford, *Northeastern and Western Yavapai*, 256–58, 260; Corbusier, "The Apache-Yumas and the Apache-Mojaves," 326–27; Mason, Response to Army letter of July 20 1874, 19; Schuyler, Apparent Response to Army letter of July 20, 1874, 13.

61. Goodwin, *The Social Organization of the Western Apache*, 156, 159; Gifford, *Culture Element Distributions*, 13; Day, Response to Army Letter of July 20, 1874, 6, 11–12.

62. Gifford, *Culture Element Distributions*, 13, 95.

63. Gifford, "The Social Divisions and Economic Life of the Western Apache," 11–15, 91–100.

64. R. S. Ewell to Commissioner of Indian Affairs, Fort Buchanan, New Mexico, April 15, 1960 (National Archives, letter, Record Group 75).

65. Lt. Col. Thomas C. Devin to Lt. Col. Roger Jones, April 1869 (National Archives, Manuscript letter, Record Group 98).

66. Gifford, *Northeastern and Western Yavapai*, 324–40.

67. Goodwin, *The Social Organization of the Western Apache*, 72.

68. Julius C. Shaw to Ben C. Cutler, Fort Wingate, New Mexico, July 24, 1865; *War Department Letters* (National Archives, Record Group 98).

69. Goodwin, *The Social Organization of the Western Apache*, 72–82.

70. Ibid., 60–61.

71. Corbusier, "The Apache-Yumas and Apache-Mojaves," 277.

72. Mason, Response to Army letter of July 20, 1874, 48; Schuyler, Apparent Response to Army letter of July 20, 1874, 18.

73. Gifford, *Northeastern and Western Yavapai*, 250–55.

74. Goodwin, *The Social Organization of the Western Apache*, 44, 50.

75. Gifford, *Northeastern and Western Yavapai*, 324–39.

76. Hammond and Rey, *Expedition into New Mexico by Antonio de Espejo*, 90, 104–8.

77. Katharine Bartlett, "Notes upon the Routes of Espejo and Farfán to the Mines in the Sixteenth Century," *New Mexico Historical Review* 17, no. 1 (1942): 21–36, 21–31, 34–35, map insert.

78. Hammond and Rey, *Expedition into New Mexico by Antonio de Espejo*, 105.

79. Alfred F. Whiting, "The Bearing of Junipers on the Espejo Expedition," *Plateau* 15, no. 2 (1942): 21–23, 21–23.

80. Hammond and Rey, *Expedition into New Mexico by Antonio de Espejo*, 105–7.

81. Ibid., 106. Schroeder, "A Brief History of the Yavapai," 111–18, 112; Albert H. Schroeder, "Documentary Evidence Pertaining to the Early Historic Period of Southern Arizona," *New Mexico Historical Review* 27, no. 2 (1952): 137–67, 148–50, 157–58.

82. Hammond and Rey, *Expedition into New Mexico by Antonio de Espejo*, 408–14.

83. Katherine Bartlett, "Oñate's Route across West Central Arizona," *Plateau* 15, no. 3 (1943): 33–36.

84. Cf. Herbert Eugene Bolton, *Spanish Exploration in the Southwest 1542–1706* (New York, 1916—"Journey of Oñate to California by Land," Zárate report, 1626.)

85. Sitgreaves, *Report of an Expedition*, 9–11 map insert.

86. Ibid.

87. Whipple, Ewbank, and Turner, *Report upon the Indian Tribes*, 14–15.

88. Whipple, *Report of Explorations*, 78–79.

89. Harold S. Colton, *The Sinagua: A Summary of the Archaeology of the Region of Flagstaff, Arizona* (Flagstaff, Ariz.: Northern Arizona Society of Science and Art, 1946).

90. Whipple, *Report of Explorations*, 79–83.

91. Ibid., 80.

92. Ibid., 84.

93. Hubert Howe Bancroft, *History of Arizona and New Mexico, 1530–1888*, vol. 17 (San Francisco, 1889), 554–57, 564–67.

94. Lansing B. Bloom, ed., "Bourke on the Southwest, III," *New Mexico Historical Review* 9, no. 2 (1934), 159–83.

95. Thomas Edwin Farish, *History of Arizona*, vol. 2 (Phoenix, 1915), 218, 219; King Woolsey to Gen. James H. Carleton, Letter written from Prescott, A. T., September 14, 1864; Thomas C. Devin to J. P. Sherburne, Letter written from Prescott, January 8, 1864; George B. Sanford (Report November 20, 1866) House Ex. Doc. no. 1, 40th Cong., 2d Sess., Serial 1324, 1867, 117–19.

96. Gifford, *Northeastern and Western Yavapai*, map insert.

97. Goodwin, *The Social Organization of the Western Apache*, 46.

98. Ibid., 44–47.

99. Gifford, *Northeastern and Western Yavapai*, 249, 250.

100. Hammond and Rey, *Juan de Oñate, Colonizer of New Mexico*, 409–10; Hammond and Rey, *Expedition into New Mexico by Antonio de Espejo*, 105–6.

101. Schroeder, "Documentary Evidence," 151–53, 157–60.

102. Sitgreaves, *Report of an Expedition*, 10–11.

103. Whipple, *Report of Explorations*, 80.

104. George M. Wheeler, "Preliminary Topographical Map" of "Explorations and Surveys South of Central Pacific R.R.," in *Explorations in Nevada and Arizona* (Washington, D.C.: U.S. Engineer Department, 1871).

105. Defendant's Exhibit G-25, Docket 22-E et al. (map showing Indian tribes in Arizona based on Wheeler Topographical Map of 1871).

106. Day, Response to Army letter of July 20, 1874, 11.

107. Navajo Land Claim Field Forms A and B. Identification of Navajo archaeological sites and commentary. Sites:

> 35: Commentary by Roscoe Wilmeth
> 44: Recorded by Maxwell Yazzie
> 45: Recorded by J. Lee Correll
> 46: Recorded by Maxwell Yazzie
> 47: Recorded by Maxwell Yazzie
> 48: Recorded by J. Lee Correll
> 68: Commentary by Roscoe Wilmeth
> 69: Commentary by Roscoe Wilmeth
> 70: Commentary by Roscoe Wilmeth
> 71: Commentary by Roscoe Wilmeth
> 72: Commentary missing.
> 73: Commentary by Roscoe Wilmeth
> 74: Recorded by Richard F. Van Valkenburgh
> 75: Recorded by Maxwell Yazzie
> 76: Recorded by Maxwell Yazzie
> 77: Recorded by Maxwell Yazzie
> 78: Recorded by Maxwell Yazzie
> 79: Recorded by J. Lee Correll

108. Goodwin, *The Social Organization of the Western Apache*, 43-44.

109. Edward H. Spicer, *Cycles of Conquest: The Impact of Spain, Mexico, and the United States on the Indians of the Southwest 1533–1960* (Tucson: University of Arizona Press, 1962), 213–31.

110. Julius C. Shaw to Ben C. Cutler, *War Department Letters*.

111. Gifford, *Northeastern and Western Yavapai*, map insert.

112. Ibid., 257, 265.

113. Goodwin, *The Social Organization of the Western Apache*, 43–48.

114. Ibid., 43–44.

115. Ibid., x.

116. Ibid., insert map facing p. 600.

117. Charles R. Kaut, *The Western Apache Clan System: Its Origins and Development* (Albuquerque: University of New Mexico Publications in Anthropology, no. 9, 1957).

118. Ibid., 53–54

119. Goodwin, *The Social Organization of the Western Apache*, 43–44.

120. Katharine Bartlett, "The Distribution of the Indians of Arizona in 1848," *Plateau* 17, no. 3 (1945): 41–45.

121. Kaut, *The Western Apache Clan System*, 130 (vi), frontispiece map.

122. Grenville Goodwin's Typescript Notes, Charlie Norman, informant, 1–7.

123. Ibid., 1.

Terrain and Ecological Conditions in the Western Apache Range

Originally published in *Apache Indian V*, American Indian Ethnohistory Series (New York: Garland Publishing, 1974), 233–60. Reprinted with permission.

1. George Peter Hammond and Agapito Rey, *Narratives of the Coronado Expedition 1540–1542* (Albuquerque: University of New Mexico Press, 1940), 74, 166, 207–8, 272.

2. Grenville Goodwin, *The Social Organization of the Western Apache* (Chicago: Chicago University Press, 1942), 12–13, 16, 156–59.

3. Hammond and Rey, *Narratives of the Coronado*, 385–415.

4. William D. Sellers, ed., *Arizona Climate* (Tucson: University of Arizona Press, 1960), 11.

5. Forrest Shreve, *The Vegetation of a Desert Mountain Range as Conditioned by Climatic Factors* (Washington, D.C.: Carnegie Institution of Washington, Publication no. 217, 1915), 11–14, 25–35.

6. Goodwin, *The Social Organization of the Western Apache*, 4–5.

7. Ibid., 157–58.

8. Ibid., 156–59.

9. Ibid., 11.

10. Ibid., shown in Tables 4 and 5.

11. United States Weather Bureau, "Climatic Summary of the United States" (Washington, D.C.: GPO, Bulletin W, sec. 25, "Northern Arizona," 1930); United States Weather Bureau, "Climatic Summary of the United States" (Washington, D.C.: GPO, Bulletin W, sec. 26 "Southern Arizona," 1933), 26–27.

12. Winfred Buskirk, "Western Apache Subsistence Economy" (Ph.D. diss., University of New Mexico, Albuquerque, 1949), 189.

13. Goodwin, *The Social Organization of the Western Apache*, 156–59, 160.

14. Buskirk, "Western Apache Subsistence Economy," 189–91.

15. Goodwin, *The Social Organization of the Western Apache*, 4–5.

16. Buskirk, "Western Apache Subsistence Economy," 190; Goodwin, *The Social Organization of the Western Apache*, 156.

17. Goodwin, *The Social Organization of the Western Apache*, 137–38.

18. Ibid., 16, 156–59.

Wildlands and Wilderness: Introduction, by Karl W. Butzer

1. Review of *The Woods and the Sea*, by Dudley Camett Lunt, and *Roadside Area*, by Paul Brooks, *Landscape* 15, no. 1 (1965): 41–42.

2. Review of *Perspectives in Conservation*, by Frederick J. Pohl, and *They All Discovered America*, by Charles Michael Boland, *Riverside Press-Enterprise* C-12 (August 12, 1962).

3. Review of *Wildlands in Our Civilization*, ed. David Brower, *Landscape* 16, no. 1 (1966): 25.

4. "People, Recreation, Wild Lands, and Wilderness" (presidential address), *Yearbook, Association of Pacific Coast Geographers*, 28, 1966 (1967), 8.

5. Review of *Wildlands in Our Civilization*, 25.

6. Ibid.

7. R. Tomaselli, "The Degradation of the Mediterranean Maquis," *Ambio* 6 (1977): 356–62.

8. J. L. Vernet, E. Badal García, and E. Grau Almero, *La Végétation Neolithique du sud-est de l'Espagne (Valencia Alicante) d'après l'analyse anthracologique* (Paris: Académie de Sciences des Paris, Comptes Rendues, 1983), 3:296, 669–72; J. L. Vernet, S. Thiebault, and C. Heinz. "Nouvelles donnes sur la végétation prehistorique postglaciaire méditerranéenne d'après l'analyse anthracologique," in *Premières Communautés Paysannes en Méditerranée Occidentale* (Paris: Centre National de la Recherche Scientifique, 1987), 87–94.

9. A. C. Stevenson and R. J. Harrison, "Ancient Forests in Spain: A Model for Land-Use and Dry Forest Management in South-West Spain from 4000 BC to 1900 AD," *Proceedings of the Prehistoric Society* 58:227–47.

10. A. M. Swain, "A History of Fire and Vegetation in Northeastern Minnesota as Recorded in Lake Sediments," *Quaternary Research* 3 (1973): 383–96. M. L. Heinselmann, "Fire and Succession in the Conifer Forests of Northern America," in *Forest Succession*, ed. D. C. West, H. H. Shugart, and D. B. Botkin (Berlin: Springer, 1981), 374–405.

11. K. W. Butzer, "The Americas before and after 1492: An Introduction to Current Geographical Research," *Annals, Association of American Geographers* 82 (1992): 345–68.

12. W. M. Denevan, "The Pristine Myth: The Landscape of the Americas in 1492," *Annals, Association of American Geographers* 82 (1992): 369–85.

13. P. L. Wagner, "Cultural Landscapes and Regions: Aspects of Communication," *Geoscience and Man* 10 (1974): 133–42.

14. Karl Butzer, "Toward a Cultural Curriculum for the Future: A First Approximation," in *Re-Reading Cultural Geography*, ed. K. Foote (Austin: University of Texas Press, 1994), 409–28.

15. M. W. Lewis, *Green Delusions* (Durham: Duke University Press, 1992).

People, Recreation, Wild Lands, and Wilderness

Originally published as the presidential address in *Yearbook, Association of Pacific Coast Geographers* 28, 1966 (1967): 1–15. Reprinted with permission.

1. Review of *Wildlands in Our Civilization*, ed. David Brower, *Landscape* 16, no. 1 (1966): 25.

2. Stephen Vincent Benét, "Nightmare, with Angels," in *The Selected Works of Stephen Vincent Benét*, vol. 2, *Poetry* (New York: Holt, Rinehart and Winston, 1963), 450–52.

3. Parts I and II (Sacramento, 1960).

4. A remarkably provocative article by Edward S. Deevey, "The Hare and the Haruspex: A Cautionary Tale," *American Scientist* 48 (1960): 415–30, is well worth examining in this connection.

5. Review of Brower, *Wildlands in Our Civilization*, 116–18.

6. *Landscape* 2 (Autumn 1952): 2–7.

People Are No Damn Good

Modified slightly from a review of *Wildlands in Our Civilization*, ed. David Brower, Sierra Club Books. Originally published in *Landscape* 16, no. 1 (1966): 25, which gave the review the present title. Reprinted with permission.

The Evolution of a Wild Landscape and Its Persistence in Southern California

Originally published in *Annals of the Association of American Geographers* 49, no. 3, pt. 2 (1959): 34–56. Reprinted with permission.

1. Philip A. Munz and David D. Keck, "California Plant Communities," *El Aliso* 2 (1949): 87–105.

2. Cf. Herbert Eugene Bolton, *Fray Juan Crespi: Missionary Explorer of the Pacific Coast, 1769–1774* (Berkeley: University of California Press, 1927); Herbert Ingram Priestley, trans. and ed., *A Historical, Political, and Natural Description of California by Pedro Fages, Soldier of Spain* (Berkeley: University of California Press, 1937).

3. W. W. Robbins, "Alien Plants Growing without Cultivation in California" (University of California Agricultural Experiment Station, Bulletin 637, 1940).

4. Homer L. Shantz, *Fire as a Tool in the Management of Brush Ranges in California* (Sacramento: California Division of Forestry, 1947), and Arthur W. Sampson, *Plant Succession on Burned Chaparral Lands in Northern California* (University of California Agricultural Experiment Station, Bulletin 685, 1944).

5. Harry P. Bailey, "Physical Geography of the San Gabriel Mountains, California" (Ph.D. diss., University of California, Los Angeles, 1950), 116–20.

6. Alfred L. Kroeber, "Handbook of the Indians in California" (Washington, D.C: *Bureau*

of American Ethnology Bulletin 78, 1925), 880–91; and "Cultural and Natural Regions in Native North America" (*University of California Publications in American Archaeology and Ethnology*, vol. 38, 1939), 131–81.

7. Kroeber, "Cultural and Natural Regions," 177–81.

8. Peveril Meigs, "The Dominican Mission Frontier of Lower California," *University of California Publications in Geography* 7 (1935). Sherburne F. Cook, "The Extent and Significance of Disease among the Indians of Baja California" (Berkeley: University of California Press, *Ibero-Americana* 12, 1937). Homer Aschmann, "The Central Desert of Baja California: Its Demography and Ecology" (Berkeley: University of California Press, *Ibero-Americana* 42, 1959).

9. Clement W. Meighan and Hal Eberhart, "Archaeological Resources of San Nicolas Island, California," *American Antiquity* 19 (October 1953): 109–25.

10. Sherburne F. Cook, "The Conflict between the California Indian and the White Civilization: I. The Indian versus the Spanish Mission" (Berkeley: University of California Press, *Ibero-Americana* 21, 1943), 161–94.

11. Sherburne F. Cook, "The Aboriginal Population of the San Joaquin Valley, California," *Anthropological Records* 16, no. 2 (1955); "The Aboriginal Population of the North Coast of California," *Anthropological Records* 16, no. 3 (1956); "The Aboriginal Population of Alameda and Contra Costa Counties, California," *Anthropological Records* 16, no. 4 (1957).

12. The most thorough effort to arrive at the aboriginal population of North America by the additive or dead reckoning method, i.e., summing the best estimates of the population of each of the resident tribes rather than dealing with numbers of people per square mile, is still James Mooney, "The Aboriginal Population of America North of Mexico" (Washington, D.C.: Smithsonian Miscellaneous Collections, vol. 80, no. 7, 1928). His figures are 1,153,000 for the whole area, of which 221,000 were in Canada, 73,000 in Alaska, and 10,000 in Greenland. Of the 849,000 Indians in the United States 260,000 were in California. Here Mooney chose to accept Merriam's estimate (C. Hart Merriam, "The Indian Population of California," *American Anthropologist*, n.s., 7 [1905]: 594–606), which was almost twice Kroeber's and, from the most recent studies, seems to be more accurate (cf. Kroeber, "Cultural and Natural Regions," 131–43).

13. David Prescott Barrows, *The Ethno-botany of the Coahvilla Indians of Southern California* (Chicago, 1900). This classic study set a pattern for ethnobotanical investigations, and for a general account of plant use by a Southern Californian Indian group it has still not been superseded.

14. Priestley, *A Historical, Political, and Natural Description*, 26–27.

15. Bolton, *Fray Juan Crespi*, 158–70.

16. This missionary, almost uniquely among the Franciscans in California, took the trouble to record a rich account of the religious and ceremonial life of the Juaneno and Gabrielino just before the native culture disintegrated completely. Geronimo Boscana, "Chinigchinich," in *Life in California* (New York, 1846), and Kroeber, "Handbook," 620–28.

17. Kroeber, "Handbook," 622–27, 712–17.

18. Cook, "Population Trends among the California Mission Indians" (Berkeley: University of California Press, *Ibero-Americana* 17, 1940).

19. Kroeber, "Handbook," Plate 57.

20. Tracy I. Storer and Lloyd P. Tevis, Jr., *California Grizzly* (Berkeley: University of California Press, 1955).

21. Barrows, *The Ethno-botany of the Coahvilla*, 25–27.

22. Ibid., 54–55.

23. Harry C. James, "Our Deserts Are Not Expendable," *Nature Magazine* 2 (November 1955): 482–84.

Publications of H. Homer Aschmann

Papers and Monographs (Refereed)

"A Geographical Approach to Some Technical Problems of Commercial Aviation." *Professional Geographer* (April 1948): 24–30.

"Kriegie Talk." *American Speech* (October–December 1948): 217–22.

"A Metate Maker of Baja California." *American Anthropologist* (October–December 1949): 682–86.

"A Fluted Point from Central Baja California." *American Antiquity* (January 1952): 262–63.

"A Primitive Food Preparation Technique in Baja California." *Southwestern Journal of Anthropology* (Spring 1952): 143–50.

"A Consumer Oriented Classification of the Products of Tropical Agriculture." *Economic Geography* (April 1952): 143–50.

"Desert Genocide." *El Museo,* n.s., 1, no. 4 (June 1953): 3–15.

"Southern California Water Problems." *Riverside Daily Press,* January 22, 1955, 3.

"Hillside Farms, Valley Ranches, Land-Clearing Costs and Settlement Patterns in South America." *Landscape* (Winter 1955–56): 17–24.

"Informe preliminar sobre investigaciones efectuadas en la Península de La Guajira, Colombia." *Boletín de la Sociedad Geográfica de Colombia,* nos. 3 and 4, vol. 14, nos. 51–52 (1956): 144–74.

"An Example of Censorship of a Scholarly Periodical." *College and Research Libraries* 18, no. 3 (May 1957): 213–16.

"The Introduction of Date Palms into Baja California." *Economic Botany* 11, no. 3 (April–June 1957): 174–77.

"Geography and the Liberal Arts College." *Professional Geographer* 10, no. 2 (March 1958): 2–6.

"Great Basin Climates in Relation to Human Occupance." Reports of the University of California Archaeological Survey, no. 42 ("Current Views on Great Basin Archaeology"), April 10, 1958, 23–40.

"The Central Desert of Baja California: Demography and Ecology." Berkeley: University of California Press, *Ibero-Americana,* 42, 1959.

"The Evolution of a Wild Landscape and Its Persistence in Southern California." In *Man, Time, and Space in Southern California,* edited by William L. Thomas, Jr. Supplement to *Annals of the Association of American Geographers* 49, no. 3, pt. 2 (September 1959): 34–56.

"Indian Pastoralists of the Guajira Peninsula." *Annals of the Association of American Geographers* 50, no. 4 (December 1960): 408–18.

"The Subsistence Pattern in Mesoamerican History." *Middle American Anthropology* 2, Social Science Monographs 10, Pan American Union, 1960, 1–11.

"The Cultural Vitality of the Guajira Indians of Colombia and Venezuela." Akten des 34. Internationalen Amerikanistenkongresses, Vienna 18–25 July 1960 (published 1962), 592–96.

"Geography in the Liberal Arts College." *Annals of the Association of American Geographers* 52, no. 3 (September 1962): 284–92.

"Evaluations of Dryland Environments by Societies at Various Levels of Technical Competence." In *Civilizations in Desert Lands,* edited by Richard B. Woodbury. University of Utah Anthropological Papers, no. 62, December 1962, 1–15.

"Proprietary Rights to Fruit on Trees Growing on Residential Property." *Man* 63, no. 84 (May 1963): 74–76.

"Historical Sources for a Contact Ethnography of Baja California." *California Historical Society Quarterly* 44, no. 2 (June 1965): 99–121.

"Comments on the Symposium, Man, Culture and Animals." In *Man, Culture, and Animals: The Role of Animals in Human Ecological Adjustments,* edited by Anthony Leeds and Andrew P. Vayda. American Association for the Advancement of Science, Washington, D.C., 1965, 259–70, Publication no. 78.

"The Head of the Colorado Delta." In *Geography as Human Ecology,* edited by S. R. Eyre and G. R. J. Jones, 231–63. London: Edward Arnold, 1966.

The Natural and Human History of Baja California. Translated and edited with introduction. Los Angeles: Dawson's Book Shop, 1966.

Foreword and introduction to *Pleistocene and Post-Pleistocene Climate Variations in the Pacific Area: A Symposium,* edited by David I. Blumenstock, vii–viii and 1–8. (Aschmann actually edited the volume but left it in Blumenstock's name as a memorial to him; Blumenstock organized the symposium.) Honolulu: Bishop Museum Press, 1966.

"Can Cultural Geography Be Taught?" In *Introductory Geography: Viewpoints and Themes.* Washington, D.C.: Commission on College Geography, Association of American Geographers, 1967, publication 5, 65–74.

"People, Recreation, Wild Lands, and Wilderness." Presidential address. *Yearbook, Association of Pacific Coast Geographers* 28, 1966 (1967), 1–15.

"Purpose in the Southern California Landscape." *Journal of Geography* 66, no. 6 (September 1967): 311–17.

Historical Accounts and Archaeological Discoveries: Working Together Two Scholarly Disciplines Enlarge Our Understanding of the Extinct Indians of Baja California (in English and Spanish). Memoria del V Simposio Anual sobre Baja California celebrado en Tijuana B.C. el Sabado 29 de Abril de 1967, Asociación Cultural de las Californias, 1968, 37–50.

"The Natural History of a Mine." *Economic Geography* 46, no. 2 (April 1970): 172–89.

"Indian Societies and Communities in Latin America: An Historical Perspective." *Proceedings of the National Conference of Latin Americanist Geographers,* Muncie, Indiana, April–May 1970, 173–91.

"Athapaskan Expansion in the Southwest." *Yearbook, Association of Pacific Coast Geographers* 32 (1970): 79–97.

"Prolegomena to the Remote Sensing of Environmental Quality." *Professional Geographer* 23 (January 1971): 59–63.

"Indian Societies and Communities in Latin America: An Historical Perspective." In *Geographic Research on Latin America: Benchmark 1970,* 124–37. Muncie, Ind.: Ball State University, 1971.

"Recovery of Desert Vegetation." In *International Geography: 1972.* Vol. 1, edited by W. Peter Adams and Frederick M. Helleiner, 631–33. Twenty-seventh International Geographical Congress, Montreal.

"Offshore Desert Islands as Centers of Development." In *Coastal Deserts: Their Natural and Human Environments,* edited by David H. K. Amiran and Andrew W. Wilson, 63–66. Tucson: University of Arizona Press, 1973.

"Distribution and Peculiarity of Mediterranean Ecosystems." In *Ecological Studies 7: Mediterranean Type Ecosystems,* 11–19. New York: Springer-Verlag, 1973.

"Man's Impact on the Several Regions with Mediterranean Climates." In *Ecological Studies 7: Mediterranean Type Ecosystems,* 363–71. New York: Springer-Verlag, 1973.

(With Leonard W. Bowden.) "Remote Sensing of Environmental Quality." In *Remote Sens-*

ing: *Techniques for Environmental Analysis,* edited by John B. Estes and Leslie W. Senger, 293–301. Santa Barbara, Calif.: Hamilton Publishing.

"La Carretera de Baja California." *Calafia* 2, no. 5 (September 1974): 36–39.

"A Late Recounting of the Vizcaíno Expedition and Plans for the Settlement of California." *Journal of California Anthropology* 1, no. 2 (Winter 1974): 174–85.

"Environment and Ecology in the 'Northern Tonto' Claim Area." In *Apache Indian V,* American Indian Ethnohistory Series, 167–232. New York: Garland Publishing, 1974.

"Terrain and Ecological Conditions in the Western Apache Range." In *Apache Indians V,* American Indian Ethnohistory Series, 233–60. New York: Garland Publishing, 1974.

"The Turno in Northern Chile: An Institution for Defense against Drought." *Geoscience and Man* 5 (1974): 97–110 (Fred B. Kniffen Festschrift).

"The Persistent Guajiro." *Natural History* 84, no. 3 (March 1975): 28–37.

"Culturally Determined Recognition of Food Resources in the Coastal Zone." *Geoscience and Man, Coastal Resources* 12 (1975): 43–47.

"Man's Impact on the Southern California Flora." *Plant Communities of Southern California,* special publication no. 2, California Native Plant Society, 1976, 40–48.

"Recuperación de la vegetación desértica." *Calafia* 3, no. 3 (October 1976): 52–57.

(With Conrad Bahre.) "Man's Impact on the Wild Landscape." In *Convergent Evolution in Chile and California: Mediterranean Climate Ecosystems,* 73–84. Stroudsburg, Pa.: Dowden, Hutchinson and Ross, 1977.

"Views and Concerns Relating to Northern Development." North Australia Research Bulletin, no. 1, September 1977, 31–57.

"Aboriginal Use of Fire." Proceedings of the Symposium on the Environmental Consequences of Fire and Fuel Management in Mediterranean Ecosystems, USDA Forest Service, General Technical Report WO-3, November 1977, 132–41.

"The Baja California Highway." In *Brand Book V,* 170–76. San Diego: San Diego Corral of Westerners, 1978.

"Desertification—A World Problem." Section E in *Drought: Our Heritage,* 9th Annual Land-Use Symposium, Albuquerque, 1978.

"Historical Development of Agriculture in Semi-Arid Regions of Winter Precipitation." Proceedings of an International Symposium on Rainfed Agriculture in Semi-Arid Regions, April 17–22, 1978, U.C. Riverside, 87–110.

"Amenities in the New Mining Towns of Northern Australia." North Australia Research Bulletin, no. 5, August 1979, 243.

"The Immortality of Latin American States." In *Geographic Research on Latin America: Benchmark 1980.* Proceedings of the Conference of Latin Americanist Geographers, edited by Tom L. Martinson and Gary S. Elbow. Vol. 8, 1981, 323–29.

"A Restrictive Definition of Mediterranean Climates." Pre-Report of the Mediterranean Bioclimatology Symposium, Montpelier, France, May 18–20, 1983, 1:1–9.

"A Restrictive Definition of Mediterranean Climates." In *Actualités Botaniques,* Bulletin de la Societé Botanique de France, vol. 131 ("Bioclimatologie Méditerranéenne"), 1985, 22–30, and comments and responses, 63, 585.

"Calendar Dates as Street Names in Asunción, Paraguay." *Names* 34, no. 2 (1986): 146–53.

"Learning about Baja California Indians: Sources and Problems." *Journal of California and Great Basin Anthropology* 8, no. 2 (1986): 238–45.

"Carl Sauer, A Self-Directed Career." In *Carl O. Sauer, A Tribute,* edited by Martin S. Kenzer, 137–43. Corvallis: Oregon State University Press, 1987.

"Human Impact on the Biota of Mediterranean-Climate Regions of Chile and California." In *Biogeography of Mediterranean Invasions,* edited by R. H. Groves and F. DiCastri, 33–41. Cambridge: Cambridge University Press, 1991.

"Paraguay: A Bilingual Country." In "Person, Place and Thing," *Geoscience and Man* 31 (1992).

Book Reviews

The Next Million Years, by C. G. Darwin. *Los Angeles Sunday (Daily) News,* February 8, 1952, 22.

Observations in Lower California, by J. J. Baegert. *Hispanic American Historical Review* (August 1952): 396–97.

Agriculture in Haiti, by Marc-Aurèle Holly. *Geographical Review* (October 1956): 587–89.

Lower California Guidebook, by Peter Gerhard and Howard E. Gulick. *Riverside Press-Enterprise,* November 25, 1956, C-8.

Atlas Aérien. Vol. 1, *Alpes, Vallée du Rhône, Provence, Corse,* by Pierre Deffontaines and Mariel Jean-Brunhes Delamarre. *Geographical Review* (April 1957): 286–88.

Man's Role in Changing the Face of the Earth, edited by William L. Thomas, Jr. *Landscape* (Spring 1957): 29–30.

Climate and Economic Development in the Tropics, by Douglas H. K. Lee. *Landscape* (Winter 1957–58): 29.

Rivers, Man, and Myths, by Robert Brittain. *Riverside Press-Enterprise,* June 8, 1958, C-8.

Lower California Guidebook, by Peter Gerhard and Howard E. Gulick. *Landscape* (Autumn 1958): 32.

The Perpetual Forest, by W. B. Collins. *Landscape* 9, no. 2 (Winter 1959–60): 34.

The White Nile, by Allan Moorhead. *Riverside Press-Enterprise,* March 5, 1961, C-15.

The Lapps, by Roberto Bosi. *Professional Geographer* 13, no. 3 (May 1961): 57.

Sahara: Desert of Destiny, by George Gerster. *Riverside Press-Enterprise,* June 11, 1961, C-15.

After the Seventh Day: The World Man Created, by Richie Calder. *Landscape* 11, no. 3 (Spring 1962): 34.

Atlantic Crossings before Columbus, by Frederick J. Pohl, and *They All Discovered America,* by Charles Michael Boland. *Riverside Press-Enterprise,* August 12, 1962, C-12.

Perspectives in Conservation: Essays on America's Natural Resources, edited by Henry Jarrett. *Landscape* 12, no. 3 (Spring 1963): 39–40.

Great Surveys of the American West, by Richard A. Bartlett. *Journal of Geography* 62, no. 5 (May 1963): 223–24.

A Guide to the Historical Geography of New Spain, by Peter Gerhard. *Annals of the Association of American Geographers* 64 (June 1964): 336.

One World Divided, by Preston E. James, and *Geography in World Society,* by Alfred H. Meyer and John D. Strietelmeir. *Landscape* 14, no. 2 (Winter 1964–65): 42–43.

The Woods and the Sea, by Dudley Cammett Lunt, and *Roadside Area,* by Paul Brooks. *Landscape* 15, no. 1 (Autumn 1965): 41–42.

The Changing Landscape of a Mexican Municipio, Villa las Rosas, Chiapas, by David A. Hill. *Hispanic American Historical Review* 45, no. 4 (November 1965): 651–52.

Warriors of the Colorado: The Yumans of the Quechan Nation and Their Neighbors, by Jack D. Forbes. *Southern California Quarterly* 48, no. 1 (March 1966): 85–88.

Die Halbinsel Baja California: Ein Entwicklungsgebiet Mexikos, by H.G. Gierloff-Emden. *Geographical Review* 56, no. 2 (April 1966): 303–4.

Wildlands in Our Civilization, edited by David Brower. *Landscape* 16, no. 1 (Autumn 1966): 25.

The Rock Paintings of the Chumash: A Study of a California Indian Culture, by Campbell Grant. *California Historical Society Quarterly* 46, no. 1 (March 1967): 78–80.

Tlapacoyan, by David Ramírez Lovoignet. *Hispanic American Historical Review* 47, no. 3 (August 1967): 414–15.

Arid Lands. A Geographical Appraisal, edited by E. S. Hills. *Science* 158, October 6, 1967, 106–7.

Campaigns Against Hunger, by E. C. Stakman, Richard Bradfield, and Paul C. Mangelsdorf. *Agricultural History* 43, no. 3 (July 1968): 273–74.

La cuenca superior del Río Grijalva, Un estudio regional de Chiapas, Sureste de México, by Karl M. Helbig. *Hispanic American Historical Review* 48, no. 4 (November 1968): 711–13.

Colombia: Social Structure and the Process of Development, by T. Lynn Smith. *Professional Geographer* 21, no. 1 (March 1969): 126.

The Autobiography of Delfina Cuero, a Diegueño Woman, as told to Florence C. Shipek, Rosalie Pinto Robinson Interpreter. *California Historical Society Quarterly* 48, no. 2 (March 1969): 177.

The Sonoran Desert: Its Geography, Economy, and People, by Roger Dunbier. *Southern California Quarterly* 51, no. 4, (December 1969), pp. 352–53.

La obra cartográfica de la provincia mexicana de la Compañía de Jesús (1537–1967), by Ernest J. Burrus, S.J. *Hispanic American Historical Review* 50, no. 2, (May 1970): 366–67.

The Indians of Los Angeles County: Hugo Reid's Letters of 1852, edited and annotated by Robert F. Heizer. *California Historical Society Quarterly* 49, no. 2 (June 1970): 175–76.

The Enduring Desert: A Descriptive Bibliography, by E. I. Edwards. *Southern California Quarterly* 52 (September 1970): 304.

The Mojave River and Its Valley, by Erma Peirson. *Southern California Quarterly* 53 (June 1973): 160–61.

The Drawings of Ignacio Tirsch, a Jesuit Missionary in Baja California, by Joyce B. Nunis, Jr. *Journal of California Anthropology* 1 (Spring 1974): 120–22.

Der Aufbau der vorspanischen Siedlungs-und Wirtschaftslandschaft im Kulturraum der Pueblo-Indianer, by Dietrich Fliedner. *Annals of the Association of American Geographers* 66 (September 1976): 482–83.

Irrigation's Impact on Society, edited by Theodore E. Downing and McGuire Gibson. *Geographical Review* 66 (October 1976): 489.

Latin American Development: A Geographical Perspective, by Alan Gilbert. *Journal of Historical Geography* 2 (October 1976): 379–80.

Royal Officer in Baja California 1768–70; Joaquín Velázquez de León, by Iris Wilson Engstrand, *Journal of San Diego History* 23 (Winter 1977): 96–97.

Geographie im Umbruch: Ein Methodologischer Beitrag zur Neukonzeption der Komplexen Geographie, by Peter Weichart. *Geographical Survey,* ca. 1977, 32–34.

Friendly Vermin: A Survey of Feral Livestock in Australia, by Tom McKnight. *Professional Geographer* 30, no. 2 (May 1978): 215–16.

Symbiosis, Instability, and the Origins and Spread of Agriculture: A New Model, by David Rindon. *Current Anthropology* 21, no. 6 (December 1980): 756.

The Natural History of Baja California by Miguel del Barco, S.J., translated by Froylan Tiscareño. *Journal of California and Great Basin Anthropology* 2, no. 2 (Winter 1980): 313–14.

The Natural World of the California Indian, by Robert E. Heizer and Albert B. Elsasser. *Journal of Historical Geography* 7, no. 3 (July 1981): 319–21.

Northern New Spain: A Research Guide, by Thomas C. Barnes, Thomas H. Naylor, and Charles W. Pelzer. *Journal of California and Great Basin Anthropology* 3, no. 2 (Winter 1981): 300–301.

People of the Magic Waters, by John R. Brumgardt and Larry L. Bowles. *Journal of California and Great Basin Anthropology* 3, no. 2 (Winter 1981): 299–300.

From Fire to Flood: Historic Human Destruction of Sonoran Desert Riverine Oases, by Henry Dobyns. *Annals of the Association of American Geographers* 72, no. 3 (September 1982): 428–29.

Die lebensräumliche Situation der Indianer im paragueyischen Chaco, by Walter Regehr. *Revista Paraguaya de Sociología,* Año 19, no. 54 (May–August 1982): 193–95.

Navajo Architecture: Forms, History, Distributions, by Stephen C. Jett and Virginia E. Spencer. *Geographical Review* 73, no. 1 (January 1983): 119–20.

Letters from Carl O. Sauer While on a South American Trip under a Grant from the Rockefeller Foundation, 1942, edited by Robert C. West. *Professional Geographer* 35, no. 2 (November 1983): 521–22.

The Letters of Jakob Baegert, 1749–1762: Jesuit Missionary in Baja California, translated by Elspeth Schulz-Bischof and edited by Joyce B. Nunis. *Journal of San Diego History* 30, no. 1 (Winter 1984): 69–71.

The Cave Paintings of Baja California, by Harry W. Crosby. *Southern California Quarterly* 69, no. 3 (Fall 1985): 322–24.

Massacre on the Gila: An Account of the Last Major Battle between American Indians and Reflections on the Origins of War, by Clifton B. Kroeber and Bernard L. Fontana. *Journal of California and Great Basin Anthropology* 8, no. 1 (1986): 138–40.

Reprinted Papers

"Geography and the Liberal Arts College." *Professional Geographer* 10, no. 2 (March 1958): 2–6.

"A Metate Maker of Baja California." *Human Geography,* Geography Department, Los Angeles State College, 1962, 182–85.

"Evaluations of Dryland Environments by Societies at Various Levels of Technical Competence." In *Land and Water Use,* edited by Wynne Thorne, 133–49. Washington, D.C.: American Association for the Advancement of Science, 1963.

The Central Desert of Baja California: Demography and Ecology. Riverside, Calif.: Hugh Manessier, 1967.

"The Evolution of a Wild Landscape and Its Persistence in Southern California." In *Physical Geography, Selected Readings,* edited by Fred E. Dohrs and Lawrence M. Sommers, 192–216. New York: Thomas Y. Crowell & Co., 1967.

"The Evolution of a Wild Landscape and Its Persistence in Southern California." In Bobbs-Merrill Reprint Series in Geography. Indianapolis: Howard W. Sams & Co., n.d.

"Historical Accounts and Archaeological Discoveries: Working Together Two Scholarly Disciplines Enlarge Our Understanding of the Extinct Indians of Baja California." *Pacific Coast Archaeological Society Quarterly* 4, no. 1, (January 1968): 46–51.

"People, Recreation, Wild Lands, and Wilderness." *Landscape* 18, no. 1 (Winter 1969): 40–44.

"People, Recreation, Wild Lands, and Wilderness." In *California: Its People, Its Problems, Its Prospects,* edited by Robert W. Durrenberger, 172–83. Palo Alto, Calif.: National Press Books, 1971.

"People, Recreation, Wild Lands, and Wilderness." In *Crisis, Readings in Environmental Issues and Strategies,* edited by Robert M. Irving and George B. Priddle, 222–33. Toronto: Macmillan of Canada, 1971.

"Purpose in the Southern California Landscape." In *California: Its People, Its Problems, Its Prospects,* edited by Robert W. Durrenberger, 104–11. Palo Alto, Calif.: National Press Books, 1971.

"Views and Concerns Relating to Northern Development." *Aluminium* (Comalco Ltd.), November 22, 1978, 5–13.

"Proprietary Rights to Fruit on Trees Growing on Residential Property." In *Whose Trees?: Proprietary Dimensions of Forestry,* edited by Louise Fortmann and John W. Bruce, 63–67. Boulder, Colo.: Westview Press, 1988.

Notes, Letters, and Reports

"The Status of Geography in the Colleges of California." Mimeographed Newsletter of the
California Council of Geography Teachers, June 1948.
"Notes on the Soils Associated with Archaeological Finds at the Avondale Site, Missouri." In
J. M. Shippee, "Archaeological Salvage at Avondale Mounds," 30–36. *Missouri Archaeolo-
gist* 15, no. 4 (December 1953): 18–39.
"Early Man in Western North America." *American Antiquity,* April 1954, 417–18.
"Hot Rock." *American Speech* (October 1954): 236–37.
"Comments on Quimby's Cultural and Natural Areas before Kroeber." *American Antiquity*
(April 1955): 377–78.
"Preliminary Report on Investigations in the Guajira Peninsula, Colombia." ONR-222 (11)
038 067-Sauer-ONR Distr., 1955.
"Miracle Mile." *American Speech* (May 1957): 156–58.
"Geography and the Liberal Arts College." Bulletin, California Council of Geography Teach-
ers, June 1957, vol. 4, no. 3, 7–10.
"Cultural Impediments to Technical Assistance Programs." *Geographical Review* 47, no. 3
(July 1957): 440–41.
(With Hugh G. J. Aitken.) "Canada's Birth Rate." *Canadian Journal of Economics and Political
Science* 24, no. 1 (February 1958): 102–3.
"Recent Developments in Geography with Reference to Latin America." Bulletin, California
Council of Geography Teachers, vol. 5, no. 2, April 1958, 13–16.
"Automobiles and Model-A: Further Comment." *American Speech* 34, no. 3 (October 1959):
227–29.
"Industrialization in a Non-Western Society." *Geographical Review* 49, no. 4 (October 1959):
564–66.
"Retreat of an Agricultural Frontier." *Geographical Review* 50, no. 4 (July 1960): 432.
"Archaeological Salvage." *Geographical Review* 53, no. 1 (January 1963): 137–39.
"Colloquia on Coastal Geography: A Report." *Professional Geographer* 15, no. 1 (January
1963): 31–33.
"Letter on Aboriginal Languages of Latin America." *Current Anthropology* 4, no. 3 (June
1963): 319.
"Comments on Pawling's Abridgment of Staszewski's Bevoelkerungsverteilung mach den
Klimagebieten von W. Koeppen." *Professional Geographer* 16, no. 1 (January 1964): 22–23.
"Geography in Liberal Arts Colleges." In *American Geography, 1960–63: Education, Employ-
ment and Other Trends.* Washington, D.C.: Association of American Geographers, 1964.
"Indians of South America." In *The New Book of Knowledge,* 201–11. New York: Grolier, 1966.
"Plant Cover and Soils, Spacecraft in Geographic Research," NAS-NRC Publication 1353,
1966, 5–6 and 48–54.
"Letter on the Brain Drain Dilemma." *Science* 155, no. 3762, February 3, 1967, 513–14.
"Comment on Origins of African Agriculture." *Current Anthropology* 9, no. 5 (1968): 494–95.
"Multiple Land Use: Arid Areas, Northern Baja California, Mexico, and Potentials of Space-
born Sensors to Collect Data of Agriculture." In *Natural Vegetation, Forestry (The Bio-
sphere), Earth Resources Surveys from Spacecraft,* NASA, vol. 1, 1968, D30–D32; C9–C11.
"Prolegomena to the Remote Sensing of Environmental Quality." Commission on Geo-
graphic Applications of Remote Sensing, Association of American Geographers, July
1970.
"Prolegomena to the Remote Sensing of Environmental Quality." Technical Report 73–O,
Commission on Geographic Applications of Remote Sensing, Johnson City, Tennessee,
September 1970, 1–13.

Comment on *Maize and the Mande Myth,* by M. D. W. Jeffreys. *Current Anthropology* 12, no. 3 (June 1971): 305–6. (Response: p. 314.)

"Climate and the Occurrence of Organisms" and "Summary Comments." Papers Presented at the International Conference on Microclimate–Tucuman, Technical Report 71–1, Chile-California Mediterranean Scrub Project, 1971, 11 pp.

Introduction to the reprinted edition of and photograph selection and captions for *The History of (Lower) California,* by Francisco Javier Clavigero, translated by Sara E. Lake and A. A. Gray, vii–ix, xvii–xxiv, and ff. Stanford: Stanford University Press, 1937. Reprint, Riverside, Calif.: Manessier Publishing, 1971.

"George McCutchen McBride, 1876–1971." *Annals of the Association of American Geographers* 62 (December 1972): 685–88.

Introduction to Section II, "Physical Geography of Lands with Mediterranean Climates." In *Ecological Studies 7; Mediterranean Type Ecosystems,* 9–10. New York: Springer-Verlag, 1973.

Introduction to Section VIII, "Human Activities Affecting Mediterranean Ecosystems." In *Ecological Studies 7; Mediterranean Type Ecosystems,* 362. New York: Springer-Verlag, 1973.

(With Leonard W. Boden, Thomas R. Lyons, and Ralph S. Solecki.) "People: Past and Present." Chap. 267 in the *Manual of Remote Sensing,* 2:1999–2060. Falls Church, Va.: American Society of Photogrammetry, 1975.

(With Conrad J. Bahre and Anthony R. Van Curen.) "Land Use in Fundo Santa Laura, Chile and Echo Valley, California." In *Chile-California Mediterranean Scrub Atlas: A Comparative Analysis,* edited by N. J. W. Thrower and D. E. Bradbury, 66–69. Stroudsburg, Pa.: Dowden, Hutchinson and Ross, 1977.

"Comments on Anderson's Review of Nava and Berger." *Journal of California Anthropology* 4, no. 1 (1977): 125–26.

"More on Early Trade," letter. *Current Anthropology* 20, no. 1 (March 1979): 243.

Comment on *Symbiosis, Instability, and the Origins and Spread of Agriculture: A New Model,* by David Rindos. *Current Anthropology* 21, no. 6 (December 1980): 765.

"A Restriction Definition of Mediterranean Climates." Pre-Report of the Mediterranean Bioclimatology Symposium, Montpelier, May 18–20, 1983, 1–9.

"In Memorium: Howard E. Gulick, 1911–1983." Baja California Symposium 22, Glendale, Calif., May 1984, 11–12.

"On Cross-Cultural Evaluation of Oneota Agricultural Production." *Current Anthropology* 32, no. 4 (1991): 445–47.

"Coromuel and Pichilingue." *Names* 40, no. 1 (March 1992): 33–38.

Contributors

Daniel D. Arreola is Professor of Geography at Arizona State University. He has published extensively in leading journals and books on topics relating to the cultural geography of Mexico and the American Southwest. His most recent book, *The Mexican Border Cities: Landscape Anatomy and Place Personality*, won a Southwest Book Award from the Border Rigional Library Association for "literary excellence and enrichment of the cultural heritage of the Southwest." He shared with Aschmann a deep understanding and appreciation of life in the Mediterranean climate of Southern California.

Conrad J. Bahre is Professor of Geography at the University of California at Davis. He is an expert on the cultural landscapes of northern Chile and northwest Mexico as examined in his books *Destruction of the Natural Vegetation of North-Central Chile* and *A Legacy of Change: Historical Human Impacts on Vegetation of the Arizona Borderlands*. His shared field experience with Aschmann in Baja California gives him unusual insight into the depopulation of that landscape after Spanish contact and what Aschmann thought about it.

Karl W. Butzer is Dickson Centennial Professor of Liberal Arts and Professor of Geography at the University of Texas, Austin. Like Aschmann, Butzer brings an understanding of several disciplines to his study of the land, a talent best demonstrated in his books *Archaeology as Human Ecology* and *The Americas before and after 1491*. He was elected to the National Academy of Sciences in 1996. Other honors include a Guggenheim Fellowship, the Busk Medal of the Royal Geographical Society, the Fryxell Medal of the Society for American Archaeology, the Stopes Medal of the Geologists' Association of London, and the Promerance Medal of the Archaeological Institute of America. He is the author of 8 books and more than 200 scientific papers.

George F. Carter is Emeritus Distinguished Professor at the Department of Geography at Texas A&M University. Carter and Aschmann shared many interests and a mutual respect since meeting in graduate school. Carter considered Aschmann an unusually perceptive observer of the land. Carter, himself a prolific author and tireless intellect, authored several books about early settlement of the New World plus 200 scholarly articles on the topics of antiquity of man in America and on the biological evidence for transoceanic carriage of plants and animals. He has received a Guggenheim Fellowship, the Honors Certificate of Merit from the Association of American Geogrphers, and the Gold Medal from the Institute for the Study of American Cultures.

William E. Doolittle, Professor of Geography at the University of Texas, Austin, specializes in human ecology, especially prehistoric and historic agriculture and water control in arid lands, the southwestern United States and northwestern Mexico in particular. He has held several positions, including Chairman of the geography department (1992–96), and Chairman of the Board of Directors of the Conference of Latin Americanist Geographers. His publications include 2 books and 100 articles and chapters, many of which focused on Latin America and relied heavily on earlier work by Aschmann.

Charles F. Hutchinson is Professor in Arid Lands Studies, among other posts at the University of Arizona. Dr. Hutchinson earned the doctoral degree from the University of California, Riverside, working in close association with Aschmann. Most of his research has focused on the applications of remote sensing and geographic information systems to agricultural and natural resource problems, largely in arid lands. He is recipient of the Gilbert F. White Fellowship at Resources for the Future. His first trip to Arizona in 1973 involved study of vegetation change at the International Biological Program test site at Silverbell, under the direction of Aschmann.

John Brinckerhoff Jackson was a pioneer in landscape studies. He adopted the American Southwest after education at Harvard and military service during World War II in Europe. Well known among geographers, planners, architects, and landscape architects, he long was a keen observer of the American landscape and a provocative and personable lecturer. In the spring of 1951 Jackson started *Landscape* magazine—a thoughtful, provocative, and challenging forum for the discussion of the changing rural landscape that he published and edited until 1968. For years Mr. Jackson taught periodically at Harvard and Berkeley. Aschmann had high regard for Jackson and the journal, and published in its pages on several occasions. Mr. Jackson was also the author of many acclaimed books, including *A Sense of Place, a Sense of Time*.

William G. Loy is Professor of Geography at the University of Oregon, Eugene, and a member of the Editorial Board of *Names* (American Names Society). He is author of *The Atlas of Oregon*, is Past President of the Oregon Academy of Science, and was a Fullbright Scholar at Trinity College, Dublin. Like Aschmann, Dr. Loy was President of the Association of Pacific Coast Geographers, and he knew Aschmann mostly in that context. Loy's presidential address "Geographic Names in Geography" discussed the importance place-names can play in identifying the cultural influences on the land, an interest Aschmann also retained throughout his career.

James J. Parsons has spent his entire academic career at the University of California, Berkeley, where he has been Emeritus Professor since 1986. He shared with Aschmann a sharp interest in landscape change in North and South America, and he is perhaps best known for his field-based studies of colonization and settlement, including prehistoric intensive "raised field" agriculture in the tropical lowlands of Columbia and Ecuador. Parsons served as President of the Association of American Geographers, and has authored many books on landscape change, most notably *Antiqueño Colonization in Western Colombia and Hispanic Lands and Peoples*. His many honors include a Guggenheim Fellowship, the David Livingston Medal from the American Geographical Society, and a dr. honorus causa from the Universidad de Antioquia. He was principally responsible for J. B. Jackson's regular visits as teacher and lecturer to the geography department at the Berkeley campus.

Martin J. Pasqualetti is Professor of Geography and Faculty Fellow for Natural Resources and the Environment at Arizona State University. He has 25 years experience researching and writing about topics of mutual interest to Aschmann, including landscape development in relationship to the energy quest. He traveled to Mexico with Aschmann on several occasions, and was personally and professionally acquainted with him for more than 20 years. He has published 4 books and more than 100 scientific papers and book chapters.

436

Index

Library of Congress Cataloging-in-Publication Data

The evolving landscape : Homer Aschmann's geography / edited by
Martin J. Pasqualetti ; with a foreword by John Brinckerhoff Jackson.
      p.   cm.
   ISBN 0-8018-5310-9 (alk. paper)
    1. Human geography. 2. Landscape assessment. 3. Landscape changes.
4. Desert biology. 5. Desert ecology. 6. Aschmann, Homer, 1920–   —
Philosophy. I. Pasqualetti, Martin J., 1945– .
GF50.E95   1997
304.2—dc20   96-27473
                                CIP